Ageing and the Built Environment in Singapore

Belinda Yuen
Editor

Ageing and the Built Environment in Singapore

Springer

Editor
Belinda Yuen
Lee Kuan Yew Centre for Innovative Cities
Singapore University of Technology
and Design
Singapore
Singapore

ISBN 978-3-030-06431-0 ISBN 978-3-319-92444-1 (eBook)
https://doi.org/10.1007/978-3-319-92444-1

Softcover re-print of the Hardcover 1st edition 2019

Cover illustration: A public housing neighbourhood in Singapore
Photo credit: Belinda Yuen

Printed on acid-free paper

This Springer imprint is published by the registered company Springer International Publishing AG part of Springer Nature
The registered company address is: Gewerbestrasse 11, 6330 Cham, Switzerland

Foreword

The Lee Kuan Yew Centre for Innovative Cities (LKYCIC) in the Singapore University of Technology and Design was established in September 2012. Shortly after, Mrs. Lee Liming, a Singapore philanthropist, agreed to gift a research grant to the Centre on the condition we work on ageing as a topic. Thus, the *Lee Liming Programme on Ageing Urbanism* became the first research programme set up in LKYCIC. The second research flagship programme, the *Future of Cities* was launched in 2014, followed by the *Chen Tianqiao Programme on Cities and Innovation*.

Taking up the topic of ageing was timely and the right thing to do. In 1997, then Prime Minister Goh Chok Tong, speaking at a Triennial Conference for grassroots leaders, raised the issue of the rapid ageing of the population. In 1998, the Inter-Ministerial Committee on the Ageing Population was formed to look into the needs and consequences of the ageing demographics. Academic interest followed. But, ageing as a research topic has gained wider attention more recently, prompted by the fact that 2030, the year when Singapore's population of those aged 65 and older would reach the projected figure of 900,000, is not far away.

Dr. Belinda Yuen working with her team completed a major survey of the ageing population to understand their needs. They verify and enrich their findings with other methodologies to get a full and rounded understanding of the requirements and aspirations of the senior population. We hope this treasure trove of data will be useful to those who have to make decisions affecting the ageing population, and for other researchers who are interested in working in the same area.

Singapore

Prof. Chan Heng Chee
Ambassador-at-large and Chairman
Lee Kuan Yew Centre for Innovative Cities
Singapore University of Technology and Design

Preface

In 2013, when I joined the Lee Kuan Yew Centre for Innovative Cities, Singapore University of Technology and Design, I was asked to develop a research programme on population ageing. Thus was born the Lee Li Ming Programme in Ageing Urbanism and the project: Understanding the Changing Needs of Singapore's Older Population. Being the first project under the programme, it reflected my continuing interest and research on housing and public spaces with an analytical lens focused on a population that I have not closely studied before: older people, their lived experiences and perceptions of growing older in these spaces.

The World Health Organization has highlighted the rapid growth of older population worldwide and the importance of the built environment including housing and outdoor spaces in the older person's health, well-being and quality of life. Yet, relatively little research has addressed the older person's experiences of these spaces. We are inspired to share our research, the methods we have used and some of the findings.

I would be remiss if I did not mention that many people (listed in the individual chapters) have contributed to the completion of the research project and this book. I duly acknowledge their contributions. In addition, I thank Mrs. Lee Li Ming for supporting our research through her generous gift.

I hope the chapter contributions in this Volume, whether you read one or all of them, will go some way towards promoting better understanding of the older person's relationship with the built environment, and will further research and consideration while planning housing and outdoor environments for an ageing society.

Singapore — Belinda Yuen

Contents

1 Surveying Older Adults' Perceptions and Aspirations 1
Belinda Yuen, Chathura Withanage and Parvathi Nair
1.1 Introduction 1
1.2 Methodology 2
1.3 Findings 4
1.3.1 Economic and Living Arrangements 4
1.3.2 Perceptions of Old Age and Growing Older 6
1.3.3 Housing Satisfaction and Aspirations 12
1.3.4 Public Spaces and Outdoor Activities 23
1.4 Conclusion 40
References 42

2 Focusing on Older Residents Views 45
Belinda Yuen and Sue Wei Cheong
2.1 Introduction 45
2.2 Methodology 46
2.3 Findings 46
2.3.1 Housing and Neighbourhood 47
2.3.2 Amenities and Transportation 53
2.3.3 Recreational and Social Spaces 56
2.3.4 The Individual and Community 60
2.3.5 Possible Differentiation Between Focus Groups 63
2.4 Conclusion 65
References 69

3 Sharing Key Informant Views 71
Belinda Yuen and Sue Wei Cheong
3.1 Introduction 71
3.2 Methodology 72

3.3 Findings . . . 73
3.3.1 Housing . . . 73
3.3.2 Neighbourhood and City . . . 86
3.3.3 Society and Individual . . . 95
3.3.4 Different Groups, Similar Concerns . . . 98
3.4 Conclusion . . . 100
References . . . 102

4 **Housing Audit** . . . 103
Belinda Yuen and Parvathi Nair
4.1 Introduction . . . 103
4.2 Methodology . . . 104
4.3 Findings . . . 106
4.3.1 HAC Age-Friendliness Matrix . . . 107
4.3.2 Component-Wise Analysis . . . 108
4.4 Conclusion . . . 121
References . . . 124

5 **Walking with Older Adults** . . . 127
Belinda Yuen and Penny Kong
5.1 Introduction . . . 127
5.2 Methodology . . . 128
5.3 Findings . . . 133
5.3.1 Respondent Usage of Space . . . 133
5.3.2 Site A: 256 Bangkit Road . . . 135
5.3.3 Site B: 434 Fajar Road . . . 139
5.3.4 Site C: 415 Fajar Road . . . 140
5.3.5 Site D: 201 Petir Road, Bukit Panjang N2 Park . . . 143
5.4 Conclusion . . . 146
References . . . 148

6 **Seeing Through the Eyes of Older Adults** . . . 149
Belinda Yuen and Emily Soh
6.1 Introduction . . . 149
6.2 Methodology . . . 150
6.3 Findings . . . 152
6.3.1 Amenities and Facilities . . . 152
6.3.2 Structured Social Spaces . . . 155
6.3.3 Incidental Social Spaces . . . 158
6.3.4 Outdoor Green Spaces . . . 160
6.3.5 Quality of Neighbourhood Infrastructure and Outdoor Street Furniture . . . 162

6.3.6 Favourite or Frequented Activity Nodes Outside of One's Neighbourhood . . . 165
6.3.7 Transportation and Travel Journeys . . . 167
6.4 Conclusion . . . 170
References . . . 173

7 Investigating Space, Activities and Social Dynamics . . . 175
Belinda Yuen and Parvathi Nair
7.1 Introduction . . . 175
7.2 Methodology . . . 176
7.3 Findings . . . 180
7.3.1 Site A—256 Bangkit Road . . . 181
7.3.2 Site B—434 Fajar Road, Void Deck . . . 188
7.3.3 Site B—Fajar Road, Hard Court . . . 192
7.3.4 Site C—415 Fajar Road, Void Deck . . . 198
7.3.5 Site D—201 Petir Road, Bukit Panjang N2 Park . . . 203
7.3.6 Site Comparison . . . 209
7.4 Conclusion . . . 216
References . . . 218

8 Real-Time Data Analysis Using a Smartphone Mobile Application . . . 221
Sumudu Hasala Marakkalage, Billy Pik Lik Lau, Sanjana Kadaba Viswanath, Chau Yuen and Belinda Yuen
8.1 Introduction . . . 221
8.2 Methodology . . . 222
8.2.1 Mobile App Development . . . 223
8.2.2 Older User Data Collection . . . 225
8.2.3 Big Data Analysis . . . 226
8.3 Findings . . . 228
8.3.1 Home Stay Duration . . . 228
8.3.2 Across Neighbourhoods . . . 229
8.3.3 Across Singapore . . . 231
8.3.4 Points of Interest (POI) . . . 233
8.4 Conclusion . . . 238
References . . . 239

9 Exploring Perceived Neighbourhood Social Capital . . . 241
Anna Lane
9.1 Introduction . . . 241
9.2 Methodology . . . 242
9.3 Findings . . . 243
9.3.1 Socio-demographics . . . 244
9.3.2 Social Capital . . . 244

9.4 Conclusion 255
References 258

10 Designing with Older Residents 259
Belinda Yuen and Sue Wei Cheong
10.1 Introduction 259
10.2 Methodology 260
10.3 Findings 261
10.3.1 Vision 263
10.3.2 Favourite Spaces 264
10.3.3 Accessibility and Connectivity 264
10.3.4 Maintenance and Management 266
10.3.5 Safety and Security 268
10.3.6 Activities and Amenities 271
10.3.7 Community Spirit and Ownership 275
10.4 Conclusion 277
References 280

Index 283

Introduction: Framing the Issue

The United Nations (2010) estimates that by 2050, 70% of the global population will be living in urban environments. Crucially, a big proportion of that population will be older than 60 years. This population segment is growing faster than any other age group as a result of both longer life expectancy as well as declining fertility rates (World Health Organization 2015). Projections are predicting an unprecedented global crossing event before the end of the decade—older people aged 65 and over will outnumber children under the age of 5 by 2020 (He et al. 2016). The dynamics of these two trends—rapid urbanisation and rapid population ageing—will undoubtedly shape the future development of cities. It is imperative that cities recognise and respond to these changing realities, now rather than later.

At the Lee Kuan Yew Centre for Innovative Cities (LKYCIC), Singapore University of Technology and Design (SUTD), we have responded to the changing reality by convening a flagship research programme, the Lee Li Ming Programme in Ageing Urbanism. The aim of the programme is to study the twin challenges of rapid urbanisation and rapid population ageing with special focus on ageing population and the urban built environment.

A key challenge facing cities is the question of how we (re)shape the built environment to accommodate the evolving needs of a fast-greying population. Specifically, how we plan and design the built environment, from housing to neighbourhoods, can determine levels of independence, dignity and self-respect in later life (Australian Local Government Association 2006; Garin et al. 2014). More than any other age groups, older people tend to spend more time at home, and if they go out, it is most frequently to the immediate neighbourhood. Getting these spaces right is a crucial factor to achieving successful, healthy ageing. At the global level, the World Health Organization has posited that an age-friendly city offers a supportive built and social environment that enables the older person to actively participate in the community (World Health Organization 2007a).

The first project under the Lee Li Ming Programme in Ageing Urbanism: Understanding the Changing Needs of Singapore's Older Population (2014–2017) seeks to investigate the lived experiences of Singapore's older population (55 years old and older) in two key spaces—housing (live) and recreation (play). Its main aim is to understand both subjective and objective older person–space relationships. The novel scope is to bring together the 'what' of knowledge with the 'how' of formulating and implementing design solution, explicitly considering the older person–space relationships using a multi-dimensional methodological approach. This approach involves quantitative and qualitative methods, traditional instrumentation as well as recent developments in mobile crowdsourcing. The project combines research with planning, with the focus to include the older person into conversation. The methods and findings are presented in this book. To our knowledge, this is the most comprehensive study using mixed methods of the two mentioned aspects in Singapore to date.[1]

Singapore is a natural laboratory for this study. With its land area of 720 sq km and population of 5.6 million, Singapore is a city-state. It is 100% urbanised and its citizen population is ageing at a rapid rate. Older Singaporeans (aged 65 and older) are expected to double to one in four by 2030. This demographic change has come about quickly. France has taken 117 years to increase its older population from 7% to 14%, Japan 24 years and Singapore is expected to take 17 years to achieve this growth (United Nations 2015). What is the impact of this demographic shift on Singapore's built environment and urban living? How can we adapt and design its housing and neighbourhoods to become more supportive of older residents' changing needs? Recognising and understanding this demographic shift is crucial. This book presents a collection of the methods and tools that may be used to help us better understand this demographic shift.

Singapore Ageing Research

Population ageing has been a topic of policy discussion in Singapore since the 1980s. Several high-level ministerial committees have been established to study ageing trends and their policy implications (see, for example the Committee on the Problems of the Aged 1984; Committee on Ageing Issues 2006; Ministerial Committee on Ageing 2016). The Ministerial Committee on Ageing released an Action Plan for Successful Ageing in August 2015, covering over 70 initiatives in 12 areas including research to help individuals to age well, foster an inclusive community and create a city for all ages (Ministry of Health 2016). This has spurred research on ageing. For example, since 2015, the National Innovation Challenge on Active and Confident Ageing has allocated S$200 million to support research and innovation on work and learning, and the study of ageing well to find ways to delay the onset of disease and disability.

[1] See other studies on Singapore's ageing society and housing such as Addae-Dapaah and Wong (2001); Bozovic Stamenovic (2012); Wu and Chan (2012); on Singapore's ageing society and outdoor spaces such as Teo (1997).

In that same year, two key national research centres were established. The first is the Centre for Ageing Research and Education at Duke-NUS Graduate Medical School Singapore to conduct research and education on ageing and health, in particular, taking a comprehensive gerontological perspective and combining the biological, social and clinical aspects of ageing to contribute to successful ageing. The second is the Geriatric Education and Research Institute to lead the development of research and education on age-related health issues to promote healthy ageing. Notwithstanding the many and growing number of research centres on population ageing in Singapore, local research has tended to largely centre on the domains of inquiry emphasised by the inter-ministerial committees of ageing like older adults' employment, health and caregiving. Analysis of the older population's experience of urban living, in particular, their attitudes, expectations and lived experiences remain an area to be further investigated (see, for example Chan 2001; Chan et al. 2010; Mathews and Leong 2014).

There is an urgent need both to identify and understand the diverse and changing ways in which the older population (and their subsets) perceive and experience the city, especially in light of the policy to promote ageing in place as the key principle of an age-inclusive built environment in Singapore (Inter-Ministerial Committee on the Ageing Population 1999; Committee on Ageing Issues 2006). The usage of methods in the present project is largely premised on this particular reasoning—if we want to improve the older residents' quality of life, we must include them in the study and ask for their wishes, suggestions, ideas and hopes. The discussion concentrates on two aspects of urban living—housing and outdoor recreational spaces.

Housing for Older People

Housing is a key focus of the present study. Housing is one of the most important aspects of age-friendly neighbourhoods, since it is a place where older people spend most of their time. According to the World Health Organization (2007b), housing is essential to the safety and well-being of older residents in the city. Housing provides the basic infrastructure for an older individual's everyday activities, communal connections and access to important services (Menec 2017). As Frochen and Pynoos (2017) observed, '[t]he condition and location of one's residence determines a great many quality-of-life outcomes resulting from structural and neighbourhood features' (p. 160).

Most, 1 in 7, older adults in Singapore live in the community within inter-generational households; less than 0.3% lived in institutional care in 2008 (Ministry of Social and Family Development 2009). Although there is a range of housing types available, the common abode is public housing; 82% of Singapore's resident population currently live in public housing, many are homeowners. About 11% (113,294) of households in public housing are headed by persons aged 65 and above while 13% (144,792) of households have head-of-household who are 55 and above (Housing and Development Board 2014). About 32% of public housing residents aged 55 and older live in 1–3-room flats and 54% in 4-room and larger flats.

There is an emerging trend of decreasing household size and increasing number of one-person households. The 2011 National Survey of Senior Citizens in Singapore reported that the average household size has been declining from 4.4 persons in 1995 to 3.3 in 2011 and more people are living in one-person households. Among the people aged 55 and above, 15% lived in one-person households in 2011 as compared to 6% in 2005. Among those aged 75 and older, the corresponding figures were 16.6% in 2011 and 4.4–7.5% in 2005 (Kang et al. 2013). The change towards smaller households is most prominent among Chinese households; one-person Chinese households increased from 8.8% in 2000 to 13% in 2010 as compared to the corresponding proportions of 4.3% and 6.5% for Malay households and 7.9% and 10.3% for Indian households (Department of Statistics 2011). Many of these one-person households are single (57%), divorced or separated (14%) or widowed (14%).

By 2030, the total number of older persons living alone could increase to 83,000 as compared to 35,000 in 2012 (*The Straits Times*, 12 April 2012). The rapidly growing number of older persons living on their own suggests that community- and home-based eldercare services as well as infrastructure and social support services will be needed even more in the future (Yuen and Soh 2017). Even while the majority of older adults are expected to remain healthy and functional, by 2030, 13% (117,000) of those aged 65 and above (more than double the current number) may be semi-/non-ambulant. The policy towards enabling the older person to age in place is necessary and urgent, especially in the context of Singapore where 5 of 6 public hospitals have over 85% bed occupancy. Rapid population ageing will put even more pressure on public hospitals.

Outdoor Recreational Spaces

Outdoor spaces are ubiquitous features of Singapore's public housing environment. On average, about 50% of public housing town are used for residential development. The remaining area is for the provision of supporting amenities and infrastructure including about 12% for open space to nurture vibrant communities. Parks are evenly distributed and allocated on a hierarchical town, neighbourhood and precinct basis. Providing open spaces in the residential neighbourhoods is equally significant for an older individual's well-being. They serve a range of functions, from visual relief and green buffer between apartment buildings to outdoor recreational spaces to support physical activities and social interaction, factors fundamental to the older resident's social, psychological and physiological health.

The World Health Organization (2010) report on physical activity for older adults suggests that compared to less active individuals, both men and women who are more active have lower rates of coronary heart disease, high blood pressure, stroke, type 2 diabetes, colon cancer, a higher level of cardiorespiratory and muscular fitness, healthier body mass and composition, a biomarker profile that is more favourable for the prevention of cardiovascular disease, among others. Marquet and Miralles-Guasch (2015) further showed in their study of Barcelona's population that living in vital urban environments contributes to building healthy mobility habits, especially when there are facilities and enough opportunities for walking nearby.

Yet, the Singapore National Health Survey 2010 (Epidemiology and Disease Control Division 2010) has revealed that physical inactivity increases with age. The data shows that 67% of Singaporeans aged between 60 and 69 did not engage in leisure time physical activities. This is not only worrying in terms of their physical health but also social and mental health. To compound matters, as stated earlier, an increasing number of older people are living in one-person households, away from their children and help or companionship may not always be readily available. The challenge is to increase the opportunities for older people to stay socially, mentally and physically active. Evidence suggests that older persons who remain both physically and socially active are diagnosed with fewer diseases and have better social health than those who remain home-locked (World Health Organization 2002). The importance of proximate outdoor spaces cannot be underestimated.

Structure of the Book

Methodology used in this project was wide ranging and combined qualitative and quantitative methods. A total of nine instruments were developed to examine older adults' perceptions and views about Singapore's housing and neighbourhoods: housing audit checklist, mobile phone application, see and snap instrument, community design workshop, interview survey, walk and talk survey, on-site observation, focus group discussion and key informant interview. Depending on the nature of enquiry, we adopted the most suitable tool in order to obtain deeper insights. Together, they provided a more complete glimpse into the older persons' lived experiences of their built environment. The research was approved by the Singapore University of Technology and Design Institutional Review Board.

Each Chapter offers a detailed description of the method used. Each method has its advantages as well as limitations. The methods are not exhaustive, but offer a broad array of some of the tools available for research into older people's lived experiences. It is but a starting point and future considerations of these methods would have to be adapted according to the research problem and question.

The Chapter following the Introduction, Chap. 1, introduces the first instrument that was used for this project, the National Survey. The aim of using this method was to provide a snapshot of older adults' experiences, perceptions, needs and aspirations for the spaces around them through personal interview. The National Survey was designed in the form of a questionnaire that covered four broad areas: demographics, home, neighbourhood and outdoor recreation spaces and respondents' life. The demographics provided answers to general questions relating to participants' age, gender, ethnicity, type of housing and living arrangements. Questions about home arrangements provided an insight into respondents' (dis)satisfaction with their home spaces, residential mobility and living arrangement preferences as they grow older. Questions about neighbourhood and outdoor recreation spaces helped us understand respondents' outdoor activities, level of satisfaction with public spaces in their neighbourhoods, and needs for facilities. Questions that inquired about respondents' life offered answers to respondents' perceptions on ageing as well as their experiences and satisfaction with growing old in Singapore.

Chapter 2 presents a qualitative research method—focus group—to offer an older user's perspective. Asked in an interactive small group setting, the broad discussion questions in the focus groups were aimed at generating opinions and discussion among the participants about their aspirations, lifestyle needs and experiences on housing and outdoor recreation spaces as they age, the innovations, barriers and opportunities in using current housing and outdoor recreation spaces, how these spaces contribute to enhancing their quality of life, and their suggestions for future development. These discussions were helpful for adding understanding—the 'why' and 'how' of issues and information collected with the survey.

Chapter 3 analyses the results from 33 Key Informant Interviews. These in-depth interviews were used to gather individual experts, community leaders and professionals' first-hand knowledge about older people's housing and outdoor recreation space design and provision. The participants came from four different sectors: government agencies and public sector, development industry, medical and healthcare sector, and caregivers, non/profit organisations, charities and service providers.

Chapter 4 introduces a Housing Audit, which delves into the housing needs and preferences of older residents. Designed as an easy-to-use checklist that can be completed by the individual or with assistance from a helper, the housing audit builds on the World Health Organization age-friendly cities guide (2007b) to identify current and potential barriers and enablers in housing design for ageing, and understand the features that are important to older people in age-friendly housing.

Chapters 5–7 move the audit activity into the outdoor spaces. Urban audit of the neighbourhood public spaces is analysed with the help of three instruments: Walk and Talk survey (Chap. 5), See and Snap study (Chap. 6) and On-site Observation (Chap. 7). Four public spaces in Bukit Panjang Town[2] were studied. The Walk and Talk and See and Snap tools were developed and used to understand the experiences of older users as well as the barriers that they face in the outdoor urban environments. The Walk and Talk survey was conducted as a mobile survey while the See and Snap took the form of a photo-taking activity. Both offer examples of participatory methods to gain insights into participants' sentiments and experiences of public spaces, neighbourhoods and the city as they journey through them on a daily basis.

The on-site observation took place on a weekday and during one weekend day, offering the researcher an opportunity to visit the site and discover its functioning—

[2]Bukit Panjang Town is located in the western region of Singapore. Built since the 1980s, Bukit Panjang's resident population in 2015 was 139,030, almost 9% of whom were residents older than age 65 (Department of Statistics 2016).

its activities, actors and processes—at different times of the day. Although time consuming and not altogether revealing about people's perceptions, motivations and feelings, this method gives a complimentary glimpse into how the public space is being used by people, not just older adults—how many, who, where and what they do while on site.

Chapter 8 presents the development and usage of an Android mobile phone application called 'City' on Google. The application was developed by the research team to capture the everyday journeys and outdoor activities of older participants in Bukit Panjang Town. The Chapter describes the methodology and development of the application as well as explains the process of data collection and analysis of big data from smart phone.

Chapter 9 investigates the social capital of older residents. Again, the Bukit Panjang Town is examined. It seeks to unveil older users' experiences of and contributions to the neighbourhood life and social outcomes. Five areas are examined: demographics of the respondents; the respondents' social connections, which are presented in terms of how many people they know in the neighbourhood, where they usually meet their friends, whom they ask for financial help and what sort of activities they are involved in; trust in respondents' neighbourhood; neighbourhood environment and the respondents' experiences of various facilities and services in their neighbourhood; and their quality of life.

The final chapter (Chap. 10), looks at the use of Community Design Workshop that was developed to work creatively together with the residents on one of the studied public spaces in Bukit Panjang Town—Bukit Panjang Neighbourhood 2 Park. The focus is on issues of identification of community concerns and opportunities for placemaking. The eight workshops explored the participants' Needs, Experiences and Vision; Accessibility and Connectivity; Park Identity and Activity Part 1 and 2; Safety and Security; Walk and Talk Urban Audit Session; Park Spirit and Sense of Place; and Community and Ownership. The workshops gave the older users (and the community) of the park an additional opportunity to voice their concerns, hopes and suggestions on how their proximate community park should look, feel and function in support of their lifestyles and quality of life. Participants' suggestions and recommendations are analysed in this Chapter.

International research suggests that there is great value in innovatively integrating research into a policy and planning agenda, especially when assessing age-friendly environment (Glicksman et al. 2014). Simply put, research on age-friendly environment is only valuable when it can contribute to the effort of creating age-friendlier neighbourhoods and public spaces, and when the agencies and organisations involved in planning the changes incorporate research's findings into their actual plans (*ibid.*). The study took that into consideration and collaborated with both the older users as well as local authorities. The Community Design Workshops (Chap. 10) are a good example of that.

Working together with the residents of Bukit Panjang Town and with the support of the local member of parliament and town council, fresh community perspectives and maps were produced as a graphic guideline for streamlined application and redesign of the community park, Bukit Panjang Neighbourhood 2 Park. In this way, research that was fuelled with direct feedback of the older users of the park got translated into planning the park's redesign and day-to-day maintenance. In the process, we gave voice to the users and enabled older adults to co-design the nearby public space.

Belinda Yuen
Špela Močnik

References

Addae-Dapaah, K., & Wong, G. (2001). Housing and the elderly in Singapore: Financial and quality of life implications of ageing in place. *Journal of Housing and the Built Environment, 16*(2), 153–178.

Australian Local Government Association (2006). *Age-friendly built environments: opportunities for local government*. ACT: Australian Local Government Association.

Bozovic Stamenovic, R. (2012). New issues in Singapore public housing from elderly perspective: Case study. *Techne: Journal of Technology for Architecture and Environment*, (4), 103–109.

Chan, A. (2001). Singapore's changing age structure and the policy implications for financial security, employment, living arrangements, and health care, Asian MetaCentre Research Paper Series No. 3. Asia Research Institute, National University of Singapore. http://www.populationasia.org/Publications/RP/AMCRP3.pdf. Accessed 18 January 2016.

Chan, A., Malhotra, C., Malhotra, R., & Østbye, T. (2010). Living arrangements, social networks and depressive symptoms among older men and women in Singapore. *International Journal of Geriatric Psychiatry, 26*, 630–639.

Committee on Ageing Issues (2006). *Report on the ageing population*. Singapore: Ministry of Community Development, Youth and Sports.

Committee on the Problems of the Aged (1984). *Report of the Committee on the Problems of the Aged*. Singapore: Ministry of Health.

Department of Statistics (2016). *General household survey 2015*. Singapore: Department of Statistics, Ministry of Trade and Industry.

Epidemiology and Disease Control Division (2010). *National health survey 2010*. Epidemiology and Disease Control Division, Ministry of Health. https://www.moh.gov.sg/content/dam/moh_web/Publications/Reports/2011/NHS2010%20-%20low%20res.pdf. Accessed 20 July 2017.

Frochen, S., & Pynoos, J. (2017). Housing for the elderly: Addressing gaps in knowledge through the lens of age-friendly communities. *Journal of Housing For the Elderly, 31*(2), 160–177.

Garin, N., Olaya, B., Miret, M., Ayuso-Mateos, J. L., Power, M., Bucciarelli, P. & Haro, J. M. (2014). Built environment and elderly population health: A comprehensive literature review. *Clinical Practice and Epidemiology in Mental Health, 10*, 103–115.

Glicksman, A., Clark, K., Kleban, M., Ring, L., & Hoffman, C. (2014). Building an integrated research/policy planning age-friendly agenda. *Journal of Aging & Social Policy, 26*(1–2), 131–146.

He, W., Goodkind, D., & Kowal, H. (2016). *An aging world: 2015*. US: Census Bureau.

Housing and Development Board. (2014). *Public housing in singapore: social wellbeing of HDB communities—HDB sample household survey 2013*. Singapore: Housing and Development Board.

Inter-ministerial Committee on the Ageing Population. (1999). *Inter-ministerial committee report on the ageing population*. Singapore: Ministry of Health.

Kang, S. H., Tan, E. S., & Yap, M. T. (2013). *National survey for senior citizens 2011*, Institute of Policy Studies, Lee Kuan Yew School of Public Policy, National University of Singapore.

Mathews, M. & Leong, T. Y. H. (2014). *Towards ageing well: aspirations, challenges and initiatives*, IPS Exchange, Institute of Policy Studies, Lee Kuan Yew School of Public Policy, National University of Singapore.

Marquet, O. & Miralles-Guasch, C. (2015). Neighbourhood vitality and physical activity among the elderly: The role of walkable environments on active ageing in Barcelona, Spain. *Social Science & Medicine, 135*, 24–30.

Menec, V. H. (2017). Conceptualizing social connectivity in the context of age-friendly communities. *Journal of Housing for the Elderly, 31*(2), 99–116.

Ministry of Health. (2016). *I feel young in my singapore: action plan for successful ageing.* Singapore: Ministry of Health.

Ministry of Social and Family Development. (2009). *State of the Elderly in Singapore 2008/2009* Release No. 3, Singapore: Ministry of Social and Family Development.

Teo, P. (1997). Space to grow old in: The availability of public spaces for elderly persons in Singapore. *Urban Studies, 34*(3), 419–439.

The Straits Times (12 April 2012) Seniors living alone may rise to 83,000 by 2030, http://app.msf.gov.sg/Portals/0/Summary/pressroom/MediaCoverage/Apr2012/120412%20ST%20SpB6%20Seniors%20living%20alone.pdf. Accessed 29 February 2016.

United Nations (2015). *World population ageing 2015*. New York: United Nations.

United Nations (2010). *Cities for all: bridging the urban divide—state of the world cities 2010/2011*. London and Sterling, VA: UN-Habitat.

World Health Organization (2002). *Active ageing: a policy framework.* Geneva: World Health Organization.

World Health Organization (2007a). *Checklist of essential features of age-friendly cities*. World Health Organization. http://www.who.int/ageing/publications/Age_friendly_cities_checklist.pdf. Accessed 18 July 2017.

World Health Organization (2007b). *Global Age-friendly Cities: A Guide*. World Health Organization. http://www.who.int/ageing/publications/Global_age_friendly_cities_Guide_English.pdf. Accessed 19 July 2017.

World Health Organization (2010). *Global recommendations on physical activity for health.* Geneva: World Health Organization.

World Health Organization (2015). *World report on ageing and health.* Luxembourg: World Health Organization.

Wu, T., & Chan, A. (2012). Families, friends, and the neighborhood of older adults: Evidence from public housing in Singapore. *Journal of Aging Research*, 1–7.

Yuen, B. (2007). Squatters no more: Singapore social housing, *Global Urban Development Magazine, 3*(1), http://www.globalurban.org/GUDMag07Vol3Iss1/Yuen.htm. Accessed 26 August 2016.

Yuen, B. & Soh, E. (2017). *Housing for older people in singapore: an annotated bibliography.* Singapore: Springer.

Chapter 1
Surveying Older Adults' Perceptions and Aspirations

Belinda Yuen, Chathura Withanage and Parvathi Nair

Abstract This Chapter presents a personal interview survey. The key instrument is a questionnaire that seeks to find out older adults' (aged 55 and older) experiences and perceptions about growing older in Singapore, their housing and neighbourhood spaces. It explores issues relating to housing satisfaction and mobility aspiration, self-reported housing modification and housing needs, frequency of outdoor activities and satisfaction with neighbourhood public spaces as well as desired facilities and service clusters.

1.1 Introduction

In this Chapter, a survey is introduced and its results discussed. The key instrument is a questionnaire used for collecting a sample population of older adults' (aged 55 and older) experiences and perceptions about the housing and neighbourhood spaces around them. The results gave a glimpse into their satisfaction, concerns and preferences for these spaces as well as their experience of growing older in Singapore.

The survey, conducted in mid-2014 to early 2015, seeks to obtain a national level snapshot of these needs, preferences and experiences. Inspired by the international literature that showed the home and its immediate neighbourhood as key places where older people spend the majority of their time (Help the Aged 2006; Bureau of Labor Statistics 2013), this is the first study to explore older people's views of their housing, neighbourhoods (outdoor public spaces) and growing old in Singapore. Past studies using survey have explored different aspects of the state of older adults in Singapore like their employment, caregiving, health and wellbeing (Ministry of Community

B. Yuen (✉) · C. Withanage · P. Nair
Singapore University of Technology and Design, Singapore, Singapore
e-mail: belinda_yuen@sutd.edu.sg

C. Withanage
e-mail: chathura@sutd.edu.sg

P. Nair
e-mail: parvathi.snjv@gmail.com

B. Yuen (ed.), *Ageing and the Built Environment in Singapore*,
https://doi.org/10.1007/978-3-319-92444-1_1

Development, Youth and Sports 1995, 1999), learning needs (Ministry of Community Development, Youth and Sports 2008), present and future living arrangements in public housing (Housing and Development Board 2014). These surveys have a different focus of study from the survey questionnaire discussed in this Chapter.

1.2 Methodology

Survey research is a common method of collecting information about a population of interest, especially when this population is very large or dispersed across a large geographic area (Andres 2012; Rea and Parker 2014; Moser and Kalton 2016). While surveys offer an excellent way to gather lots of information from a large target population through standardised survey questions, survey administration can be both a time- and manpower-consuming activity. It has the challenge of inflexibility because of the structured process of asking standardised survey questions.[1] Numerous textbooks have been written about the complex but important operation of survey research design and procedures (e.g. Buckingham and Saunders 2004; De Vaus 2014; Blair et al. 2014). This bears no repetition except a reminder to study them closely before embarking on survey research. Survey research has many key features including a predefined questionnaire and sampling.

The survey in the present study was designed as a structured, personal interview with the help of a printed questionnaire. Only Singaporeans or Singapore permanent residents who were 55 years and older were invited to participate in the survey. A total of 3025 older adults (above 55 years old) were interviewed.[2] In view of the feminisation of ageing, a slightly higher proportion of women were interviewed.[3] Instead of a household-based sample, older adults of different age groups and ethnicity were surveyed in a range of public places across Singapore (e.g. markets, hawker centres, parks, public housing void decks,[4] senior activity centres, community centres, transportation and recreational hubs, among others). Since many of those residing in one- and two-room HDB[5] flats were interviewed at senior activity centres, this had resulted in a slightly higher proportion of Chinese respondents.

---

[1]To address this, focus group discussions were conducted to offer respondents an opportunity to further discuss their personal feelings, perceptions and views. See Chap. 2.

[2]Based on the Department of Statistics (2012) there are approximately 865,000 Singaporeans and Singapore permanent residents aged above 55 years. With a 95% confidence level and ±1.8% sampling error, we arrived at a target sample of 3000.

[3]At national population level, female residents outnumber male with a sex ratio of 963 male per 1000 female.

[4]This is a uniquely Singaporean common space in public housing. It refers to the ground floor of public housing blocks that has been purposefully left open (void) since the 1970s as a sheltered space for residents' common use, e.g. as a venue to meet friends or for block parties, funerals or weddings. See National Heritage Board (2013) for further details about its history, importance, common features and programming.

[5]HDB is the abbreviation for Housing and Development Board, the public housing authority in Singapore. Public housing is popularly referred to as HDB housing/block/flats.

The survey was administered in English, Malay, Tamil, Mandarin and key Chinese dialects in light of Singapore's multi-ethnic population.[6] There was no incentive for participation. The questionnaire comprised some 40 questions in four sections covering,

- Respondents' demographics—questions relating to participants' nationality, age, gender, ethnicity, type of housing and living arrangements;
- Understanding respondents' home—questions relating to their home, levels of importance and satisfaction with home spaces, residential mobility and plans for adapting home spaces, preferences for future home and concerns about housing needs as they age;
- Understanding respondents' neighbourhood and outdoor recreation spaces—questions relating to their frequency of outdoor activities, levels of importance, satisfaction and experiences with public spaces in their neighbourhoods, preferences for near-home facilities and services to maintain independence and active lifestyle;
- Understanding respondents' life—questions relating to their perceptions on ageing, their experiences and satisfaction with various aspects of life and growing old in Singapore.

The questions were mainly closed-ended. They included a couple of dichotomous (2-point questions, e.g. yes or no) followed by contingency questions (this routine avoids asking respondents questions that are not applicable to them), many multiple choice, often including a response option on 'Others' to give respondents the opportunity to go beyond what is provided in the question as well as a number of scaled questions using a five-point Likert scale[7] and matrix questions. All participants' responses were voluntary and anonymous. All responses were analysed as a group, not identified by the individual. As it was a pen-and-paper survey, data checking, cleaning and entry took some time to complete. A total of 2942 questionnaires were admitted to data analysis using SPSS Statistics. The four key demographic (explanatory) variables—age, gender, ethnicity and housing type—of respondents were considered.[8] The sample used for analysis was weighted by ethnic ratios to approximate national proportions.

Age-based segmentation was the primary demographic variable in the analysis to identify the different experiences, attitudes and preferences of generational cohorts of older people. The assumption was that people of the same generation, e.g. baby boomers 1946–1964, traditionalists 1925–1945, by and large go through the same societal contexts and circumstances, and norms and behavioural expectations. Several studies on age and values have shown that values tend to vary with age (Rokeach 1968; Lascu et al. 1996; Bengtson et al. 2009).

[6]The ethnic composition of Singapore's resident population includes 74.3% Chinese, 13.4% Malay, 9.1% Indian and 3.2% Others. See Department of Statistics (2016).

[7]Likert scaling is a widely used bi-polar scaling method to measure the intensity of respondents' feelings for a given item. See Likert (1932), Carifio and Perla (2007).

[8]Even though the survey dealt with all four demographic variables, only age will be considered in this Chapter. A more comprehensive analysis that will include the remaining variables will be published separately.

1.3 Findings

The findings are organised under four broad sections, introducing respondents' economic and living arrangements, their perceptions of old age and growing old in Singapore, and their experiences and perceptions of home and neighbourhood public spaces. It attempts to explore how different older age groups in Singapore: the emerging old (age 55–64), young old (age 65–74), middle old (age 75–84) and oldest old (above age 85) perceive ageing and their surrounding built environment. The importance of this discussion lies in its focus on the views of older people themselves.

1.3.1 Economic and Living Arrangements

Reflecting national residential pattern, most respondents (81.9%) lived in public housing. The majority are homeowners (80.6% among public housing respondents and 90.4% among those residing in private housing). There is a seeming decline in home ownership with age while renting and other housing arrangements (e.g. staying on their own in a housing unit owned by children) increase with advancing age (Table 1.1). Renting appears more prevalent among public housing residents (12% as compared to 3.2% renting among private housing respondents). There is an increase in the percentage of respondents selecting 'other' housing tenure with increasing age. The majority of them elaborated that they are staying with their children or in housing units owned by their children (children living elsewhere).

The majority of respondents have lived in their present dwelling for periods of 11 or more years with the oldest old staying the longest—over 20 years in their housing unit (Table 1.1). The results suggest a general preference to age in place. Marginally more among the public housing respondents reported longer length of residence—70.1% have stayed in their public housing units for 11 or more years as compared to 61.6% among private housing respondents.

On employment status of respondents, there is a trend towards retirement and having no earned income with increasing age. The majority of the oldest old are either retired (51.2%) or economically inactive (36.2%). Only 1.3% of oldest old remain in employment while 8.8% said they are unemployed. In contrast, 60.5% of the emerging old respondents are working (in a range of occupations from administrative/managerial, sales/services and professional to clerical, production, cleaners and labourers). About one in five of emerging old is retired while 6.3% are unemployed. About 2% of all respondents take part in voluntary work (1.7% of male and 2.4% of female respondents).

An increasing trend towards widowhood can be discerned with age. About 50% of the oldest old are widowed and about 60% of them are living in one or two-persons households. Many of them are women. The majority (70.3%) of the emerging old respondents are living in households of three or more people. Many are living with spouse and/or children. Living with spouse (59.2%) and/or children (49.9%)

Table 1.1 Housing tenure

		Respondent age								All respondents	
		Emerging old (55–64)		Young old (65–74)		Middle Old (75–84)		Oldest old (85 and older)			
		N	%	N	%	N	%	N	%	N	%
House ownership	Own	1252	91.1	822	79.8	286	65.0	43	60.3	2404	82.4
	Rent	68	4.9	131	12.7	91	20.6	15	20.5	304	10.4
	Others[a]	55	4.0	75	7.3	64	14.5	14	19.3	208	7.1
Length of stay	<1 year	39	2.8	17	1.6	8	1.9	1	1.3	65	2.2
	1–5 years	196	14.1	149	14.3	60	13.5	6	8.8	411	14.0
	6–10 years	213	15.4	144	13.9	76	17.1	12	16.5	445	15.1
	11–20 years	468	33.7	267	25.8	80	18.0	17	22.9	831	28.3
	>20 years	470	33.9	459	44.2	219	49.4	37	50.5	1185	40.3

Note [a]E.g. stay on own in house owned by children (children living elsewhere)

are the most common living arrangements among respondents. A small number of respondents indicated other living arrangements such as living with tenants.

There is, however, an increasing likelihood towards living alone with age—about one in four of those aged 75 and older reported living alone. The trend towards smaller household size with age could increase the number of older people living alone in the future. At the national population level, the Department of Statistics has observed an increase of living alone households from 10.1% to 11.9% over the past decade (2005–2015), especially prevalent among households with heads aged 65 and older (26.6%) (Department of Statistics 2016).

1.3.2 Perceptions of Old Age and Growing Older

The majority of the respondents do not think of themselves as old. When asked about their self-perceived (how old do you feel you are) and desired old ages (at what age would you consider yourself elderly), the majority would prefer to be and feel younger than their actual chronological age (Table 1.2). The means of the responses to these two questions significantly increased between age groups with advancing age (t-test, $p < 0.05$). This suggests that age seems to influence people's perceptions of their own age: the gap between actual and perceived ages is widening with age. Self-perceived age should not be ignored as international evidence suggests that it can relate to future expectations and to individual envisioned trajectories and pathways of ageing (Furstenberg 2002).

Across the age groups, respondents' self-perceived age (how old people feel) is generally lower than their real age (5–17 years). The mean age at which respondents perceive that old age starts is around 73.6 years with a standard deviation of 10.23 (Table 1.2). The definition of when old age begins varies among individual respondents. As a group, the emerging old tend to perceive old age as starting at 71 years while for the oldest old it is 83 years. Men tend to perceive old age as starting slightly later than women of their age: the mean for men is 74 while for women it is 73. The greatest observed gender difference is among the 75–84 where old age starts at the mean age of 80 for men and 77 years for women. Married respondents, especially among the middle old and oldest old, tend to perceive old age as starting later than others of their age.

Upon closer examination, those with higher incomes seem more inclined to report old age as starting later (one to six years). Those with no income and those with less than S$1000 monthly incomes generally tend to perceive old age as starting earlier (one to six years and one to eight years respectively) than wealthier counterparts of their age. The difference tends to increase with age and becomes greatest among the middle old age group. Respondents who are not employed tend to perceive old age as starting earlier than their employed counterparts. An exception is observed among those retired: the emerging old and oldest old retirees are inclined to give a later beginning date for old age than their employed counterparts.

Table 1.2 Perception of age and ageing

	Respondent age									
	Emerging old (55–64)		Young old (65–74)		Middle old (75–84)		Oldest old (85 and older)		All respondents	
	Mean	Standard deviation	Mean	Standard deviation	Mean	Standard deviation	Mean	Standard deviation	Mean	Standard deviation
How old do you feel you are?	50	9	58	10	67	10	76	11	55.85	11.73
At what age do you consider yourself as elderly?	71	9	75	10	78	10	83	13	73.59	10.23

Private housing respondents generally tend to think that old age starts later than their counterparts in public housing across age groups, apart from the oldest old where the trend appears reversed (old age starts at 82 and 83 years for private housing and public housing respondents respectively). On the whole, for the majority of the respondents, especially among those who are older, married, wealthier and working, old age appears to start later than the official older person age definition of 65 years old adopted in Singapore. This finding seems to lend support to international studies on older people feeling younger than their actual age as they grow older (Pew Research Centre 2009). Perceptions of ageing are an important area for further study as they are at the very core of old age definition, identity and behavioural outcomes (Levy 2003; Sarkisian et al. 2005; Sneed and Whitbourne 2005).

We turn next to self-assessed good health. The results indicate a declining trend with age. The majority (84.6%) of the emerging old respondents consider themselves to be of good/very good health as compared with just 50% among the oldest old (Table 1.3). Less than 1% of emerging old respondents perceive themselves as having poor health. It is possible that respondents were inclined not to reveal the actual condition of their health. When self-perceived age and self-perceived health are examined, it appears that those who feel younger have better self-perceived health, suggesting that self-perceived age might be a potential predictor of one's state of health. Further research will be required to test this postulation.

Self-perceived health was found to corroborate with respondents' satisfaction with their current state of health. The majority (above 70%) of respondents express satisfaction with their present state of health. Younger respondents are generally more satisfied with their state of health. Dissatisfactions with self-assessed health tend to increase as respondents get older: 11.3% among the emerging old respondents, 11.6% among the young old, 14.8% among the middle old and 28.4% among the oldest old are dissatisfied with their current state of health. In addition to health, respondents were asked to rate several other aspects of their subjective quality of life including how satisfied they are with their current state of happiness, family's emotional support, circle of friends and social network, level of participation in social activities, involvement with community, familiarity with technology, and life in general.

Respondents across the age groups are generally satisfied with their life (over 85%) and various other dimensions, especially happiness (over 80%), family support (over 80%), circle of friends (over 70%) (Table 1.4). Higher self-assessed happiness is recorded among the middle old (84.8%) and young old (82.9%) than the emerging old (81.3%) and oldest old respondents (77.9%). More among the younger respondents are satisfied with their circle of friends and social network (over 70% among the emerging old and young old as compared to 64.9% among the middle old and 63.9% among the oldest old) and with their family support (over 80% among the emerging old, young old and middle old as compared to over 75% among the oldest old).

More older respondents are satisfied with their level of participation in social activities and involvement with community than their younger counterparts (over 50% for the older age groups as compared to 46–47% among the emerging old). In the area of their familiarity with technology, the trend is towards increased dissatisfaction

Table 1.3 Self-perceived health

		Respondent age								All respondents	
		Emerging old (55–64)		Young old (65–74)		Middle old (75–84)		Oldest old (85 and older)			
		N	%	N	%	N	%	N	%	N	%
How would you rate the state of your current health?	Very good	308	22.2	153	14.7	32	7.2	4	5.1	497	16.9
	Good	865	62.4	653	62.9	276	62.3	33	45.0	1827	62.1
	Not too good	203	14.7	218	20.9	129	29.0	32	43.5	581	19.8
	Poor	10	0.7	15	1.4	6	1.4	5	6.4	36	1.2

Table 1.4 Satisfaction with quality of life

		Respondent age								All respondents	
		Emerging old (55–64)		Young old (65–74)		Middle old (75–84)		Oldest old (85 and older)			
		N	%	N	%	N	%	N	%	N	%
Your state of health	Very dissatisfied	18	1.3	17	1.6	12	2.6	6	8.1	52	1.8
	Dissatisfied	139	10.0	110	10.6	54	12.2	15	20.3	318	10.8
	Neither satisfied nor dissatisfied	240	17.4	164	15.8	70	15.8	12	16.4	486	16.5
	Satisfied	801	57.8	647	62.2	268	60.3	31	42.5	1746	59.4
	Very satisfied	187	13.5	102	9.8	40	9.1	9	12.7	338	11.5
Your state of happiness	Very dissatisfied	6	0.4	5	0.4	3	0.6	1	1.8	14	0.5
	Dissatisfied	45	3.2	37	3.6	9	2.1	6	8.8	97	3.3
	Neither satisfied nor dissatisfied	209	15.1	137	13.2	56	12.5	8	11.4	410	13.9
	Satisfied	874	63.1	689	66.4	316	71.1	46	62.7	1924	65.4
	Very satisfied	252	18.2	171	16.5	61	13.7	11	15.2	495	16.8
Your family's emotional support	Very dissatisfied	11	0.8	6	0.6	5	1.2	3	3.8	26	0.9
	Dissatisfied	37	2.7	38	3.7	24	5.4	4	5.1	103	3.5
	Neither satisfied nor dissatisfied	199	14.4	137	13.2	54	12.1	11	15.2	401	13.6
	Satisfied	860	62.2	664	64.0	295	66.5	43	59.5	1862	63.4
	Very satisfied	275	19.9	193	18.6	66	14.8	12	16.5	545	18.6
Your circle of friends and social network	Very dissatisfied	6	0.5	4	0.4	2	0.4	1	1.3	13	0.4
	Dissatisfied	34	2.5	45	4.4	22	5.0	6	8.8	108	3.7
	Neither satisfied nor dissatisfied	347	25.1	248	23.9	132	29.8	19	26.1	745	25.4
	Satisfied	822	59.4	607	58.4	247	55.7	41	56.3	1716	58.4
	Very satisfied	174	12.6	134	12.9	41	9.2	5	7.6	355	12.1

(continued)

Table 1.4 (continued)

		Respondent age								All respondents	
		Emerging old (55–64)		Young old (65–74)		Middle old (75–84)		Oldest old (85 and older)			
		N	%	N	%	N	%	N	%	N	%
Your level of participation in social activities	Very dissatisfied	17	1.2	19	1.8	5	1.1	0	0.0	40	1.4
	Dissatisfied	175	12.6	157	15.1	88	19.8	10	13.9	430	14.6
	Neither satisfied nor dissatisfied	533	38.5	336	32.4	122	27.5	26	35.5	1017	34.6
	Satisfied	554	40.0	428	41.3	200	45.0	31	43.0	1212	41.3
	Very satisfied	108	7.8	97	9.4	29	6.6	6	7.6	239	8.1
Your involvement with community	Very dissatisfied	20	1.5	13	1.3	9	2.1	1	1.8	44	1.5
	Dissatisfied	183	13.2	129	12.4	67	15.1	12	16.0	391	13.3
	Neither satisfied nor dissatisfied	543	39.2	375	36.1	139	31.3	20	27.8	1077	36.6
	Satisfied	511	36.9	435	41.9	202	45.4	36	49.3	1184	40.3
	Very satisfied	127	9.2	86	8.3	27	6.0	4	5.1	243	8.3
Your familiarity with technology	Very dissatisfied	38	2.7	71	6.9	60	13.6	17	24.0	187	6.3
	Dissatisfied	174	12.6	216	20.8	112	25.2	22	29.9	524	17.8
	Neither satisfied nor dissatisfied	373	26.9	307	29.6	133	30.1	22	30.4	835	28.4
	Satisfied	695	50.1	399	38.4	125	28.2	9	11.9	1227	41.7
	Very satisfied	106	7.6	46	4.4	13	3.0	3	3.8	167	5.7
Your life	Very dissatisfied	4	0.3	0	0.0	2	0.4	1	1.3	7	0.2
	Dissatisfied	31	2.2	19	1.9	12	2.8	3	3.8	65	2.2
	Neither satisfied nor dissatisfied	151	10.9	111	10.8	46	10.4	9	12.6	317	10.8
	Satisfied	911	65.9	727	70.2	319	71.9	48	65.9	2005	68.4
	Very satisfied	286	20.7	177	17.1	64	14.5	12	16.5	539	18.4

with age. Only 15.7% of the oldest old respondents are satisfied with their current familiarity with technology as compared with 57.7% among the emerging old, 42.8% among the young old, 31.2% among the middle old.

Respondents' experiences of ageing are further measured with a twelve-item scale where they are prompted to think of old age and their own ageing experience and state the extent to which they agree or disagree with each of the twelve statements (Table 1.5). The response scale is a five-point Likert scale, ranging from strongly agree to strongly disagree. Most respondents offered an opinion, either agreeing or disagreeing with these statements. Less than a third of respondents gave no view—neither agreeing nor disagreeing. The two statements with the highest ambivalent values are: "When I think of old people, I think of them as generally unhappy" (28.3% of respondents returned neither agreeing nor disagreeing) and "Old people do not get respect in Singapore society" (27.4% of respondents returned neither agreeing nor disagreeing). More among the younger respondents appear undecided.

Most respondents, especially the younger ones, do not think of themselves as old: 68.9% among emerging old, 63.8% among young old, 50.8% among middle old and 50.1% among oldest old. A similar trend is observed for the statement relating to: "As I grow older, I feel that I am not important to my friends". The majority—74.0% among the emerging old, 73.8% among young old, 70.4% among middle old and 58.8% among oldest old—do not feel that they will become not important to their friends with age. Overall, ageing is viewed to be a positive experience for the majority of the respondents.

Almost half of the respondents (especially among the young old and middle old) disagree that old people are not respected in Singapore society. The majority (66.2%) of the respondents state that they could still contribute meaningfully to society, as they grow older. This perception is strongest among younger respondents: 72.3% among emerging old, 67.0% among young old, 49.8% among middle old and 36.7% among oldest old.

1.3.3 Housing Satisfaction and Aspirations

Regarding how suitable their house is to their personal needs, respondents were asked to rate their satisfaction with various aspects of the dwelling (e.g. the number and size of rooms, its layout, location, etc.) on a five-point Likert scale, ranging from least satisfied to most satisfied (Table 1.6). Admittedly, housing satisfaction and housing needs are likely to change over the course of the individual's life cycle. Notwithstanding, the results indicate that the respondents are generally satisfied with their present dwelling. This finding is consistent with other surveys on residential satisfaction (see e.g. Housing and Development Board 2014).

The three housing aspects with the most respondents expressing satisfaction are: the sense that this is my home (87.9%), the location of my house (81.5%) and the number of rooms in my house (80.5%). These are also the aspects with the least

Table 1.5 Experience of growing older

		Respondent age								All respondents	
		Emerging old (55–64)		Young old (65–74)		Middle old (75-84)		Oldest old (85 and older)			
		N	%	N	%	N	%	N	%	N	%
As I grow older, I feel that I am not important to my family	Strongly disagree	270	19.5	165	15.9	60	13.6	9	12.7	504	17.1
	Disagree	767	55.4	636	61.2	268	60.4	39	53.7	1710	58.2
	Neither agree nor disagree	216	15.6	126	12.2	52	11.7	12	17.2	407	13.8
	Agree	120	8.7	104	10.0	58	13.1	12	16.4	294	10.0
	Strongly agree	12	0.9	8	0.8	5	1.2	0	0.0	26	0.9
As I grow older, I feel that I am not important to my friends	Strongly disagree	183	13.2	115	11.1	44	9.9	6	8.9	348	11.9
	Disagree	841	60.8	651	62.7	269	60.5	36	49.9	1797	61.1
	Neither agree nor disagree	264	19.1	171	16.5	80	18.1	14	19.7	530	18.0
	Agree	87	6.3	97	9.3	47	10.7	14	19.0	245	8.3
	Strongly agree	9	0.6	5	0.4	4	0.8	2	2.5	19	0.6
As I grow older, I feel less stressed and worry less	Strongly disagree	53	3.8	22	2.1	8	1.8	3	4.4	86	2.9
	Disagree	329	23.8	198	19.0	90	20.2	17	24.0	634	21.6
	Neither agree nor disagree	316	22.9	194	18.7	74	16.6	5	6.3	589	20.1
	Agree	619	44.8	551	53.0	247	55.6	40	55.1	1457	49.6
	Strongly agree	65	4.7	74	7.1	26	5.9	7	10.1	172	5.9
Old age is a very depressing stage of life	Strongly disagree	110	8.0	67	6.5	25	5.7	4	5.1	206	7.0

(continued)

Table 1.5 (continued)

		Respondent age								All respondents	
		Emerging old (55–64)		Young old (65–74)		Middle old (75-84)		Oldest old (85 and older)			
		N	%	N	%	N	%	N	%	N	%
	Disagree	581	42.0	487	47.2	191	43.1	24	32.9	1283	43.8
	Neither agree nor disagree	351	25.4	203	19.6	93	20.9	17	23.5	663	22.6
	Agree	322	23.3	260	25.2	130	29.4	27	37.2	739	25.2
	Strongly agree	19	1.4	16	1.6	4	0.8	1	1.3	39	1.3
As I grow older, I expect to become more lonely	Strongly disagree	100	7.2	83	8.0	18	4.1	5	6.3	206	7.0
	Disagree	621	44.9	505	48.7	198	44.7	30	41.7	1354	46.1
	Neither agree nor disagree	281	20.3	145	14.0	61	13.7	14	19.0	501	17.0
	Agree	357	25.8	288	27.8	155	34.8	20	27.8	819	27.9
	Strongly agree	26	1.9	16	1.5	12	2.6	4	5.1	57	1.9
As I grow older, I expect to be able to do the things I've always done	Strongly disagree	23	1.7	16	1.5	6	1.4	3	3.8	48	1.6
	Disagree	242	17.5	226	21.8	126	28.3	25	34.3	618	21.1
	Neither agree nor disagree	298	21.5	168	16.1	85	19.1	14	18.9	563	19.2
	Agree	718	52.0	545	52.5	207	46.6	28	37.9	1497	51.0
	Strongly agree	101	7.3	84	8.1	21	4.6	4	5.1	209	7.1
As I think of old people, I think of them as generally unhappy	Strongly disagree	55	4.0	34	3.3	7	1.7	2	2.6	99	3.4
	Disagree	566	40.9	446	42.9	192	43.3	33	45.6	1236	42.1

(continued)

Table 1.5 (continued)

		Respondent age								All respondents	
		Emerging old (55–64)		Young old (65–74)		Middle old (75-84)		Oldest old (85 and older)			
		N	%	N	%	N	%	N	%	N	%
	Neither agree nor disagree	421	30.5	282	27.1	111	25.0	17	24.4	831	28.3
	Agree	327	23.6	265	25.5	130	29.4	18	24.9	740	25.2
	Strongly agree	13	1.0	12	1.1	3	0.6	2	2.6	30	1.0
As I grow older, I worry that my health will get worse	Strongly disagree	44	3.2	35	3.4	4	0.8	3	3.8	86	2.9
	Disagree	237	17.1	205	19.8	89	20.1	18	24.6	549	18.7
	Neither agree nor disagree	230	16.6	120	11.6	42	9.5	7	10.1	399	13.6
	Agree	774	55.9	609	58.7	278	62.8	40	55.1	1700	57.9
	Strongly agree	99	7.1	69	6.6	30	6.8	5	6.4	202	6.9
I don't think of myself as old	Strongly disagree	18	1.3	15	1.5	16	3.7	3	3.8	53	1.8
	Disagree	159	11.5	169	16.3	106	24.0	24	32.9	457	15.6
	Neither agree nor disagree	252	18.3	190	18.4	95	21.5	10	13.2	547	18.7
	Agree	812	58.9	556	53.6	207	46.9	33	45.0	1608	54.9
	Strongly agree	138	10.0	106	10.2	17	3.9	4	5.1	265	9.0
Old people do not get respect in Singapore's society	Strongly disagree	48	3.4	40	3.9	12	2.7	5	6.4	105	3.6
	Disagree	565	40.8	434	41.9	201	45.5	26	36.5	1226	41.8
	Neither agree nor disagree	402	29.0	280	27.1	103	23.2	19	26.4	803	27.4

(continued)

Table 1.5 (continued)

		Respondent age								All respondents	
		Emerging old (55–64)		Young old (65–74)		Middle old (75-84)		Oldest old (85 and older)			
		N	%	N	%	N	%	N	%	N	%
	Agree	324	23.5	254	24.6	122	27.5	22	30.7	722	24.6
	Strongly agree	45	3.2	26	2.5	5	1.0	0	0.0	76	2.6
Retirement is a time of leisure	Strongly disagree	20	1.5	12	1.1	1	0.2	0	0.0	33	1.1
	Disagree	232	16.8	156	15.1	43	9.6	6	8.9	437	14.9
	Neither agree nor disagree	285	20.6	192	18.5	101	22.8	15	21.0	593	20.2
	Agree	723	52.4	595	57.4	275	61.8	48	66.3	1641	55.9
	Strongly agree	121	8.8	82	7.9	25	5.6	3	3.8	230	7.8
As I grow older, I can still contribute meaningfully to society	Strongly disagree	17	1.2	9	0.9	12	2.7	3	3.8	41	1.4
	Disagree	101	7.3	123	11.8	86	19.3	23	31.7	333	11.3
	Neither agree nor disagree	265	19.1	211	20.3	125	28.2	20	27.9	621	21.1
	Agree	826	59.6	578	55.6	193	43.4	24	32.9	1620	55.1
	Strongly agree	176	12.7	119	11.4	28	6.4	3	3.8	325	11.1

Table 1.6 Housing satisfaction

		Respondent age								All respondents	
		Emerging old (55–64)		Young old (65–74)		Middle old (75–84)		Oldest old (85 and older)			
		N	%	N	%	N	%	N	%	N	%
Number of rooms in my house	Least satisfied	8	0.6	4	0.4	4	1.0	0	0.0	16	0.5
	Less satisfied	58	4.2	23	2.2	17	3.8	4	5.1	101	3.5
	Neither satisfied nor dissatisfied	231	16.8	152	14.7	64	14.4	7	10.1	455	15.5
	Satisfied	721	52.4	595	57.3	265	59.8	45	61.5	1626	55.5
	Most satisfied	358	26.0	264	25.4	93	21.0	17	23.3	732	25.0
Size of rooms in my house	Least satisfied	13	1.0	4	0.4	4	1.0	0	0.0	21	0.7
	Less satisfied	67	4.9	39	3.7	15	3.5	3	3.8	124	4.2
	Neither satisfied nor dissatisfied	326	23.8	195	18.8	110	24.9	11	15.2	642	22.0
	Satisfied	656	47.8	555	53.6	227	51.4	43	59.0	1481	50.7
	Most satisfied	309	22.5	244	23.5	85	19.3	16	22.0	654	22.4
Layout of the house	Least satisfied	6	0.4	5	0.4	5	1.2	0	0.0	15	0.5
	Less satisfied	67	4.9	33	3.2	18	4.1	3	3.8	121	4.2
	Neither satisfied nor dissatisfied	331	24.1	243	23.5	109	24.6	15	21.0	698	23.9
	Satisfied	680	49.5	524	50.6	216	49.0	42	57.4	1462	50.0
	Most satisfied	290	21.1	232	22.3	93	21.2	13	17.7	628	21.5
Provision of elder friendly features in my house	Least satisfied	63	4.6	33	3.2	17	3.7	2	2.6	114	3.9
	Less satisfied	222	16.2	117	11.3	50	11.2	6	7.7	394	13.5
	Neither satisfied nor dissatisfied	556	40.4	340	32.8	128	28.8	17	23.9	1040	35.6
	Satisfied	414	30.1	411	39.7	190	42.9	35	49.2	1050	35.9
	Most satisfied	121	8.8	134	12.9	59	13.2	12	16.7	325	11.1

(continued)

Table 1.6 (continued)

		Respondent age								All respondents	
		Emerging old (55–64)		Young old (65–74)		Middle old (75–84)		Oldest old (85 and older)			
		N	%	N	%	N	%	N	%	N	%
Degree of flexibility of my house for ageing needs	Least satisfied	42	3.0	13	1.2	7	1.7	2	2.5	64	2.2
	Less satisfied	153	11.1	75	7.3	30	6.8	4	5.1	262	8.9
	Neither satisfied nor dissatisfied	523	38.0	377	36.4	166	37.5	26	36.2	1092	37.3
	Satisfied	536	39.0	440	42.6	188	42.4	32	43.5	1196	40.9
	Most satisfied	122	8.9	129	12.5	52	11.7	9	12.7	312	10.7
Degree of adaptability of my house for ageing needs	Least satisfied	29	2.1	13	1.3	6	1.5	2	2.5	51	1.7
	Less satisfied	146	10.6	80	7.7	31	7.1	3	3.8	260	8.9
	Neither satisfied nor dissatisfied	593	43.2	424	41.0	197	44.4	30	41.3	1244	42.5
	Satisfied	482	35.1	424	41.0	169	38.1	31	42.3	1106	37.8
	Most satisfied	123	8.9	93	9.0	40	9.0	7	10.1	263	9.0
Ease of maintenance within the home	Least satisfied	13	0.9	7	0.7	3	0.8	1	1.3	24	0.8
	Less satisfied	101	7.3	56	5.5	18	4.0	3	3.8	178	6.1
	Neither satisfied nor dissatisfied	437	31.8	284	27.5	134	30.2	19	26.9	873	29.9
	Satisfied	696	50.6	537	52.1	243	54.9	39	53.9	1515	51.9
	Most satisfied	128	9.3	147	14.3	45	10.2	10	14.1	331	11.3
Location of my house	Least satisfied	8	0.6	7	0.7	3	0.8	0	0.0	19	0.7
	Less satisfied	51	3.8	37	3.6	22	4.9	4	5.1	113	3.9
	Neither satisfied nor dissatisfied	228	16.8	110	10.8	53	12.1	10	13.9	402	13.9
	Satisfied	675	49.7	503	49.2	228	51.8	39	53.8	1445	49.9
	Most satisfied	396	29.1	365	35.7	134	30.4	20	27.2	914	31.6

(continued)

Table 1.6 (continued)

		Respondent age								All respondents	
		Emerging old (55–64)		Young old (65–74)		Middle old (75–84)		Oldest old (85 and older)			
		N	%	N	%	N	%	N	%	N	%
Ease of access to key services	Least satisfied	13	0.9	12	1.2	2	0.5	0	0.0	27	0.9
	Less satisfied	93	6.8	45	4.4	27	6.2	8	11.4	174	5.9
	Neither satisfied nor dissatisfied	264	19.2	144	14.0	62	14.1	5	6.9	475	16.3
	Satisfied	673	49.0	517	50.0	244	55.0	41	56.4	1475	50.5
	Most satisfied	331	24.1	315	30.5	107	24.2	18	25.3	771	26.4
Having close ties with community	Least satisfied	22	1.6	10	1.0	3	0.6	0	0.0	35	1.2
	Less satisfied	162	11.8	93	9.0	40	9.1	4	5.1	299	10.2
	Neither satisfied nor dissatisfied	505	36.8	308	29.8	131	29.7	19	26.2	962	33.0
	Satisfied	522	38.0	453	43.9	201	45.6	35	48.2	1211	41.5
	Most satisfied	163	11.9	168	16.3	66	15.0	15	20.5	412	14.1
Interaction with my neighbours	Least satisfied	20	1.5	8	0.8	1	0.2	1	1.3	30	1.0
	Less satisfied	98	7.2	54	5.2	30	6.8	5	6.4	187	6.4
	Neither satisfied nor dissatisfied	443	32.3	242	23.5	96	21.7	16	23.0	797	27.3
	Satisfied	624	45.4	509	49.3	240	54.2	31	42.9	1403	48.1
	Most satisfied	187	13.6	218	21.2	75	17.0	19	26.4	500	17.1
Proximity to family and friends	Least satisfied	15	1.1	13	1.2	8	1.9	0	0.0	36	1.2
	Less satisfied	129	9.4	79	7.7	48	10.8	7	10.1	263	9.0
	Neither satisfied nor dissatisfied	383	27.8	293	28.4	127	28.7	20	27.4	822	28.1
	Satisfied	638	46.3	463	44.8	187	42.4	32	43.5	1320	45.1
	Most satisfied	212	15.4	185	17.9	71	16.2	14	19.0	483	16.5

(continued)

Table 1.6 (continued)

		Respondent age								All respondents	
		Emerging old (55–64)		Young old (65–74)		Middle old (75–84)		Oldest old (85 and older)			
		N	%	N	%	N	%	N	%	N	%
The sense that this is my home	Least satisfied	6	0.4	2	0.2	3	0.8	0	0.0	11	0.4
	Less satisfied	36	2.6	20	1.9	8	1.9	2	2.6	66	2.3
	Neither satisfied nor dissatisfied	118	8.6	96	9.2	56	12.6	5	7.7	275	9.4
	Satisfied	602	43.8	447	43.1	211	47.5	38	52.6	1297	44.3
	Most satisfied	612	44.5	472	45.6	165	37.2	27	37.2	1276	43.6

number of respondents registering neither satisfied nor dissatisfied. The items with the most ambivalent responses (neither satisfied nor dissatisfied) are: the degree of adaptability of my house for my ageing needs (42.5%), the provision of elder-friendly features in my house (35.6%), and having close ties with the community (33.0%).

The three housing aspects with the least number of respondents reporting satisfaction are: the degree of flexibility of my house for my ageing needs (51.6%), the provision of elder-friendly features in my house (47.0%), and the degree of adaptability of my house for my ageing needs (46.8%). The levels of satisfaction received for these three housing dimensions tend to increase with age. Over 65% of oldest old respondents state satisfaction with the provision of elder-friendly features in my house as compared to 38.9% among the emerging old. A number of emerging old (nearly one in five) are dissatisfied with this provision.

Aside from housing satisfaction, the importance of these housing aspects was rated on a five-point Likert scale. Five aspects were rated important/very important by over 80% of respondents in their housing consideration for personal needs: the sense that this is my home (93.1%), ease of access to key services (92.4%), the location of my house (92.2%), provision of elder-friendly features in my house (81.7%) and ease of maintenance within the home (79.9%). These results suggest the importance of pride of home ownership, issues of accessibility and convenience (and ultimately, the value of property) and in-dwelling comfort among respondents.

The oldest respondents (85+) more than any other age groups highlighted the importance of elder-friendly features and ease of access to key services. Younger respondents appear more concerned with location, ease of maintenance and sense that this is my home. The sense that this is my home also appears to be important among many of the oldest old. Importance on the layout, number and size of rooms in my house tends to decline with age while having close ties with the community, interaction with my neighbours, and proximity to family and friends seem to become more important as one grows older.

When asked about their residential mobility aspirations, most (86.3%) of the respondents said that they were very unlikely to move from their present house within next three to five years, indicating a desire to age in place. The percentage of respondents unlikely to move is significantly associated with age (chi-square, $p<0.001$), peaking with the oldest-old (92.4%). Among those indicating that they were very likely to move, younger age respondents are more likely to move.

To a further question about whether they would continue living in their current house as they age, almost three-quarters (73.8%) of respondents indicated that they would continue living there. This inclination appears to increase with age (chi-square, $p<0.001$), and is at its strongest among the oldest old (83.6%) as compared to 68.6% among the emerging old. About 18% of respondents, especially among younger respondents are not sure if they would continue living in their current house as they age. Interestingly, with age, the percentage of respondents who indicated "no" and "not sure" appears to decline.

When asked what is the one most important reason for wanting to move, the most frequently mentioned reasons are: house is too small (16.5%), house is too big (13.6%), financial considerations (11.0%) and to be nearer to family and friends

(9.6%). Apart from the oldest old, the other age groups had more reasons for moving with the widest range coming from the young old and emerging old. The key reasons for moving among both these age cohorts revolved around the appropriate size of house (too small or too big), financial considerations, lifestyle change and empty nest. The diversity of reasons indicates the heterogeneous nature of lifestyles and needs, making it impossible to have a single housing solution applicable to all.

Regarding respondents' reasons for not moving, the two reasons that garnered the most responses, especially among younger respondents, are: I own this house (27.3%) and good location (20.8%). House ownership and good location would appear to be important pull factors influencing people's decision to stay on. But, both reasons seem to lose their appeal with age while the reasons relating to good neighbours, like the neighbourhood, too hard to move, feel safe and secure, easy access to services appear to become more often mentioned as one ages. Design and amenities did not feature prominently among respondents' reasons for not moving; they received the lowest votes—house design/layout supports my ageing (2.1%) and easy access to services (3.5%).

On respondents' self-report about the building changes that have been carried out and will be made to their house as they age, the majority, especially among the younger respondents, said they have made no changes to their housing unit (61.6%) and did not intend to make changes in the future (47.4%). These intents, however, seem to weaken with age. Where changes have been made, these are primarily to the bathroom (29.3%) and kitchen (16.8%). About one in five respondents has installed hand rails/grab bars while 11.5% have carried out stairs/ramps/floor modifications. These are also the modifications commonly carried out by the agency in public housing home improvement programme for ageing in place (Yuen and Soh 2017).

Where future changes are planned, most future housing modifications are anticipated to take place in the bathroom (28.8%), especially among the emerging old. The leading future works are to install hand rails/grab bars (34.4%) and to make stairs/ramps/floor modifications (24.9%). About their housing needs in the future, respondents, especially older respondents, are most concerned about access to health care (35.9%) and safety and security (27.9%). Over 40% of oldest old are concerned about these issues while 26.4% and 33.4% of emerging old are concerned about safety and security and access to healthcare respectively.

Transport and related infrastructure (13.0%) and home modification costs (9.4%), especially among younger respondents, are the other frequently mentioned concerns about future housing needs. A small proportion were concerned about other needs such as affordability and maintenance of the household to suit their changing needs. Among the gender, more male respondents express concern with access to healthcare, home modification costs, and adaptable housing design while more women seem concerned over transport and related infrastructure, and safety and security.

When asked what is the most appropriate form of housing for older persons, the preference is for the older person to continue to live in own house (67.0%). This preference seems to increase with age: 62.5% among the emerging old, 68.9% among young old, 74.6% among middle old and 81.0% among oldest old respondents. About one in five respondents (17.2%) considers retirement village as a possible housing option while 7.2% mentions elderly studio apartment.[9] Both of these preferences appear to decline with age. The majority of respondents who indicated other options preferred to live with their children, as they grow older.

When asked to select their preferred housing option if they or their spouse became dependent on others and needed assistance, the majority would opt to continue to live in own house, either with outside help (34.8%) or hire a live-in domestic helper (17.3%). Less people are interested in moving to age-specific boarding houses (2.0% of respondents) though the idea appears to have more currency among the young old (2.8%) and oldest old (2.5%) than the emerging old (1.8%) and middle old (0.8%).

1.3.4 Public Spaces and Outdoor Activities

Regarding respondents' frequency of participation in a range of outdoor activities, it is possible that there may be under/over-reporting as the question is based on recall. Also, respondents may not fully wish to reveal their actual levels of participation in any of the activities. Notwithstanding, the results offer a glimpse into some of the outside home and social activities of the respondents. As indicated in Table 1.7, the three most frequently participated activities are: shopping (56.5%), sports (55.8%) and dining out (45.6%).

Perhaps to be expected, these activities would be at their lowest frequency among the oldest old respondents. Shopping seems to be a favourite activity, especially among the middle old (75–84) before declining among the oldest old: 58.4% for old reported participating frequently in shopping as compared to 55.6% among the emerging old, 57.8% for young old and 41.3% for oldest old. The numbers who do not participate in this activity at all tend to increase with age: 2.9% among the emerging old, 5.5% for young old, 9.7% for middle old and 24% for oldest old respondents.

Another is sports/exercise. The number who do not participate at all in sports/exercise would increase with age (17.5% among the emerging old compared with 44.4% for oldest old) even though many reported frequent participation in these activities: 49.5% among the emerging old, 60.1% for young old, 67.0% for middle old and 46.7% for oldest old respondents. Although many among the oldest old (46.7%) would frequently participate in sports and exercise, many more of them equally do not participate in this activity at all (44.4%) or participate infrequently (5.1%) and neither frequently or infrequently (3.8%).

At the national level, the National Sports Participation Survey 2011 has reported a 37–41% sports participation rate among older people (aged 55 and older) as compared to about 50–84% sports participation rate among the under 29 (Singapore Sports Council 2011). The national ambition is to encourage older people to lead healthy,

[9]Even though retirement village is a common housing typology for older people in Western developed countries (e.g. UK, USA, Australia), there is no development in Singapore at time of writing. The elderly studio apartment is one housing option that has been developed by Singapore's public housing agency in recent decades. See Yuen and Soh (2017).

Table 1.7 Frequency of outdoor activity

		Respondent age								All respondents	
		Emerging old (55–64)		Young old (65–74)		Middle old (75–84)		Oldest old (85 and older)			
		N	%	N	%	N	%	Count	%	N	%
Chatting with my friends/neighbours at void deck	Not at all	385	27.8	267	25.7	121	27.3	20	27.9	793	27.0
	Very infrequent	272	19.7	160	15.5	41	9.2	5	7.6	479	16.3
	Infrequent	157	11.3	93	9.0	29	6.7	4	5.1	284	9.6
	Neither infrequent nor frequent	187	13.5	91	8.7	27	6.2	6	8.1	311	10.6
	Frequent	274	19.8	220	21.2	96	21.6	12	17.2	602	20.5
	Very frequent	110	7.9	206	19.9	128	29.0	25	34.1	469	16.0
Sports/exercise	Not at all	242	17.5	193	18.6	77	17.4	32	44.4	545	18.5
	Very infrequent	159	11.4	84	8.1	23	5.2	4	5.1	270	9.2
	Infrequent	134	9.7	58	5.6	28	6.2	0	0.0	220	7.5
	Neither infrequent nor frequent	165	11.9	77	7.5	18	4.1	3	3.8	264	9.0
	Frequent	426	30.7	283	27.2	103	23.1	5	6.3	816	27.7
	Very frequent	260	18.8	342	32.9	195	43.9	29	40.4	826	28.1
Shopping/retail	Not at all	40	2.9	57	5.5	43	9.7	17	24.0	158	5.4
	Very infrequent	119	8.6	95	9.2	49	11.1	9	12.7	273	9.3
	Infrequent	220	15.8	134	12.9	42	9.5	6	8.8	402	13.7
	Neither infrequent nor frequent	235	17.0	153	14.7	49	11.2	10	13.2	448	15.2
	Frequent	596	43.0	468	45.1	198	44.8	24	33.7	1287	43.8
	Very frequent	175	12.6	132	12.7	60	13.6	5	7.6	373	12.7
Banking	Not at all	240	17.4	314	30.4	203	45.6	47	65.2	804	27.4
	Very infrequent	384	27.8	319	30.9	126	28.3	14	19.6	843	28.7

(continued)

Table 1.7 (continued)

		Respondent age								All respondents	
		Emerging old (55–64)		Young old (65–74)		Middle old (75–84)		Oldest old (85 and older)			
		N	%	N	%	N	%	Count	%	N	%
	Infrequent	317	22.9	186	18.0	61	13.8	6	8.9	571	19.5
	Neither infrequent nor frequent	253	18.3	126	12.2	34	7.7	5	6.3	419	14.3
	Frequent	173	12.5	75	7.3	17	3.9	0	0.0	266	9.1
	Very frequent	15	1.1	13	1.3	3	0.7	0	0.0	32	1.1
Meeting friends at hawker centre/coffee shops	Not at all	210	15.2	176	16.9	102	23.1	33	45.6	521	17.7
	Very infrequent	246	17.8	164	15.7	73	16.4	8	11.4	490	16.7
	Infrequent	208	15.1	145	13.9	49	11.0	5	6.3	406	13.8
	Neither infrequent nor frequent	220	15.9	132	12.7	34	7.6	4	5.1	389	13.2
	Frequent	375	27.1	275	26.5	116	26.3	14	19.0	780	26.5
	Very frequent	124	9.0	149	14.3	69	15.6	9	12.7	351	12.0
Religious activities	Not at all	317	22.9	222	21.4	104	23.5	22	29.9	665	22.6
	Very infrequent	175	12.6	179	17.3	69	15.4	16	21.5	438	14.9
	Infrequent	179	12.9	124	11.9	66	15.0	4	5.1	373	12.7
	Neither infrequent nor frequent	138	10.0	93	8.9	28	6.2	4	5.6	263	8.9
	Frequent	505	36.5	355	34.3	148	33.2	26	35.5	1034	35.2
	Very frequent	69	5.0	64	6.2	30	6.7	2	2.5	165	5.6
Volunteering	Not at all	681	49.2	560	54.0	285	64.3	63	86.5	1589	54.1
	Very infrequent	195	14.1	114	11.0	46	10.4	1	1.3	356	12.1
	Infrequent	111	8.0	72	7.0	29	6.5	1	1.3	213	7.3

(continued)

Table 1.7 (continued)

		Respondent age								All respondents	
		Emerging old (55–64)		Young old (65–74)		Middle old (75–84)		Oldest old (85 and older)			
		N	%	N	%	N	%	Count	%	N	%
	Neither infrequent nor frequent	149	10.7	78	7.6	7	1.6	0	0.0	234	8.0
	Frequent	197	14.2	150	14.5	44	9.9	7	9.7	398	13.5
	Very frequent	53	3.8	62	6.0	32	7.3	1	1.3	148	5.0
Visiting family/friends	Not at all	109	7.9	134	12.9	91	20.4	29	40.4	363	12.4
	Very infrequent	265	19.2	254	24.4	135	30.4	14	19.0	668	22.7
	Infrequent	322	23.3	224	21.5	77	17.3	9	12.2	631	21.5
	Neither infrequent nor frequent	271	19.6	154	14.9	53	11.9	8	11.4	486	16.5
	Frequent	373	27.0	220	21.2	72	16.3	11	14.5	677	23.0
	Very frequent	43	3.1	52	5.0	16	3.6	2	2.5	113	3.9
Community centre/social club activities	Not at all	612	44.2	391	37.7	191	43.3	32	44.1	1226	41.7
	Very infrequent	219	15.8	128	12.3	42	9.5	6	9.0	396	13.5
	Infrequent	119	8.6	102	9.8	31	7.0	4	5.2	255	8.7
	Neither infrequent nor frequent	142	10.3	79	7.6	25	5.7	3	3.8	249	8.5
	Frequent	215	15.5	208	20.1	60	13.6	12	16.2	495	16.9
	Very frequent	77	5.5	130	12.5	93	21.0	16	21.7	315	10.7
Taking educational courses/classes	Not at all	847	61.2	732	70.6	378	85.1	70	96.2	2027	69.0
	Very infrequent	230	16.6	115	11.0	17	3.8	1	1.3	362	12.3
	Infrequent	97	7.0	48	4.6	10	2.4	1	1.3	156	5.3
	Neither infrequent nor frequent	123	8.9	68	6.5	8	1.9	0	0.0	198	6.8

(continued)

Table 1.7 (continued)

		Respondent age								All respondents	
		Emerging old (55–64)		Young old (65–74)		Middle old (75–84)		Oldest old (85 and older)			
		N	%	N	%	N	%	Count	%	N	%
	Frequent	75	5.4	62	5.9	24	5.5	1	1.3	161	5.5
	Very frequent	12	0.9	13	1.3	6	1.5	0	0.0	32	1.1
Dining out	Not at all	59	4.3	106	10.2	72	16.3	17	23.3	254	8.6
	Very infrequent	181	13.0	177	17.0	100	22.4	15	20.3	472	16.0
	Infrequent	216	15.6	160	15.4	81	18.3	16	22.3	474	16.1
	Neither infrequent nor frequent	194	14.0	147	14.1	53	12.0	7	10.1	401	13.7
	Frequent	554	40.0	317	30.5	98	22.1	15	20.3	983	33.5
	Very frequent	180	13.0	133	12.8	40	8.9	3	3.8	355	12.1
Theatre/cultural activities	Not at all	670	48.3	553	53.3	287	64.6	59	81.5	1569	53.4
	Very infrequent	346	25.0	260	25.0	87	19.6	7	9.6	700	23.8
	Infrequent	128	9.2	94	9.0	26	5.8	4	5.1	251	8.5
	Neither infrequent nor frequent	156	11.2	67	6.4	22	5.0	2	2.5	247	8.4
	Frequent	75	5.4	53	5.1	13	3.0	1	1.3	143	4.8
	Very frequent	11	0.8	12	1.1	9	2.0	0	0.0	32	1.1
Medical/health appointment	Not at all	199	14.3	83	8.0	23	5.2	6	8.8	311	10.6
	Very infrequent	636	45.9	496	47.9	218	49.1	27	36.7	1376	46.9
	Infrequent	191	13.8	191	18.4	104	23.4	17	23.6	503	17.1
	Neither infrequent nor frequent	279	20.1	214	20.6	75	16.9	15	20.3	582	19.8
	Frequent	66	4.8	41	3.9	21	4.7	5	7.6	133	4.5
	Very frequent	15	1.1	11	1.1	3	0.7	2	3.1	32	1.1

active lives under Singapore's Action Plan for Successful Ageing, and overcome the prevalent misconceptions and thoughts about being too old for sports or had to be good at a particular sports in order to participate in it (Ministry of Health 2016; Singapore Sports Council 2011).

Notwithstanding and which may well perhaps be expected, there is a trend for participation frequency in several of the activities to decline with age. For example, for dining out: 53.0% among the emerging old respondents has frequently participated in this activity contrasted with 43.3% for young old, 31.0% for old and 24.1% for oldest old. A similar trend can be discerned in the activities of banking, religious activities, visiting family/friends, volunteering and theatre/cultural activities.

Among the many diverse outside home activities, meeting friends at hawker centre/coffee shops[10] seems to be a favourite activity of the young old and middle old respondents before declining among the oldest old: 36.1% among the emerging old frequently participates in this activity contrasted with 40.8% for young old, 41.9% for middle old and 31.7% for oldest old. Yet, there are some activities where participation seems to increase with age. An obvious example is medical/health appointment: 5.9% among the emerging old frequently participates in this activity contrasted with 5.0% for young old, 5.4% for middle old and 10.7% for oldest old. Others include: community/social club activities and chatting with friends/neighbours at void deck, suggesting the importance of near-home spaces and social activities for the older person.

A third or more of the respondents would frequently participate in social activities. Prime among these include meeting friends at hawker centre/coffee shop (38.5%), religious activities (40.8%), visiting family/friends (26.9%), chatting with my friends/neighbours at the void deck (36.5%). As alluded to earlier, the latter seems to be a favourite pastime of older respondents: 51.3% of oldest old as compared to 27.2% among emerging old would frequently participate in this activity. Perhaps because of the shrinking circle of friends and resources with age, 45.6% of oldest old do not meet friends at hawker centre/coffee shop at all while 17.7% do this infrequently and 5.1% neither frequently nor infrequently.

Compared to the other age groups, the oldest old respondents register the highest proportions of 'not at all' participation in many of the outdoor activities. Overall, comparatively fewer people participate on a frequent basis in theatre/cultural activities (5.9%), medical/health appointments (5.2%), taking educational courses (6.2%) and banking (10.4%). About 10–15% of respondents reported not having any medical/health appointments (higher among the younger cohorts) while two-thirds had infrequent medical/health appointment and 20.1% neither frequently nor

[10]This is another unique local amenity. Located all over Singapore, hawker centre is established in every public housing and near large private housing estate as a hygienic alternative to street food hawking. It provides a one-stop indoor venue of multiple cooked food stalls, up to some 200 stalls each. There are over 100 hawker centres in Singapore. Coffee shops or *kopi tiam* in local parlance exist among the ground floor shops of many public housing blocks, providing drinks (e.g. coffee (*kopi*), tea (*teh*), soft drinks and sometimes even beer) and some 6–8 food stalls (like noodle, rice, breakfast items such as kaya toast and soft-boiled eggs). The bigger *kopi tiam* may have more food stalls. See Kong (2007), Lai (2010).

infrequently. As this is based on self-report, it is possible that these results may not represent the full frequency.

Regarding the reasons that prevent respondents from doing the abovementioned outside home activities more frequently, a majority of the respondents opined that they arc happy with their present levels of participation while many others express simply "not interested" to do more or already "do enough" under the other reasons category. Some older respondents pointed out that family members help them with their activities like shopping, banking, etc. Where reasons were offered, a lack of time appears to be a common reason, especially among the younger respondents even though many are not in employment.

For medical/health appointments, dining out, theatre/cultural activities, banking, lack of finance is a major barrier to more frequent participation while lack of company is another hindrance mentioned in the activities of meeting friends at hawker centre/coffee shops, chatting with friends and neighbours at void deck, and dining out.

Respondents were asked to consider and evaluate their neighbourhood outdoor spaces along a number of dimensions with the use of five-point Likert scales. Respondents were asked to indicate: (a) the extent to which public spaces in their local neighbourhood support their outdoor life, and (b) the extent to which local public spaces affect their overall experience and satisfaction of living in their neighbourhood/town. To the first perceived assessment, over 40% of respondents indicated much/very much support while 30.8% opined some support, 19.5% a little and 7.1% not at all.

There are no sharp differences among the age groups and gender though more among the oldest old respondents considered these public spaces to be of little (24.0% compared to 19.4% among the emerging old) or not at all (12.7% compared to 6.5% of emerging old) support to them. The oldest old are also the group that saw these spaces to be of very much support to them (16.5%) compared to the other age groups (10.7% among the emerging old), suggesting that perhaps they more than the other groups are dependent on these nearby outdoor spaces.

To the second question, about one in three respondents indicates that local public spaces affect their overall experience and satisfaction of living in their neighbourhood/town much/very much though there appears to be an increasing trend with income (35.1% among no earned income and 33.5% among <S$1000 pm as compared to 40.2% among S$3000–3999 and 43.7% among >S$4000), and a declining trend with age (21.5% of oldest old as compared to 39.0% of emerging old rated much/very much).

Respondents were then asked to indicate their degree of agreement (using a five-point Likert scale) with a set of ten quality statements about public spaces in their neighbourhood, giving a glimpse into their perceived assessment of these spaces (Table 1.8). The majority (over 65%) of respondents appear to have favourable views about these public spaces, especially in terms of accessibility, safety, good connectivity between public spaces and considering needs of people with reduced mobility. The older respondents, especially the oldest old, are inclined to view the public spaces positively.

Table 1.8 Aspects of neighbourhood public spaces

		Respondent age								All respondents	
		Emerging old (55–64)		Young old (65–74)		Middle old (75–84)		Oldest old (85 and older)			
		N	%	N	%	N	%	N	%	N	%
Different parts of the city are easily accessible through public transport and safe pedestrian connections	Strongly disagree	15	1.1	7	0.7	3	0.6	0	0.0	25	0.8
	Disagree	77	5.6	49	4.7	24	5.3	5	6.3	154	5.2
	Neither agree nor disagree	199	14.3	121	11.6	54	12.2	5	6.3	378	12.9
	Agree	916	66.1	705	67.9	313	70.3	57	78.5	1991	67.7
	Strongly agree	179	12.9	157	15.1	51	11.6	6	8.8	393	13.4
Construction/renovation of public spaces consider the needs of people with reduced mobility	Strongly disagree	16	1.2	8	0.7	5	1.0	0	0.0	28	1.0
	Disagree	124	9.0	99	9.5	29	6.6	8	11.4	260	8.9
	Neither agree nor disagree	311	22.4	188	18.1	84	18.9	8	11.5	591	20.1
	Agree	744	53.7	599	57.8	280	63.1	48	65.8	1671	56.8
	Strongly agree	191	13.8	143	13.8	46	10.4	8	11.4	389	13.2
There is good connectivity between public spaces	Strongly disagree	13	1.0	3	0.3	2	0.4	0	0.0	18	0.6
	Disagree	103	7.4	68	6.6	25	5.7	5	7.6	201	6.9
	Neither agree nor disagree	304	22.0	213	20.6	101	22.8	7	10.1	626	21.3
	Agree	819	59.2	626	60.4	266	59.8	55	76.0	1765	60.1
	Strongly agree	144	10.4	127	12.2	50	11.3	5	6.3	325	11.1
Public spaces promote chance meetings and socializing	Strongly disagree	13	1.0	9	0.8	3	0.6	0	0.0	25	0.8
	Disagree	130	9.4	95	9.2	60	13.5	4	5.1	289	9.8
	Neither agree nor disagree	370	26.7	230	22.2	118	26.6	26	35.6	744	25.3
	Agree	770	55.6	618	59.6	244	55.0	35	48.0	1667	56.8

(continued)

Table 1.8 (continued)

		Respondent age								All respondents	
		Emerging old (55–64)		Young old (65–74)		Middle old (75–84)		Oldest old (85 and older)			
		N	%	N	%	N	%	N	%	N	%
	Strongly agree	101	7.3	85	8.2	19	4.2	8	11.4	213	7.3
Public spaces are intergenerational places where old and young can meet	Strongly disagree	15	1.1	4	0.4	2	0.4	0	0.0	21	0.7
	Disagree	200	14.4	141	13.6	82	18.6	12	17.0	436	14.8
	Neither agree nor disagree	318	22.9	208	20.1	85	19.0	18	24.8	628	21.4
	Agree	730	52.6	591	57.1	249	56.0	38	51.8	1607	54.7
	Strongly agree	124	9.0	92	8.9	27	6.0	5	6.3	247	8.4
Public spaces are comfortable places	Strongly disagree	13	0.9	7	0.7	1	0.2	0	0.0	21	0.7
	Disagree	114	8.3	86	8.3	38	8.7	6	9.0	245	8.3
	Neither agree nor disagree	417	30.1	266	25.7	108	24.3	16	22.6	807	27.5
	Agree	740	53.4	577	55.8	268	60.5	43	59.4	1628	55.4
	Strongly agree	102	7.3	99	9.6	28	6.3	6	9.0	236	8.0
Public spaces are safe places	Strongly disagree	8	0.6	3	0.3	2	0.4	1	1.3	14	0.5
	Disagree	103	7.4	59	5.7	12	2.8	4	5.1	178	6.1
	Neither agree nor disagree	304	22.0	190	18.4	66	14.8	11	15.2	572	19.5
	Agree	841	60.6	658	63.7	323	72.7	51	70.9	1873	63.8
	Strongly agree	131	9.4	123	11.9	41	9.3	6	7.6	300	10.2
Public spaces are vibrant spaces	Strongly disagree	11	0.8	6	0.6	2	0.4	0	0.0	19	0.6
	Disagree	205	14.9	179	17.2	88	19.8	13	18.3	485	16.5
	Neither agree nor disagree	495	35.8	319	30.7	132	29.8	24	32.5	969	33.0

(continued)

Table 1.8 (continued)

		Respondent age								All respondents	
		Emerging old (55–64)		Young old (65–74)		Middle old (75–84)		Oldest old (85 and older)			
		N	%	N	%	N	%	N	%	N	%
	Agree	613	44.3	473	45.7	202	45.5	31	42.9	1319	45.0
	Strongly agree	58	4.2	60	5.8	20	4.6	5	6.3	142	4.9
Public spaces are designed to cater for a diverse range of activities and uses	Strongly disagree	16	1.2	8	0.7	2	0.4	0	0.0	26	0.9
	Disagree	169	12.2	113	10.9	36	8.1	10	13.4	328	11.1
	Neither agree nor disagree	351	25.3	237	22.9	112	25.1	26	35.9	725	24.7
	Agree	766	55.2	617	59.5	275	61.8	32	44.3	1689	57.5
	Strongly agree	84	6.1	63	6.0	20	4.6	5	6.4	172	5.8
Public spaces can create/foster a sense of belonging	Strongly disagree	14	1.0	12	1.2	3	0.6	0	0.0	29	1.0
	Disagree	102	7.3	85	8.2	21	4.8	3	4.4	212	7.2
	Neither agree nor disagree	429	30.9	303	29.2	142	31.9	21	28.6	895	30.4
	Agree	718	51.8	529	51.0	253	56.9	42	58.2	1541	52.4
	Strongly agree	124	8.9	108	10.4	26	5.7	6	8.8	264	9.0

In two areas—public spaces promote chance meetings and socialising, and public spaces are intergenerational places where old and young can meet—there are more younger respondents agreeing with these statements: 62.9% among emerging old as compared with 59.4% for oldest old in the 1st statement and 61.6% among emerging old as compared to 58.1% for oldest old in the 2nd statement. The statement: public spaces are designed to cater for a diverse range of activities and uses, seems to resonate more among the young old (65.5%) and middle old (66.4%) as compared with 61.3% for emerging old and 50.7% among oldest old. The statement: public spaces are vibrant spaces, has the least votes of agreement (49.9%) and the most neither agree nor disagree responses (33%) while 17.1% disagree with this statement.

Other aspects with ambivalent views of more than one in five who neither agreed nor disagreed include: public spaces can create/foster a sense of belonging (30.4%), public spaces are designed to cater for a diverse range of activities and uses (24.7%), public spaces are safe places (19.5%), public spaces are comfortable places (27.5%), public spaces are intergenerational places where old and young can meet (21.4%), public spaces promote chance meetings and socialising (25.3%) and there is good connectivity between public spaces (21.3%).

Respondents further rated their neighbourhood public spaces on nine bipolar adjective scales (including safe vs. dangerous, clean vs. dirty) to help identify the connotative meaning (suggestive significance) of local public spaces. Most respondents, especially among older respondents, see these spaces as safe (81.5%), clean (63.7%) and enjoyable (68.5%). Fewer viewed these spaces as unique places (18.0%). Factor analysis of the bi-polar adjective pairs did not seem to show significant difference across age groups.

Even so, some interesting affective reactions may be observed: the mean values from the young old and middle old age groups for the traits of passive-active (mean = 3.36, SD = 0.90) and dangerous-safe (mean = 4.01, SD = 0.82) are marginally higher than the means of the emerging old group, suggesting their different qualitative judgement and affective responses to these characteristics, towards more positive attitude among older age groups (Fig. 1.1). The mean value from the oldest old for the weak-strong (mean = 3.69, SD = 0.85) and dirty-clean (mean = 3.83, SD = 0.90) adjective pairs is highest among the age cohorts, indicating the older respondent's concern with these characteristics. In contrast, the adjective pair of standard-unique generated a slightly larger mean among the emerging old group than the other age groups. Most scores pointed towards the positive adjectives (+3.3) with the exception of 3 pairs: quiet-noisy (mean = 2.63, SD = 0.97), easy-difficult (mean = 2.35, SD = 0.85), standard-unique (mean = 2.66, SD = 0.93).

Respondents' perceived satisfaction with local public spaces was rated on a five-point Likert scale across a range of eleven physical and social attributes of public spaces in their neighbourhood (Table 1.9). Generally, one in two respondents is satisfied with these attributes with the exception of the dimension on self-realisation—it enables self-realisation; I can choose to do things that I have never done before, which registered the lowest satisfaction score (39.1%). More among the young old (42.9%) and middle old (44.2%) express satisfaction with this quality (compared to 37.4% for the emerging old and 29.1% for oldest old. More people (45.8%) are

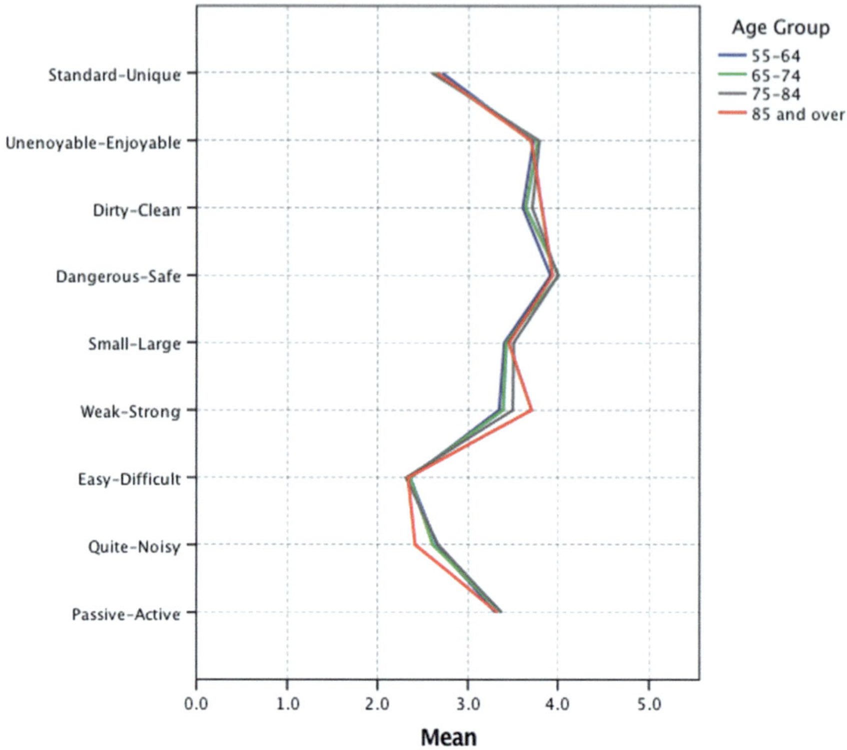

Fig. 1.1 Semantic profile of respondent age groups

neither satisfied nor dissatisfied while 14.1% state dissatisfaction with this attribute, the second highest dissatisfaction score after "I get to make new friends" (15.2% are dissatisfied). More among the young old (55.7%) and middle old (54.3%) respondents seem satisfied with this attribute (compared to 45.2% for the emerging old and 48.6% for oldest old).

For the rest of the attributes, about one third of the respondents are neither satisfied nor dissatisfied. The highest satisfactions are recorded for these attributes: I can come and go as I wish; I do not have difficulty getting here (70.4%), People here get along with one another (62.0%), It allows me to have control of the things I would like to do (62.0%), and There are enough parks, gardens and open spaces for me (60.8%). In most cases, satisfaction seems to increase with age. In particular, the oldest old respondents express greater satisfaction than younger cohorts in the domains on control (It allows me to have control of the things I would like to do, I can decide how I like to use this space and decide what I want to do), comfort and amenity (I feel welcomed and accepted when I am here, People here get along with one another, Seating areas are comfortable and well designed, There are enough parks, gardens

Table 1.9 Satisfaction with physical and social attributes of neighbourhood public spaces

		Respondent age								All respondents	
		Emerging old (55–64)		Young old (65–74)		Middle old (75–84)		Oldest old (85 and older)			
		N	%	N	%	N	%	N	%	N	%
I can come and go as I wish; I do not have difficulty getting here	Least satisfied	8	0.6	6	0.6	5	1.0	2	2.5	20	0.7
	Less satisfied	52	3.8	30	2.9	15	3.4	5	7.6	103	3.5
	Neither satisfied nor dissatisfied	348	25.2	263	25.4	119	26.8	14	19.5	744	25.4
	Satisfied	744	54.0	535	51.7	240	54.2	39	53.9	1558	53.2
	Very satisfied	226	16.4	200	19.3	64	14.5	12	16.5	503	17.2
It allows me to have control of the things I would like to do	Least satisfied	11	0.8	7	0.7	3	0.6	1	1.3	22	0.8
	Less satisfied	76	5.5	48	4.6	25	5.6	8	11.5	156	5.3
	Neither satisfied nor dissatisfied	487	35.3	352	34.1	153	34.8	16	22.4	1009	34.5
	Satisfied	636	46.1	470	45.5	208	47.0	38	53.3	1352	46.2
	Very satisfied	170	12.3	155	15.1	53	12.0	8	11.6	387	13.2
I can decide how I like to use this space and decide what I want to do	Least satisfied	14	1.0	15	1.4	3	0.6	2	3.1	34	1.2
	Less satisfied	150	10.9	99	9.7	50	11.4	8	11.5	308	10.6
	Neither satisfied nor dissatisfied	552	40.0	387	37.6	155	35.1	24	34.1	1118	38.3
	Satisfied	534	38.7	401	39.0	183	41.6	30	42.3	1148	39.3
	Very satisfied	129	9.4	127	12.3	50	11.3	6	9.0	312	10.7
It gives me pleasure to enjoy the things that I do	Least satisfied	6	0.5	7	0.7	3	0.6	2	2.6	18	0.6
	Less satisfied	114	8.3	70	6.7	44	10.0	7	10.2	235	8.0
	Neither satisfied nor dissatisfied	497	36.1	340	33.0	143	32.4	27	37.9	1008	34.5
	Satisfied	590	42.9	444	43.1	196	44.4	28	39.0	1259	43.1
	Very satisfied	169	12.3	169	16.4	56	12.6	7	10.3	401	13.7

(continued)

Table 1.9 (continued)

		Respondent age								All respondents	
		Emerging old (55–64)		Young old (65–74)		Middle old (75–84)		Oldest old (85 and older)			
		N	%	N	%	N	%	N	%	N	%
It enables self-realisation; I can choose to do things that I have never done before	Least satisfied	15	1.1	10	0.9	4	0.9	2	3.1	31	1.1
	Less satisfied	181	13.1	134	13.0	57	12.8	9	12.6	381	13.0
	Neither satisfied nor dissatisfied	667	48.4	447	43.2	186	42.1	40	55.2	1339	45.8
	Satisfied	419	30.4	343	33.2	163	37.0	17	22.8	942	32.2
	Very satisfied	97	7.0	100	9.7	32	7.2	5	6.3	233	7.9
I feel welcomed and accepted when I am here	Least satisfied	15	1.1	9	0.9	2	0.4	0	0.0	25	0.9
	Less satisfied	116	8.4	71	6.8	46	10.3	5	7.6	237	8.1
	Neither satisfied nor dissatisfied	495	35.9	311	30.1	114	25.7	21	29.6	941	32.2
	Satisfied	593	43.0	476	46.1	223	50.4	37	51.4	1330	45.5
	Very satisfied	160	11.6	165	16.0	58	13.1	8	11.4	392	13.4
People here get along with one another	Least satisfied	6	0.5	10	1.0	1	0.2	0	0.0	18	0.6
	Less satisfied	91	6.6	48	4.7	37	8.4	3	3.8	180	6.2
	Neither satisfied nor dissatisfied	489	35.6	286	27.9	116	26.4	18	25.3	910	31.2
	Satisfied	635	46.2	512	49.9	214	48.6	40	54.4	1401	48.1
	Very satisfied	153	11.1	169	16.4	72	16.3	12	16.4	405	13.9
I get to make new friends	Least satisfied	11	0.8	11	1.0	2	0.4	1	1.3	24	0.8
	Less satisfied	198	14.4	132	12.8	77	17.4	12	17.2	419	14.4
	Neither satisfied nor dissatisfied	546	39.6	314	30.5	123	27.9	24	32.9	1007	34.5
	Satisfied	505	36.6	427	41.5	188	42.7	28	38.4	1147	39.3

(continued)

Table 1.9 (continued)

		Respondent age								All respondents	
		Emerging old (55–64)		Young old (65–74)		Middle old (75–84)		Oldest old (85 and older)			
		N	%	N	%	N	%	N	%	N	%
	Very satisfied	118	8.6	146	14.2	51	11.6	7	10.2	323	11.1
Seating areas are comfortable and well designed	Least satisfied	27	2.0	12	1.1	4	0.8	1	1.3	44	1.5
	Less satisfied	132	9.6	89	8.6	35	7.9	5	7.6	262	8.9
	Neither satisfied nor dissatisfied	524	38.0	372	35.9	153	34.6	19	26.1	1068	36.5
	Satisfied	602	43.7	444	42.9	214	48.4	43	58.7	1303	44.5
	Very satisfied	92	6.7	118	11.4	37	8.3	5	6.3	252	8.6
There are enough parks, gardens and open spaces for me	Least satisfied	31	2.2	17	1.6	12	2.8	1	1.3	61	2.1
	Less satisfied	156	11.3	116	11.2	50	11.4	9	12.6	332	11.3
	Neither satisfied nor dissatisfied	358	25.9	268	25.8	113	25.5	17	23.6	756	25.8
	Satisfied	678	49.1	499	48.1	227	51.3	39	53.6	1443	49.2
	Very satisfied	157	11.4	137	13.2	40	8.9	6	8.9	339	11.6
There are enough opportunities to interact with people of my age	Least satisfied	25	1.8	6	0.6	5	1.2	1	1.3	37	1.3
	Less satisfied	152	11.0	75	7.2	37	8.3	7	9.5	271	9.2
	Neither satisfied nor dissatisfied	543	39.3	371	35.8	149	33.7	20	27.4	1083	36.9
	Satisfied	551	39.9	445	43.0	204	46.2	35	48.0	1235	42.1
	Very satisfied	110	8.0	138	13.4	46	10.5	10	13.9	305	10.4

and open spaces for me), and social interaction (There are enough opportunities to interact with people of my age).

As with housing, respondents are asked to rate the importance they attach to abovementioned list of physical and social attributes of neighbourhood public spaces. The levels of importance and satisfaction across the physical and social attributes received high levels of ratings from all age groups. The attributes that received the most votes of importance are also those where respondents have most satisfaction with (chi-square, $p<0.001$): I can come and go as I wish; I do not have difficulty getting here (82.8%), People here get along with one another (78.6%), It allows me to have control of the things I would like to do (73.2%), and There are enough parks, gardens and open spaces for me (78.7%). It gives me pleasure to enjoy the things I do (71.3%), I feel welcomed and accepted when I am here (72.1%) and seating areas are comfortable and well designed (71.6%) were felt to be important as well to many.

The attributes relating to individual control and self-fulfilment (like I can come and go as I wish; I do not have difficulty getting here, I can decide how I like to use this space and decide what I want to do, It gives me pleasure to enjoy the things I do) are considered more important among the younger cohorts, especially the emerging old as compared with the oldest old. For the latter, social interaction appears more important: I feel welcomed and accepted when I am here, People here get along with one another, There are enough opportunities to interact with people of my age.

The quality of: It enables self-realisation; I can choose to do things that I have never done before, is voted important by the least number of respondents (51.0%), and by most respondents to be of least important (14.7%) and neither important nor unimportant (34.2%). Other attributes recording high ambivalent values (neither important nor unimportant) are: People here get along with one another (30.3%), and There are enough opportunities to interact with people of my age (29.1%). There did not appear to be major differences among the gender though more female respondents seem to rate the items important (difference of 0.3–7.2%) while more men rated neither important nor unimportant (difference of 0.6–6.4%).

The space attributes relating to individual control and self-fulfilment: I can come and go as I wish; I do not have difficulty getting here, It allows me to have control of the things I would like to do, and It gives me pleasure to do enjoy the things that I do, have more higher income and also private housing respondents rating them as important. Four others about space ambience and provision have more low income respondents (those earning <$1000 pm) rating them as important: I feel welcomed and accepted when I am here, People here get along with one another, I get to make new friends, Seating areas are comfortable and well designed, and There are enough opportunities to interact with people of my age.

Regarding respondents' views on the things that would encourage/discourage them to use their local neighbourhood public spaces, the most frequently mentioned one thing that would encourage their use is the company of family and friends (23.2%). The other 'encouraging' factors are: pleasant environment (16%), barrier free access (14.0%), social activities (13.5%), good transport connectivity (10.5%), and safety and security (9.7%). Two of the factors seem to recur more among older respondents: Company of family and friends and barrier free access are particularly important to the oldest old (mentioned by 33.4 and 26.5% of oldest old respectively). The majority of the respondents who indicated 'others' mentioned that they would like facilities that would challenge them physically and intellectually. The factors of: presence of social activities, good facilities, safety and good security, pleasant environment, appear more frequently among younger respondents, especially among the emerging old.

Among the group of things that would discourage their use of local neighbourhood park, poor accessibility (21.7%), unsafe place (13.9%), lack of social activities (11.9%) and poor transport (11.6%) are most frequently mentioned, resonating with the responses received for the things that would encourage their use of neighbourhood public spaces. The majority of the respondents who indicated other 'discouraging' factors mentioned the lack of time and the weather. The lack of sheltered spaces to protect them from the weather conditions also discourages them. In addition, the presence of teenagers and noisy children is highlighted as a deterrent by 10.9% of the respondents, especially among older respondents. More older respondents, especially among the 75–84 and women, would mention poor accessibility. Younger respondents more than older respondents included the lack of social activities, lousy facilities, unsafe place and bad design as deterrents to neighbourhood outdoor public space use.

On desired provision, respondents were asked to select one new community space or facility (from a list of nine including an 'others' option where they can elaborate) they would like to have near to their house that would contribute to their personal wellbeing and to give the reason for their choice. Most respondents (one in four) chose health and wellness centre (25.1%) and public park (22.4%). The selection is particularly strong among the younger age cohorts, perhaps suggesting their inclination towards staying healthy and active. The three most common reasons for selecting the particular new facility are: Because it helps me to keep fit and healthy (25.6%), Because I can go there by myself at any time (18.8%), and Because I want to be close to nature (14%). These reasons are particularly important to the younger cohorts (emerging old and young old).

Respondents were further asked to indicate their choice of a cluster of three most desired facilities and services to have provided near to their house to help maintain their independence and active lifestyle. Of the many choices, the three most popular clusters of three services (ranked according to weighted averages) in ascending order are: health facilities, transport facilities and cooked food (hawker centre, food court, for instance). Most respondents rank the following facilities as their topmost preferred facility selection: dry/wet market (19.4%), health facilities (16.6%), transport facilities (15.6%) and cooked food (14%), and the following as within their top two priorities: health facilities (36.1%), cooked food (28.7%), transport facilities (28.6%) and dry/wet market (27.9%). The following services received the least votes in descending order: leisure facilities, personal grooming services and financial facilities, being the least preferred. Examination of the data on the top preferred facility showed that almost half (44.4%) of oldest old have voted health facilities as their number one priority while younger respondents, especially the emerging old, placed transport facilities (44.3%) and cooked food (37.9%) as their topmost choice.

1.4 Conclusion

The survey using personal interview with the help of a questionnaire has provided a channel to investigate the lived experiences of the older person. The results offer a glimpse into the older respondents' experiences and perceptions of their home and neighbourhood spaces, old age and growing older in Singapore. The focus has been on the views of older people themselves, describing their age identity, their activities and aspirations, their assessments and expectations about their home and neighbourhood public spaces. Such knowledge can contribute to understanding the older person's feelings about their proximate environments and spatial preferences, which is a gap in the existing literature.

Our analysis shows that:

Economic and living arrangements—Although the majority of the respondents are homeowners, there is a trend towards declining home ownership and rising renting with age. Crucially, there is a trend towards living alone with old age even though there is a general desire to age in place (this preference appears to strengthen with age). Many have lived in their present dwelling for periods of 11 or more years. Economically, the likelihood towards retirement and having no earned income increases with age while 7.5% of respondents describe themselves as 'unemployed'. From a policy perspective, it is important to further study and understand the needs of these older adults, the implications for quality of life and opportunities in later years in view of increasing longevity.

Perceptions of old age and growing old—Generally, growing old is viewed to be a positive experience among the majority of the respondents. Most respondents do not think themselves as old, lending support to international studies. The mean age at which respondents consider old age to start is 73.6 years. Importantly, those who feel younger tend to report better self-assessed health and satisfaction with life. This perception is a major finding that requires further research on self-perceived age as a correlate of self-perceived health and quality of life.

Housing satisfaction—Most respondents are generally satisfied with most aspects of their present dwelling, resonating with other local residential satisfaction studies. The high satisfaction accorded to: the number of rooms in my house, location of my house, and the sense that this is my home, seems to motivate a level of attachment to support their desire to age in place. Areas requiring attention as glimpsed from their dissatisfaction ratings include the degree of flexibility of my house for my ageing needs, the provision of elder-friendly features in my house, and the degree of adaptability of my house for my ageing needs. More among the younger respondents expressed dissatisfaction with these aspects. Expression of housing dissatisfaction, no matter how minor, prompts further research in view of the potential lock-in effect of housing and relationship to general subjective wellbeing.

Housing aspirations—Majority of the respondents aspire to continue to live in their own home as they age. This aspiration is strongest among the oldest old respondents. Good location of the house and pride of home ownership appear to be the key pull factors to continue to dwell in place while house size (too small or too big), financial considerations, and desire to be nearer to family and friends are the major push factors in respondents' preference to move from present dwelling. Besides good location and pride of home ownership, most respondents consider ease of access to key services, provision of elder-friendly features in my house, and ease of maintenance within the home to be important when assessing the suitability of their house for their ageing needs. Most, especially among the younger respondents, indicated that they have not made any change to their housing and did not intend to make changes in the future. These intents, however, seem to weaken with age. Where changes have been made, they are mostly to the bathroom and kitchen, indicating the need to pay attention to these spaces when future proofing our housing stock to meet ageing needs.

On their housing needs in the future, most respondents, especially older respondents are most concerned about access to healthcare, safety and security. About one in four respondents chose either a health and wellness centre or a public park as their one desired new facility to have in their neighbourhood as they were concerned about keeping fit and healthy, they can go there by themselves at any time, and to be close to nature. Most respondents elected health facilities, transport facilities, and cooked food (e.g. hawker centre, food court) as their preferred cluster of three desired facilities to have in their neighbourhood to support their personal wellbeing as they age.

Outdoor activities and public spaces—Although most respondents are generally satisfied with their own level of participation in outdoor activities, there is a trend for the frequency of outdoor activity participation to reduce with age. Lack of time appears to be a common reason for not participating more frequently in outdoor activity. The significance of neighbourhood public spaces is revealed by respondents' self-perceived function of these spaces in their life space. A majority of the respondents reported that their neighbourhood public spaces supported their outdoor life while one in three respondents indicated that their local public spaces affected their overall experience and satisfaction of living in a neighbourhood. The oldest old are more likely to rate local public spaces to be of very much support to them compared to the other age groups, suggesting the importance of these spaces in their everyday life sphere.

The company of family and friends, and the presence of a pleasant, barrier-free, well connected environment with social activities are seen as the key factors influencing the use of local public spaces, particularly for the oldest old. Poor accessibility, unsafe place, lack of social activities, and poor transport are important deterrents to respondents' use of these spaces. Along with the respondents' satisfaction and importance ratings of the different qualities of public spaces, these results remind us of what older adults see and hope for in their local public spaces. They are definitely worth exploring further. International evidence underscores the importance of relating the quality of public spaces to usability (which closely depends on people's needs and perception) when public spaces are successful (Carr et al 1992; Gehl and Svarre 2013).

Acknowledgements We gratefully acknowledge the many people who have helped us with the survey presented in this chapter: Mayor Dr. Teo Ho Pin, Northwest Community Development Council (CDC), for facilitating access to various community and grassroots seniors events and activities; Member of Parliament (MP) Dr. Amy Khor, Senior Minister of State in the Ministries of Health and of Manpower, for facilitating access to seniors groups in Hong Kah North constituency; MP Desmond Lee, Minister of State for National Development, for facilitating access to seniors community clubs in Jurong Spring constituency; MP Sam Tan, Minister of State, Prime Minister's Office and Ministry of Culture for facilitating access to the Radin Mas Active Ageing Council and seniors community groups in Radin Mas constituency; MP Dr. Lily Neo, Tanjong Pagar GRC, for facilitating access to Senior Activity Centres in the Kreta Ayer/Chinatown area; Mrs. Yu-Foo Yee Shoon for introducing us to Active Contributive Engaged Seniors (ACE Seniors); Siti Habibah Siraj, Director, Malay Heritage Foundation for connecting us to the elderly Malay community; Prabha Seth and Thomson Community Centre Indian Activity Executive Committee for helping us to connect with their local Indian community and administering the Survey in Tamil, where needed; Southeast Community Development Council, Community Outreach Programme for the Elderly (COPE), various Voluntary Welfare Organisations (VWOs) who work with seniors to reach the older population within the Southeast Community Development Council; Eurasian Association for assisting us to connect with the Eurasian older community; Staff of Sunlove Senior Activity Centres, NTUC SilverAce Senior Activity Centres, Tung Ling Community Services, Goodlife! Family Services, Kreta Ayer/Chinatown Senior Activity Centres, Presbyterian Community Services Sarah Seniors Activity Centre, Radin Mas Active Ageing Council, Active Contributive Engaged Seniors (ACE Seniors), Jia Ying Community Services Society, Jalan Kukoh Senior Activity Centre, Jamiyah Ar-Rabitah Mosque, Joo Seng Neighbour Ring Community Services, Telok Blangah Seniors Activity Centre, Thye Hua Kuan Moral Charity Senior Activity Centres for granting us access to conduct the survey; constituency directors and staff of the various constituencies and community centres for facilitating access to the seniors communities in their constituencies/centres; the many individuals who facilitated access to private housing estates, country/community clubs, elderly interests, qigong, wellness, theatre and cultural groups; our colleagues, Debbie Loo, Laura Jasmine, our student interviewers, seniors interviewers from ACE Seniors Club and other adult interviewers for helping us conduct such an extensive number of interviews; Dinusha Senanayake for data compilation and digitising the completed questionnaires; Cheong Sue Wei, student intern Lim Aunn Ning and other part-time helpers for data entry and verification; Cheong Sue Wei and Emily Soh Yongxu for their inputs to the writing of the "Summary of Key Findings"; all residents who participated in our interview and generously shared with us their responses and experiences of their ageing, homes and neighbourhoods.

References

Andres, L. (2012). *Designing and doing survey research*. London: Sage.

Bengtson, V. L., Gans, D., Putney, N. M., & Silverstein, M. (Eds.). (2009). *Handbook of theories of ageing*. New York: Springer.

Blair, J., Czaja, R. F., & Blair, E. A. (2014). *Designing surveys: A guide to decisions and procedures*. London: Sage.

Buckingham, A., & Saunders, P. (2004). *The survey methods workbook: From design to analysis*. Cambridge: Polity Press.

Bureau of Labor Statistics, United States Department of Labor. (2013). *American time use survey summary, bureau of labor statistics 2013*. http://www.bls.gov/tus/. Accessed September 29, 2015.

Carifio, J., & Perla, R. J. (2007). Ten common misunderstandings, misconceptions, persistent myths and urban legends about Likert scales and Likert response formats and their antidotes. *Journal of Social Sciences, 3*(3), 106–116.

Carr, S., Francis, M., Rivlin, L. G., & Stone, A. M. (1992). *Public Space*. New York: Cambridge University Press.

De Vaus, D. (2014). *Surveys in social research*. New York: Routledge.

Department of Statistics, Singapore (2012). *2012 population trends*. Singapore: Department of Statistics. http://costofhealthyliving.com/wp-content/uploads/2013/11/population2012.pdf. Accessed July 14, 2017.

Department of Statistics, Singapore. (2016). *Population trends 2016*. Singapore: Department of Statistics.

Furstenberg, A. L. (2002). Trajectories of aging: Imagined pathways in later life. *International Journal of Aging and Human Development, 55,* 1–24.

Gehl, J., & Svarre, B. (2013). *How to study public life*. Washington, D.C.: Island Press.

Help the Aged. (2006). *Older people*. Decent Homes and Fuel Poverty, London: Help the aged.

Housing and Development Board. (2014). *Sample household survey 2013*. Singapore: Housing and Development Board.

Kong, L. (2007). *Singapore hawker centres: People, places, food*. Singapore: National Environment Agency.

Lai, A. E. (2010). *The Kopitiam in Singapore: An evolving story about migration and diversity*. Asia Research Institute Working Paper Series No. 132. Singapore: Asia Research Institute.

Lascu, D., Manrai, L. A., & Manrai, A. K. (1996). Value differences between Polish and Romanian consumers: A caution against using a regiocentric marketing orientation in East Europe. *Journal of International Consumer Marketing, 8,* 145–167.

Levy, B. R. (2003). Mind matters: Cognitive and physical effects of aging self-stereotypes. *Journals of Gerontology. Series B, Psychological Sciences and Social Sciences, 58,* 203–211.

Likert, R. (1932). A technique for the measurement of attitudes. *Archives of Psychology, 140,* 5–55.

Ministry of Community Development, Youth and Sports (1995). *1995 National survey of senior citizens*. Ministry of Community Development, Youth and Sports. https://www.msf.gov.sg/publications/Documents/Materials_SC_NationalSurvey1995.pdf. Accessed July 14, 2017.

Ministry of Community Development, Youth and Sports, Singapore. (1999). *1999 Survey of transitions in health, wealth and welfare of elderly Singaporeans*. Singapore: Ministry of Community Development, Youth and Sports.

Ministry of Community Development, Youth and Sports, Singapore (2008). *2008 Survey on the learning needs of seniors in Singapore*. Singapore: Ministry of Community Development, Youth and Sports. https://www.moh.gov.sg/content/dam/moh_web/Publications/Reports/2008/MCYS%20Survey%20on%20the%20Learning%20Needs%20of%20Seniors%20in%20Singapore.pdf. Accessed July 14, 2017.

Ministry of Health, Singapore. (2016). *I feel young in my Singapore! Action plan for successful ageing*. Singapore: Ministry of Health.

Moser, C. A., & Kalton, G. (2016). *Survey methods in social investigation*. New York: Routledge.

National Heritage Board. (2013). *Void decks, community heritage series III*. Singapore: National Heritage Board.

Pew Research Centre. (2009). *Growing old in America: Expectations vs. reality*. Washington D.C.

Rea, L. M., & Parker, R. A. (2014). *Designing and conducting survey research: A comprehensive guide*. New York: Wiley.

Rokeach, M. (1968). *Beliefs, attitudes and values*. San Francisco: Jossey Bass.

Sarkisian, C. A., Prohaska, T. R., Wong, M. D., Hirsch, S., & Mangione, C. M. (2005). The relationship between expectations for aging and physical activity among older adults. *Journal of General Internal Medicine, 20,* 911–915.
Singapore Sports Council. (2011). *National sports participation survey 2011, final report*. Singapore: Singapore Sports Council.
Sneed, J. R., & Whitbourne, S. K. (2005). Models of the aging self. *Journal of Social Issues, 61,* 375–388.
Yuen, B., & Soh, E. Y. S. (2017). *Housing for older people in Singapore: An annotated bibliography*. Springer Briefs in Ageing. Cham: Springer.

Chapter 2
Focusing on Older Residents Views

Belinda Yuen and Sue Wei Cheong

Abstract Focus Group Discussions are conducted to yield a more qualitative and in-depth perspective about the ageing community and their everyday urban spaces. It is used to explore the range of views and feelings that cannot be captured in the more quantitative, closed-ended questionnaire of the survey discussed in Chap. 1. The discussions shared the participants' perspectives on their housing and neighbourhood, amenities and transportation, recreational and social spaces as well as their individuality and community connection as they grow older. Their views revealed some of the key desired characteristics of age-friendly house and neighbourhoods.

2.1 Introduction

The Focus Group Discussion (FGD) is conducted to yield a more qualitative and in-depth perspective about the ageing community and their everyday urban spaces. It is used to explore the range of views and feelings that cannot be captured in the more quantitative, closed-ended questionnaire of the survey discussed in Chap. 1. The focus group discussion is an established method for engaging people in a non-threatening small group setting, to learn about their opinions and collect insights into how the gathered group thinks about a designated topic. It is similar to needs assessments survey but seeks to obtain what participants are thinking or feeling, offering insights, nuance and variety to the discussion that would be less accessible without the interaction found in a group environment (Lindlof and Taylor 2002; Liamputtong 2011).

This Chapter outlines the study methodology, key findings and recommendations from these discussions.

B. Yuen (✉) · S. W. Cheong
Singapore University of Technology and Design, Singapore, Singapore
e-mail: belinda_yuen@sutd.edu.sg

S. W. Cheong
e-mail: suewcheong@gmail.com

B. Yuen (ed.), *Ageing and the Built Environment in Singapore*,
https://doi.org/10.1007/978-3-319-92444-1_2

2.2 Methodology

The FGD was designed to provide an opportunity for older people to further share views on age-friendly housing and neighbourhoods. Five FGD sessions were conducted between November 2014 and December 2014 at local community centres to facilitate residents' participation. All but one session was conducted in English. The remaining session was conducted in Mandarin. The proceedings of each session, approximately 1.5 h, were voice-recorded with the consent of all participants within that focus group, then transcribed and analysed.

The discussions were conducted with groups of six to ten participants around a uniform set of predetermined open-ended, relatively broad and qualitative questions. The discussion questions were aimed at fostering a free-flowing discussion among participants, in particular, their lifestyle needs, experiences, attitudes and feelings about housing and outdoor recreation spaces as they age, the innovations, barriers and opportunities in using current housing and outdoor recreation spaces, how these spaces contribute to enhancing their quality of life, and their suggestions for future provision.

A total of 41 respondents participated in the FGD, comprising survey respondents who indicated consent for re-contact and desired groups of participants gathered by snowball sampling. Each FGD was a one-off session, comprising participants of different demographic profiles (50+ of Chinese, Malay, Indian and Eurasian ethnicities), one moderator and one note-taker. All participants were Singaporeans or Singapore Permanent Residents. The composition of the five FGD included:

- 1 Mandarin-speaking group of mixed ages (>50 years old): Mandarin group;
- 1 mixed gender group in the older age range (>70 years old): older old group;
- 1 mixed gender group in the young old age range (55–70 years old): young old group;
- 1 all-female group of mixed ages (>50 years old) and different ethnicity: all-female group;
- 1 all-male group of mixed ages (>50 years old) and different ethnicity: all-male group.

Even though this study has used the FGD to complement and explore the meanings of survey findings that cannot be explained statistically, FGD can also be carried out prior to designing survey questionnaires (Krueger and Casey 2015).

2.3 Findings

Recurrent themes in participants' discussion points and textual descriptions of certain situations and issues were content analysed. They will be examined under the following issues:

- Housing and Neighbourhood;
- Amenities and Transportation;
- Recreational and Social Spaces;
- The Individual and Community.

Since the results are from facilitated discussion among groups of carefully selected participants, the findings will not be representative of the views of the entire population. Many of the participants' views relate to local areas and happenings. Notwithstanding, they give a glimpse of the everyday concerns and experiences of the older person as he/she goes through the neighbourhoods and Singapore. The findings are highlighted with the use of participant verbatim comments, where relevant. As spoken English and sometimes Singlish (colloquial Singaporean English), they may not always reflect correct grammar.

2.3.1 Housing and Neighbourhood

Universal Design An initial response to what age-friendly housing and age-friendly neighbourhoods mean to respondents across all FGD groups often related to the 'hardware' or physical provisions. Specifically, this referred to universal design—infrastructural design features that enabled the ageing or decreasingly ambulant individual to move safely, with ease and confidence within the home and immediate surroundings. Universal design in these spaces included barrier-free access (BFA) doorways, corridors and ramps, grab rails, non-slip tiles, sufficiently sized bathrooms, appropriately heighted cabinets and lift access on every floor. In the neighbourhood (this is a crucial space for individual mobility), the desired provisions consisted of covered linkways, BFA ramps, toilets, assistive technology at traffic crossings (e.g. longer green man crossing), suitably located sheltered resting areas with seating, and lift access for overhead bridges. As one respondent shared,

> To me…age-friendly housing essentially conforms with what BCA [Building and Construction Authority] calls universal housing and this is basically design concepts which are friendly to the ageing and the aged. (Older old group male participant)

In discussing Singapore's efforts for achieving universal design and age-friendliness in housing and neighbourhoods, several challenges and successes were highlighted. These included:

Housing Estates Upgrades While there were laudable efforts in implementing BFA in newer public housing towns[1] and in new housing within older public housing estates (e.g. Bukit Batok Town), older units built between 1960s and 1980s had designs that were not easily retrofitted to include age-friendly features. One Mandarin group participant shared that her four-storey walk-up apartment block was unlikely to be upgraded with a lift under current guidelines. The young old group cited examples

[1]Since 1960s, the public housing authority, the Housing and Development Board has developed 23 new towns and several public housing estates across Singapore.

such as narrow corridors and doorways, which were hard to rectify for wheelchair access; the problem of steps before a main entry, which after conversion into a ramp, might still be unsuitable for wheelchair access due to a steep ramp gradient. Respondents recognised the need to build BFA compliant ramps and the importance of not creating new barriers, but space constraints did not always allow it.

Successes of estate upgrading were discussed. The Mandarin group participants expressed satisfaction and shared how they benefited from their estate improvements and home enhancement subsidies (e.g. EASE—Enhancement for Active Senior Programme[2] by the Housing and Development Board, public housing authority). Their rubbish chutes were upgraded, leading to better hygiene and less pests like cockroaches. Other upgraded flat features included pipes (e.g. metal pipes changed to plastic pipes), handrails, BFA toilets and flooring (non-slip coating, for instance), particularly for flats that were 30 years and older. However, not all participants were aware of these home enhancement subsidies. Those that had learnt about EASE during the FGD were enthusiastic about the new knowledge and the prospect of upgrading their home.

The uneven distribution of upgrading programmes between areas was highlighted by the groups. The Mandarin group noted the discrepancy between different public housing towns while the older old and all male groups identified it as a public versus private housing issue. They claimed that private and low-density estates (Serangoon Gardens private housing estate, for instance) were often neglected or delayed in receiving maintenance or upgrading (e.g. covered linkways). An older old participant living in landed property, though wishing for infrastructural conveniences in his neighbourhood was empathetic to why he thought the government had prioritised upgrading in public estates. He said,

> I think government focus continues rightfully to be for the 81%, which makes sense, you know what I mean? 81% people stay in [these] apartment… I would very much like to stay in a HDB[3] because the facilities are all there… (Older old group male participant)

An all-male group participant was less understanding when he observed,

> So if … the government is really talking about accessibility and making more amenities for seniors, they cannot separate between private and public… [Do] you mean in private, the seniors would have more leeway or more things to do? When you [have] two seniors with dementia or walking disability, they are the same whether you are in private or you are in public housing… Why have this discrimination? (All male group participant)

The older old and all-male groups observed that seniors residing in landed properties often walked longer distances to public transport and amenities, and without the shade of covered linkways. Contrastingly, for public housing residents, in addition

[2]Enhancement for Active Senior Programme (EASE) offers subsidized home modification to older residents in public housing to help them age in place. The improvements include slip resistant treatment to toilet and bathroom flooring, installation of grab bars and ramps. See Yuen and Soh (2017).

[3]HDB is the abbreviation for Housing and Development Board, the public housing authority in Singapore. Public housing is popularly referred to as HDB housing/block/flat.

to covered linkways, the older resident could take cover across void decks[4] along the way. Without good connectivity or walkability, the older person might stay 'in their little corner', negatively affecting their quality of life. The all-male and older old groups concluded that public housing estates, as a whole, were more age-friendly than private estates. Some considered moving from private to public housing for better accessibility to public transport, medical facilities and other amenities as they age.

Other comments on the design of age-friendly housing and neighbourhood included a suggestion of surveying the needs and interests of residents in each geographical area/estate. This would allow for suitable, context-specific upgrades and avoid features/provision that do not work (mentioned by young old, all-male and older old groups). The all-male group stressed that a wider spectrum of user needs—from the older person to their caregivers—must be considered. The young old group suggested that the older, underserved community members and low-income estates including the older estates of Queenstown and Commonwealth should be prioritised in having their needs surveyed and receiving estate improvement.

Challenges of Relocation The groups discussed the challenges of relocating when their current residence became unsuitable in meeting their needs as they aged. Comparing upgrading with resettling, the all-female group shared that the former was 'easier' than trying to repair the negative effects that resettlement (HDB Selective En bloc Redevelopment Scheme[5], for instance) could have on the older person's psychological and emotional wellbeing. The sense of belonging and identity that comes with living in the same residence for a long period of time should not be neglected. As one female participant voiced,

> Why can't I live happily in the same home that I have been living in the last 40, 50 years? To me, that's the bigger concern… After all this hard work of building a home for yourself, you're then told that: Oh, your home is no longer suitable for you, please go and downsize, or please go and live in an old aged village… (All-female group participant)

The all-female group highlighted housing relocation programmes as a key driver in the erosion of the concept of a life-long home, running counter to the notion of ageing in place. There was a general sense that the popular viewpoint was towards new being considered better than the existing. As another female participant shared,

> When [Singaporeans] build for housing, I don't know why [they] always think, "I have to upgrade." Actually when you build a house, you should assume that the person will live in the house until they pass away. Their whole life. (All-female group participant)

A young old group participant shared his experience of moving to a new flat in Pasir Ris town to prepare for old age. The participant noted that while the new residence

---

[4]These are typically found on the ground floor of public housing apartment blocks. The space has been purposefully left vacant (open space) and typically used for communal activities.

[5]Introduced in 1995, the scheme seeks to renew older low-density public housing estates and towns. Small precincts within these estates/towns are selected for demolition and redevelopment. Affected residents are offered new flats close by, giving them the opportunity to continue to live in the neighbourhood. See Phang and Kim (2013) for further details.

was in the area where his children lived, it was far away from where he grew up. He shared the pain and adjustment needed,

> This is my coming [of] age… So I'm shifting for my benefit, for my family's benefit… I was born in Bukit Panjang, I grew up [at] Fajar [Secondary] School, the school field [was] my house, 15 rooms… I stayed there [for] 29 years then I shifted over here, so I had a pain also to leave the place. But then, you see the needs is there, not the wants… So we have to shift to adjust things. If thing cannot adjust, we actually adjust ourselves. (Young old group male participant)

Similarly, a Mandarin-speaking participant shared her woes about moving. Ideally, she would like a lift installed in her four-storey walk-up flat so she did not have to move with increasing frailty. However, she opined that her apartment block was unlikely to receive such an upgrade and relocating would mean moving away from her caretaker daughter (if she moved, she would like to move to live close to her daughter). To some others, for example, an all- male group participant who lived in a landed house, moving or downsizing for necessity was viewed as 'downgrading'. He said,

> [The questions is] whether you can adapt to that downgrading and then live on, which I think I can. Because I was once… also a heartlander. But, then is this really what Singaporeans are heading for? We were told to work hard and you know, get something that you want, but, in the end, you have to go back to the basics again. (All-male group participant)

Other participants from the all-male group regarded relocating for old age pragmatically. They shared that downsizing allowed access to equity and contributed to sustaining an existing standard of lifestyle during retirement. Additionally, they believed in 'necessity trumping sentimentality', which differed from an all-female group participant's emphasis on comfort, familiarity and attachment to one's current home,

> Basically, it's to move to a place that fits our lifestyle and fits our budget. (All-male group participant)

> I think it doesn't matter where we stay because Singapore is such a small place and is so easily connected, it doesn't matter. Personally, it's the place where you go, where most important, you feel comfortable, you like it… as you age… (All-female group participant)

The young old and all-female groups offered ideas other than retrofitting or relocating to encourage ageing in place. The all-female group brought up the idea of flexible homes with features such as reconfigurable partition walls within the homes to reduce the need to move with age. Young old group respondents suggested age-friendly features (such as BFA ramps) to be designed and implemented at the inception of the home design and construction to anticipate future needs.

Housing Alternatives for Older Residents Housing alternatives for the ageing population such as retirement villages and elderly studio apartments were suggested and discussed. The groups had mixed opinions about these housing typologies but agreed on the need for a variety of housing options to suit the respective older age subgroups, and a flexibility to choose them (this idea will be discussed further under the final issue, Sect. 2.3.4). Another was the need for flexibility in housing repayment plans.

The Mandarin group shared that retirement villages were a current and pressing need. The Mandarin and young old groups were concerned with the idea of 'graceful ageing', which they opined could be achieved in retirement villages. Seniors could enjoy leisure activities and services specifically catered to them, as well as companionship and support. There was a perception amongst the young old participants that retirement villages were costly, but the Mandarin group shared that it made more sense to rent in retirement villages even if the monthly rental was higher for the advantage of having companions and conveniences. To them, the upfront payment for an elderly apartment was too high and that money was better on hand than tied down in a 30-year apartment lease.

The young old group noted that some seniors who lived with their children lacked decision-making powers in the household. Similarly, the all-male group commented that having several generations under one roof was a source of friction. Hence, some preferred (the independence in) retirement villages. As one participant shared,

> I can't decide because it's my son's house! I can't decide, it's my daughter's house… Of course, I prefer to be in a retirement village! … I press the bell, someone will attend to me, got doctor, you know… But, then again, I got no money! … How can I ask my son or my children to put me there? They also got no money! So, I live with them. So, when I live with them, I can't say anything. (Young old group female participant)

Contrastingly, the all-female group opined that a retirement village demographic where everyone was uniformly old was unnatural. Segregating the older person from the rest of the community would create an isolated community, and would go against a natural desire to be a part of society. The all-female group shared that there should not be age limits (in housing), which only reinforced the stigma of being old and being old as defined by a set number (this, to the group, was illogical). Both the all-female and young old group believed in the need for a 'beautiful mix' of young and old. As one female participant said,

> I don't want to be thrown into a place where all I see are fellow aged people…Yes, we are older. Yes, we cannot climb the stairs. Yes, please make things a lot more convenient for our use, but those are just the physical aspects of it. The emotional aspect of it, the feeling that we want to be a part of the community, you know? (All-female group participant)

The Mandarin group participants highlighted some challenges faced in elderly studio apartments.[6] One participant currently living in an elderly studio apartment complained about a water leakage problem from her bathroom into the bedroom. Some participants blamed it on the barrier-free design, which eliminated kerbs between rooms while others felt it was a workmanship issue. When the participant filed a complaint, she did not receive the help she expected to get. Further design challenges included unsuitable window placement that affected sunlight entry and the drying of laundry. Cabinets were inappropriately heighted. Overall, participants opined that the design of elderly apartments was insufficiently resolved and there was much room

---

[6]The first of these apartments were developed in the 1990s. See Yuen and Soh (2017).

for improvement. The participants lamented about the quality of new developments in general, which in their views, were getting worse and worse. Another participant opined that her feedback on age-friendly improvements to the housing stock was unlikely to be realised in time to benefit her generation, hence, living in a retirement village might be a better solution.

Privacy and Security Both the Mandarin and older old groups spoke about the importance of privacy in their living environment. The former shared that older units were often aligned orthogonally—a design flaw—offering little privacy and was difficult to amend. She did not want neighbours to look into her apartment and neither did she want to see the goings-on in adjacent units. Although newer housing estates had higher block density, their angled orientation provided a degree of privacy for each unit. Similarly, an older old respondent living in landed property added that, since one could not choose one's neighbours, it was important that the environment was designed to respect individual privacy.

The Mandarin group shared examples of security features that could make a housing environment safer. Several participants suggested that in selected public estates (e.g. those with older population), for a nominal increase in service and conservancy charges, measures such as a security guard or gate with card-access could be implemented for added security. Another Mandarin group participant suggested the idea of a key deposit service for older residents, alike a hotel concierge. This could lower the risk of losing keys, the trouble, time and cost for engaging a locksmith each time the lost happens.

Maintenance and Management The Mandarin group participants were concerned with cleanliness and hygiene in their public housing estates. They grieved over anti-social behaviour such as urination in lifts and indiscriminate dumping of waste (e.g. food) in the void decks. Another participant bemoaned the misuse of the common corridor as a storage space for unwanted household items—corridor clutter posed a danger during an emergency evacuation and a hindrance to those with disabilities and walking aids. Such misdemeanours were viewed as detrimental to the communal environment and could draw vermin (e.g. flies, rats, cockroaches were mentioned). Respondents blamed this on the breakdown of 'software' in society, or people's civic responsibility. One participant opined that employing foreign labour for estate maintenance perpetuated the problem, absolving culprits of their offence and obligation.

According to the Mandarin group respondents, many newer housing developments had poor building quality, workmanship and management. They identified issues such as broken tiles and water ponding along the corridors. Apparently, these flooring defects had been addressed several times but the problem remained unresolved. Such challenges in the housing environment could promote unnecessary stress, particularly on older residents, lowering their quality of life.

2.3.2 Amenities and Transportation

Amenities A key theme that emerged on age-friendly neighbourhoods was the availability of amenities, and critically, their proximity and accessibility from residential areas. Types of amenities most desired by the participants included: wet markets, hawker centres, food courts, grocery shops, hospitals, grooming services, and polyclinics. Libraries were mentioned as a good place to 'read and nap' though a young old group respondent lamented that sleeping was discouraged in public spaces, even at senior centres.

Proximity to essential amenities was stressed by all groups. It affected whether they felt motivated to leave their house and travel out. An older old group participant suggested a distance not beyond three to four kilometres from residential areas,

> …As you get older, you probably have [more] reluctance to move too far away from where you stay, I notice. I'm a typical example that I would always try and look at the closest point, a closest library, a closest polyclinic and all that and when provided with alternatives that are much further I would just not go… I will just forego. (Older old group male participant)

An all-female group participant described the significance of accessible amenities to her husband's daily life,

> It's really very, very important. I see it with my husband… My husband's expectation is walking out, meeting friends, go to *kopitiam* [coffee shop], have breakfast, and exercise. They literally go down to Chinatown, just to sit around and in the air con, in the afternoon to while away the time, have dessert before he comes home. (All-female group participant on her husband)

Whether the journey to amenities was sufficiently sheltered was a cause for concern. The Mandarin group shared that there were large unsheltered distances between home and shops in Pasir Ris Town while the all-male group noted a similar trend in the West Coast area. This omission could lead to a lower level of accessibility to amenities, in turn preventing older people from venturing out of their homes. Accessible amenities affect those taking care of the older adult as well. This was stressed by a young old group respondent,

> Our rule says we have to send [our father] to the nearest hospital, which is Khoo Teck Puat… But there's also one little gripe of course, Khoo Teck Puat is not so accessible as Tan Tock Seng… For us to go and visit him, for anyone to visit him, [it's about] accessibility. Khoo Teck Puat is a bit out of the way. I think when they planned Khoo Teck Puat Hospital, they should have built it near a MRT station. (Young old participant)

The young old and all-female groups favoured the concept of 'decentralised' amenities or having them spread out across the public housing blocks (e.g. *mamak* shops,[7] coffee kiosks in void decks) rather than concentrated in a central area.

---

[7] '*Mamak*' is a Tamil word, meaning 'uncle' or 'elder'. These are small, one-man operated convenience stores (approximately 9 sq m), traditionally owned and operated by Indians, offering friendly and personal service. See Yeoh and Kong (1995).

> … Facilities should be more nearby… don't concentrate all the facilities in the town centre, then you have to go so far to buy your things. At least now my place, at least not too bad, close by. But the new town… not so good… my place at least got hawker centre, got wet market, everything… within walking distance… nowadays the new town… very hard to find food. Must walk very far… (All-female group participant on Punggol New Town)

Not all new towns were planned in this manner. A young old respondent praised Sengkang New Town for its decentralised spread of amenities,

> …They have a mini-mart, a small coffee shop, even have clinic and the amenity… within the precinct itself…So you don't need to travel to go down. If you're sick, you just go downstairs, the most you go is a few blocks away, you can access that. You no need to really take a bus or to have to travel a lot. (Young old group male participant on Sengkang New Town)

Transportation and Infrastructure. Age-friendly features in transport services and infrastructure were discussed. These included 'hardware' such as 'kneeling buses' (buses are lowered at bus-stops to facilitate wheelchair access), priority seats on buses and trains and lifts in transport stations. A young old group participant expressed satisfaction with current improvements,

> Public transport [is] … now much more handicap friendly and [accessible]. Like my mum, she's on wheelchair, she can go out with a maid on buses and most of the bus are handicap [accessible]. And she used to travel almost any other day, if she want to go, she just get her caregiver and they just [go] out themselves. So, I think we have improved the environment for that, handicapped or ageing requirement. (Young old group participant)

Apart from the hardware, 'software' initiatives including the MyTransport.SG app[8] and senior citizen concession card by the Land Transport Authority were identified as useful senior's mobility features. According to the older old group, transport cost could deter travel for the income-less older person. The all-male group noted that cheap travel was helpful and would work in tandem with concession entry into events and local attractions to encourage older adult's participation and visitation.

Good connectivity within the transport system itself was also vital in motivating older people to move out of their houses and neighbourhood. One participant from the young old group shared his experience in Bukit Panjang Town, which had two forms of localised transport service (bus and light rail transit) and yet, did not adequately connect residents within the town,

> Bukit Panjang is a tale of two cities…The old Bukit Panjang where Petir Road is, near the bus interchange. The new Bukit Panjang is where Segar Road is, all the hills. So lately, they cut down all the hills and they build more inner roads and they build more flats. But, there's no access for these flats to the main road. It's bounded by Kranji Expressway… down to Woodlands, then BKE [Bukit Timah Expressway]. The only access is through the Bukit Panjang Road. So, they've been cramming all the flats there and they don't put buses there. The reason being they have the LRT [light rail transit] there. But the problem with the LRT is that between one station and another station, it's three bus stops. And, if your house is

[8]This is one-stop mobile platform on MyTransport.SG, a portal developed by the Land Transport Authority of Singapore to provide information and e-services for all land transport users. Some of the information on the MyTransport.SG portal includes interactive maps to get around Singapore, latest public transport campaigns and initiatives, real time information on bus arrival timings.

> here, and your school is here, you have to walk two bus stops [to your house], two bus stops to the school. So in other words, you have a LRT in front of your block, but you cannot use that to go to school because you will have to go there, and then you have to walk back. So they took away all the buses. So, for 18 years, we have no buses. (Young old group male participant)

Some young old group respondents were displeased that express buses were full when they arrived in their estate from other estates—a result, in their view, of inadequate bus services and the centralised nature of amenities in the region. The Bukit Panjang resident went on to share that many of the facilities—sports stadium, polyclinic and HDB office for Bukit Panjang residents—were located and shared with larger neighbouring estates, hence, the crowding when the bus arrived in his estate.

Some older old participants revealed that they were unfamiliar with public transport, having commuted using private cars most of their lives. Similarly, these participants noted that those living in low-density areas missed out on improvements made to public transport. To this, they suggested transport advances like autonomous cars to improve age-friendliness of the environment. However, they pointed out that the technology might not be implemented in time to benefit the current ageing group.

Observations were made about the increased use of motorised mobility aids, in particular, motorised wheelchairs and the hazard they posed (especially to pedestrians). The young old group shared that such wheelchairs might not be suitable for all, particularly those capable of higher speeds. Participants also foresaw themselves using these aids but noted the gap in interface—homes and public infrastructure had not been designed to cater to the increasing use of this transport type,

> ...How do we manoeuver into my three-room flat when the bathroom is so small? I have to stop outside and then get in. Are they going to make the door, the wall... more comfortable? ... We are pioneers in that sense, we are also pioneers in this thing, you know? So if we're living by ourselves, I don't want to get stuck between the walls and start calling my son, "Son! I'm stuck between walls! I cannot move! What am I going to do? (Young old group participant)

The young old and Mandarin groups highlighted that the co-existence of bicycles, motorised mobility aids and new electrically powered vehicles on the same pathways presented many challenges and needed immediate attention. The Mandarin group suggested licensing of electrical vehicles as a possible step forward. Opinions were divided amongst the young old participants as to whether dedicated lanes for the different forms of transport could be executed due to space constraints in Singapore,

> They can [widen the road] if they want to, look at the PIE [Pan Island Expressway], they upsized the PIE. Remember? Two extra lanes... what is another two metres of space for cyclists or motorised wheelchair...?
>
> You think it's possible?
>
> Of course! It's whether they want to or they don't. (Young old group participants' exchange)

2.3.3 Recreational and Social Spaces

A variety of recreational and social spaces for seniors were discussed. They included both indoors and outdoors, informal and formal, private and public such as Community Centres, Residents Committees, Senior Citizens Corner, void decks, open public spaces, parks and community gardens. Like the amenities and transport, accessibility to these spaces was emphasised by all groups. Understandably, when planning social activities for older adults, location and distance should be carefully considered. Activities that were organised faraway and required commuting might tire seniors out too much to enjoy the activity upon arrival. Auxiliary facilities in these spaces such as toilets, shelters, seats (sheltered), covered linkways were also desired so seniors can maintain a level of independence. The older old and all-female groups further highlighted the need for social spaces to cater to both organised activities as well as for ad hoc or spontaneous interaction.

Planning and Design. A participant shared that his neighbourhood park in Ang Mo Kio Town was one kilometre away (a good feature). He did not expect this everywhere but emphasised the need for outdoor spaces to be within 'reasonable distance' from one's home—this would encourage the older person 'to move out of their house into open space'. Open spaces were a necessary counterpoint to the small living confines of many homes. These spaces would benefit the older adults' psycho-emotional wellbeing (discussed further below) and were crucial in reversing their reduced social interactions. The all-female group cited the example of how the introduction of a new park in her Woodlands (public housing) neighbourhood sparked off community-based activities, which grew rapidly in participation numbers and drew all age groups. This organic development helped to widen the residents' social circles and provided regular exercise opportunities. She shared,

> …With the building of the park, there were activities that were organised… then people would automatically come… Grandparents bringing their young or couples taking evening walks… And then very interestingly… soon followed all the community activities. This particular group that I had was 'fun-dance'… And it soon catches on. It started as a free, free club but soon because the number [it got] too big, 100, 200, so therefore she [the volunteer instructor] started charging… But word gets around. So these are things, people are important. Word of mouth… (*All-female group participant*)

An all-male group respondent shared that new outdoor spaces such as park connectors benefited cyclists and those physically fit but provided no additional networking opportunities for those less fit and ambulant. He believed that more inclusive design was needed and that surveying the interests and requirements of users in each neighbourhood would support this objective.

Space Preservation While new outdoor spaces were being introduced in some areas, in other areas, they were being taken away. A young old group participant bemoaned the levelling of topography and removal of vegetation that once surrounded his area of residence in Bukit Panjang Town.

> … All the hills are cut, all the forests are gone. And in place, they have built up blocks of flats, facing one another, very clustered together. So much so there is no space for even to get out of the estate. It's all hemmed in. (Young old group participant)

An older old group participant echoed this sentiment. He felt that increasing housing stock in place of open space was not the best way forward,

> So every little open space is being used for building, I think that's the wrong thing. These open spaces should be left because that [is]...part of the concept of having a nice environment in the town. (Older old group participant)

Auxiliary Facilities To create more conducive outdoor spaces for older people, an older old group participant suggested 'sufficient cool shelters', which did not mean air-conditioning but shelters shaded with strategically grown vegetation. In regard to safety, participants of the older old group discussed an emergency call button in parks. The idea was adopted from golf courses where aid arrived on the press of the button. The button location had to be clearly marked out or it would not be utilised. Other participants opined that apps would cover this need better.

Informal Gathering Spaces Like the older old group, the Mandarin and all-female groups highlighted the need for more shelters and seating—these would provide informal gathering spaces, especially at void decks and more commercial areas. Again, air-conditioning was unnecessary. If seating were well placed in shaded areas with a view of other activities (people-watching), it would be welcome. People naturally drew other people; infrastructure could be simple, the area could be small and still work. Such facilities were suggested to be in decentralised clusters but active areas (so older people could enjoy them in their neighbourhoods and not just in Chinatown). As several respondents shared,

> I saw at my hawker centre, a lot of old people sitting there because... they can see things happening there, that's why they sit there.

> You look at Chinatown... just tables and chairs, you know people would play chess there, Chinese chess, and people would crowd around to observe. (Exchange among All-female group participants)

> I think we should have more activity corner because like in Bangkit [area in Bukit Panjang Town]... there's always the pavilion where the senior citizens would play dum [checkers], play chess. And because it's near to the market, near to the hawker centre and very centrally located and near to the POSB [Post Office Savings Bank] bank, so it's like a meeting point. (Young old group participant)

Currently, free public seating around commercial areas was felt to be lacking. According to the all-female group, the lack of seats was tied to the very economic motivation of commercial establishments. As one respondent said,

> They want you to buy *kopi*[9] and sit down, you have to pay money to... (All-female group participant)

The Mandarin and all-female groups identified that the lack of public seating, especially in newer estates, was also to prevent sleepers. There was a need to reassess the design of facilities when they were not used in the 'intended manner', instead of simply eliminating them. They shared that the removal of seats led to the deprivation of seats for the rest of the community. For example,

[9]Local Chinese dialect word for 'coffee'.

> One person's misbehaviour, everyone's punished. (All-female group participant)

Another example given of unconstructive action for the purpose of changing public behaviour was the banning of 'youngster' football games at void decks. They shared that space for play would always be needed by the youth and it was counter-intuitive to suppress these activities. As one female participant observed,

> That means the design is wrong. The design is wrong. They need some wall to hit the [football]. It's like, the Orchard Youth Park? For the skating thing? They need a space like that so you have to design a space like that, otherwise where would they do it? On the road! You see? So firstly, the observation. Then, the designers have to come in, what do people need? They design for what people need, and not "I design it and you use it." (All-female group participant)

Programme Respondents from the various groups reviewed the use of elderly communal spaces and programmes. The following accounts show that the interests of the older person differ from community to community.

A young old group participant shared his challenge in running a Senior Citizens Corner: initially named Cosy Corner (in Bukit Batok Town), young couples took to the name and used the space for themselves instead. Subsequent attempts were made to redesign the space as a clubhouse for older residents with a provision of games like dum (checkers) and carom, but the take-up remained slow. Participants from the Mandarin group shared that in their Community Centre, popular activities included karaoke, dancing, mahjong and rummy-O (keeps mental facilities active) while outdoor sports facilities were not well used by older residents. To draw the older person outdoors, the participants suggested that community gardens or small tea-houses could work.

Difference in interests occurred between genders as well. The older old group cited examples of women preferring line dancing (90% women) and men preferring board games or mahjong. It was observed that men generally engaged less than women in community activities. As one female participant noted,

> I realise that men, they [don't] join in the activities… Don't talk about the married men, those who are retirees, widowers, divorced, whatever. Actually, in fact whatever activities we are involved in, mostly women. That's why we are always asking where have all the men gone to? (Older old group female participant)

The same participant suggested that what was needed might be a Men's Club, as the dominant presence of females in these activities and spaces might be a cause of discomfort for males who want to participate. Another male participant joked about naming it the OMCA (Old Man's Christian Association) instead of YMCA (Young Man's Christian Association).

Respondents discussed the incompatibility between the community programmes implemented and what older people want. Community-based solutions were needed. According to the all-male group, those in public office did not fully understand people's needs and (when they do) changes were implemented too slowly. Bureaucracy or the emphasis on 'key performance indicators' and other rigid regulations posed obstacles to the organic development of activities. They suggested that the running of activities be left to the community members: there should be a level of freedom to organise their own activities on an ad hoc basis. As one participant suggested,

> … I think the party in power, the senior civil servants, the CCs [Community Centre], the RCs [Residents' Corner] and the PA [People's Association] and all these people, they need to realise that they don't have all the solutions. And they need to sort of open up and let the residents do their own thing and you know, organise themselves into exercise clubs or whatever it is, without being worried about whether they're going to get fined for playing cards at the void deck. (All-male group participant)

An older old group participant shared a similar sentiment. As a board member of an Anglican nursing home for older people, he realised that a centralised location and top-down decision-making process were out-dated methods. Hence, he had taken a step back and begun surveying community needs,

> I found that running a Home, alone in a community in Potong Pasir [public housing estate] may not be the most efficient way of doing things in terms of facilities, so I am introducing wellness centres, active ageing and all that to the CC [Community Centre] so that the Home in itself becomes secondary. But, it becomes a fulcrum, a centre of activities for Sennett Estate… so I'm working with the MPs [Members of Parliament] and all these people there to see… So [with] NUS [National University of Singapore], I'm working out a needs assessment social. (Older old group male participant)

Supporting Psycho-Social Wellbeing Recreational and social space provision for the older population extends past meeting material needs and comforts. Equally, if not more importantly, is its support to the older individual's psycho-social wellbeing. As one respondent observed,

> It goes beyond the provision of food and a roof. It is what is in the mind and in the heart that we have to meet. (Older old group male participant)

The older old group shared that as people grew older, there were fewer social opportunities. Mobility was reduced and reliance on others was greater. Hence, in designing age-friendly spaces, these spaces should be envisioned as reliable, 'go-to' centres where the older person can expect to get the necessary interaction and support. As one participant explained,

> Always my recommendation to the Minister is that we must provide psycho[logical], social, and emotional support. There's no point in Lions Club, Rotary Club going there, *hoh kwa, boh hoh jiak* [local Chinese Hokkien dialect, which translates to 'look good but cannot eat: no substance'], "Here Uncle, here's a hamper." What the hell do they want a hamper for? They want somebody to come and talk to them, 20 min, half an hour, four times a week. They look forward to a young person coming, you know, and talking with them. (Older old group male participant)

Some described how group fitness classes in community centre, swimming pools and gyms allowed the older person to keep physically active. They provided the added advantage of being socially engaged when the older person exercised with friends or met and developed social network with these group activities.

2.3.4 The Individual and Community

In the above three themes, of focus was environmental design in age-friendly housing and neighbourhoods. In this final theme, less tangible factors related to the individual and the community are considered.

Sense of Ownership Sense of ownership was identified by the groups to contribute to the sustainability and success of age-friendly environments. The Mandarin group referred to this as the 'software' of society, and that it was currently problematic. There was a tendency for a lack of such a value in certain demographic groups, affecting the liveability of a place, e.g. when people do not take ownership of their civic spaces. A direct outcome of this was poor condition of physical environment. The all-female group referred to the environment and society as *macro* concerns, which could be addressed externally whereas matters concerning the self were *micro* concerns that could only be addressed by the individual through personal desire and conviction. As one participant observed,

> …It's up to us, it's really up to us. How we feel about ourselves, our self-worthiness, self-esteem, and whether you really want to go out and have your next phase, and how you want it to be. Your perception of things is really a lot about yourself. There's no point in having expectations on others or what others think. (All-female group participant)

Volunteerism and Ground-Up Approaches To develop a sense of ownership for the community, the all-female group suggested volunteerism. Volunteering in the community could be thought of as a way of 'blessing others' while keeping occupied without expecting returns. It could lead to a happier life. Volunteerism could encourage people to develop a bigger stake in their communities when they contribute time and energy without financial motivation, and can become monitors in their own community.

Ground-up or people-centric approaches where members created and led their own community's activities were another way to cultivate ownership. As one participant described,

> You need facilitators, somebody to facilitate. But, after that you need to give ownership. That means the people, they feel that this is theirs, their own, so, then they will continue to do it. Not the government give them this. You think they will go and do it? (All-female group participant)

According to the older old group, grassroots programmes could be more time- and cost-effective than relying and waiting on the government to implement them. Additionally, as examined earlier (see Sect. 2.3.3, Issue 3: Recreational and Social Spaces—Programme), bottom-up initiatives could better meet the needs of the community. The older old group shared that Family Service Centres assessed the demographics of their own communities in order to organise relevant activities. An example was the Fei Yue Family Service Centre that worked on the classification of different social behaviours within the community; they came up with the metaphors of '*kopi* [coffee] and *toast*' to describe the extroverted residents taking a lead in encouraging introverted others to come out of their home and mingle.

Other participants shared that extroverts or natural leaders tended to emerge during these ground-up initiatives or community demographic assessments. Often, these were the same people who would also help in perpetuating future community activities. As one participant said,

> Because if you have sufficient number of willing leaders in the community they will draw out the best of people and that's, I think, very under exploited in Singapore. Leadership shouldn't just be at the government level, it can be at different levels and this, I think, we have not fully exploited this quality of leadership in Singapore. (Young old group participant)

Choice and Control Older people's profiles and their needs would continue to evolve with society. Providing the older person with the flexibility of choice and a sense of control was a progressive way in which to address changing needs. On retirement age and housing options, for example, for seniors who wanted to continue working beyond the retirement age, the all-male group suggested creating better job options and not limiting them to the roles of only cleaners or servers. The Mandarin group desired flexibility in housing type and repayment plans (smaller down payments, or pay-as-you-go in retirement villages). They emphasised that ultimately, older people wanted the opportunity to live gracefully and not be preoccupied with retirement finances (where the government seemed to have its focus). They opined that the government, in its effort to cater for the older population, had often ended up stigmatising them.

On a psychological level, sense of control was related to enabling and empowering older adults to decide for themselves, especially when faced with increasing constraints from health-related issues and societal stigmas, etc. As a couple of the participants shared,

> I don't think I want people doting over me because I'm old. I think what I like people to do is make me aware of what's available in these open spaces, keep on reminding me and I will then choose when I want to be there. Because this whole issue is about, you know, "Let's get these old folks together, let's bring them here." That's out. (Older old group participant)

> I think for some people, independence, in this sense [is]: I decide when I want to go out. I decide where I want to go. I can even take a bus to go to Chinatown, stay there for half the day, then come back when it is dinner time… Independence is also your children saying, "Okay, Mommy, go ahead and do whatever you want," or "Mommy, I'm so happy that you have all your activities planned out, so we don't worry that you are sitting at home, alone, or not having friends." So, I get the sense that mostly it's about self-decision, self-control on the decision, self-reliance. Even also the fact that you can do what you like to do. (All-female group participant)

Technology and social media were a way in which some group participants felt they could stay connected, be informed and yet have the freedom to choose whether they want to participate. As some said,

> That's why, of course the Facebook is quite important. Facebook is quite good. Now MND [Ministry of National Development], URA [Urban Redevelopment Authority] start to feature it, then you look at nice pictures, then you'd want to go and do it. (All-female group participant)

> …Technology needs to be used. Because as we get old, we have a memory problem. We need to be reminded…we all have smart phones…We like to decide in the morning when we get up… I got this SMS [short message service], eh, I thought, I'm free today so let's do it…So don't bother us with appointment, don't lock us in, every morning tai chi from this time to this. I am not interested. Just keep me informed and I will just breeze in when I like it. (Older old group participant)

The discussions surfaced the theme of community and communal support as a facet to address the social needs of those growing older.

Kampung Spirit Groups raised and explored the idea of reviving the *kampung* (local Malay word for 'village') spirit. To the young-old group, the *kampung* spirit was about living in a community (retirement village cited as a possibility) where they could age gracefully together, where residents provided mutual support and encouragement for one another. As one offered,

> …How can we help each other to grow old graciously? So if you [talk about] somebody who is [growing] old and [does] not know what to do… This is where we as a community, supposing that we all live together with one another and we all live in that retirement village… can actually make use of that village to motivate each and everyone of us, to lend that practical support, and it can be emotional, it can be spiritual, it can be physical; it can be the total aspects of our being… (Young old group participant)

Young old participants also emphasised the importance of relations with neighbours as they could provide immediate assistance when needed. Knowing at least one or two neighbours was essential. To participants, the *kampung* spirit had to start with themselves, not others. As some shared,

> …When you live in flats, you don't have that kind of kampung spirit anymore, but actually it can be revived… How do we do it? We start ourselves first, because if we do not start… Other people will never… do anything… Because everybody will be saying, no it's not my business, no, no, no. But if we take the initiative ourselves and we see as a vision that can be fulfilled, that something like this is possible, it's really, really possible. (Young old group participant)

> This generation is thinking differently relative to what aged living is, relative to the next generation who'll become aged in 30 years. They have grown up not in the kampung spirit. Kampung spirit is us. (Older old group participant)

One older old participant noted that *kampung* spirit in the past was not an integrated one (i.e. *kampungs* were either Chinese *kampungs* or Malay *kampungs*, not usually racially integrated), whereas the aim of its revival now in today's communities had very much to do with integration of individuals and communities, regardless of creed.

Changing Attitudes and Needs Integration of communities had to start with changing attitudes towards the older person and the ageing population. According to the all-female group, ageing was a gradual process rather than an abrupt change upon reaching a specific age. Assumptions made based on the person's age rather than on how active or mobile they are were unconstructive. This did not sit well for those who were still physically and mentally healthy. Attitudes in society needed to change. As one female participant argued,

> You see, I don't want you to treat me as your nanny… or when you want a break you come to me. No, I am part and parcel of your community and I think that is an idea that many people don't realise, and we are being sold this idea of, oh, you're old, so now we give you special treatment, we build the house specially age-friendly for you because you're old… (All-female group participant)

The Mandarin group shared similar insights. They argued that it was crucial to redefine words such as 'old' or 'pioneer'. Today, for the baby-boomer generation, "80 is the new 60"; today's seniors are more educated, have better health and live longer than previous cohorts; the women have less children due to family planning policies. As one respondent observed,

> It's a totally different generation from my mother. So, I think that word 'old', you know, 'pioneer' or whatever, it has to be re-defined because these are very active, useful… (Mandarin group participant speaking in English)

The profile of the next generations (X, Y, Z and so on) would continue to evolve and policies had to evolve with them. This is important, as one respondent highlighted,

> …because if that is not well-defined …The administration cost will be higher later on… (Mandarin group participant speaking in English)

Some emphasised the need for greater tolerance and understanding between different generational and demographic groups. As one participant shared,

> …The caution is, I think, we need to also not have a sense of entitlement. There is this… There are enough stories out there on public transport about how old people start the fight on the trains because they demand the seat when… they should just ask nicely! You know? I think most people, they are probably all now engaged in phones, so they're not looking up, which is a bad thing, but you don't compound one bad thing with another by being demanding. And that's one thing. So I think that there's also danger of labelling young people as being disrespectful. Actually, I'm seeing, in the work I'm doing, a lot of young people going out and doing community work, and they're in hospitals with needy children, cancer-stricken children, these are young people. And they are doing it, and a lot of them are actually doing it now as careers, which don't pay well. So, I think we need to make sure we don't create a new divide between old and young people. So, that's what I mean by inclusivity. (All-male group participant)

The group further added that older people had a responsibility to do their part as well in an inclusive society—take a first step and help set the tone and expectations of those around them. As one respondent said,

> …I find that when I stepped into society, there was an expectation but, because you wanted to move on, so, therefore, you, you yourself have to take the first step to meet that need. And, then from there, the first step is very crucial. (All-female group participant)

2.3.5 Possible Differentiation Between Focus Groups

Discussions from the five FGDs suggested that there might be more evident differences in the way gender, educational level and socio-economic groups perceived ageing and the city than between age groups (e.g. between young old—baby boomers and older old—mature generation).

Gender Differences The key differences included male and female participants' sentiment towards housing relocation and participation in social activities. Female participants (there were three from the all-female group who distinctly) expressed their dissatisfaction towards relocating to meet ageing needs. They emphasised a sense of attachment to the current residence, old friends and neighbours, desire for familiarity and freedom of choice to stay put. The two female participants who aspired moving into a retirement home did so because one wished for companionship (she lived independently) and the other because she had little decision-making powers when living with her children.

In contrast, the male participants were more receptive to the idea of moving, naming practical issues like access to amenities, maintenance/upgrades (when moving from landed house to public housing) and age-friendly features. One male participant said that he missed the area where he grew up but had resigned to life's necessities when he moved (of being closer to his children and to prepare for old age). The one male participant who bemoaned the idea of moving did so because giving up his landed property for an elderly studio apartment had felt like a compromise and affected his pride.

There was a general sentiment that more female participated in community activities than male. This was highlighted both by female and male participants from the older old group. The female participant shared that in her community centre, attendance was usually 90% women. We observed a similar trend in the FGD sessions (23 female and 18 male participants), Community Design Workshops and See and Snap sessions.[10] This could be attributed to a combination of factors: type of activity, men not preferring organised activities and men preferring the company of other men, especially in the absence of women. As observed in our See and Snap and Onsite Observation surveys, self-organised activities like chess games and bird-singing competitions in public spaces were popular among the older male and drew other male audiences. Hence, the importance of informal gathering spaces and demographic surveys to better identify the needs of each community.

Age Groups The different groups seemed to raise similar issues about the built environment, perhaps because of shared experiences of the city. Differences were more in terms of nuances, informed by individual circumstances and interpretations. For example, the older old participants appeared less familiar with public transport services than the young old while the old stressed the need for social spaces to enable social interaction and recreation. The older old participants did not show any evident inferiority in engagement in community, society goings-on, open space or park usage and technology usage than the young old group (an older old participant said that he received text notifications from his local library notifying him of activities that he could attend).

[10]See Chaps. 6 and 10.

2.4 Conclusion

The discussions revealed two main approaches in understanding and designing for ageing in the city—through the hardware or physical environment, and through the software or society and individual. The concept of age-friendliness was consistently and reflexively discussed in both spheres. Age-friendliness in the physical environment was largely seen as an enabler of continued mobility, independent living, social interaction (both organised and spontaneous), physical and mental wellbeing and better quality of life. Age-friendliness in the society was seen as a compassion, civic responsibility and inclusivity; in the individual, it was related to self-sufficiency, sense of ownership and freedom of choice. To FGD respondents, these latter intangible attributes underpinned the success and sustainability of age-friendly urban environments. They stressed that external material factors could help them age better insofar as there was also a wholesome civic, familial and personal conviction to bolster the endeavour.

The findings reinforced what age-friendly environments should support and respond to:

- *Ageing individual's continued independence*[11] *and freedom* (materially and philosophically including mobility, lifestyle choices, housing type and social participation) is of utmost importance. It means enabling older people to maintain their normal way of life without relying or relying little on aid or others for as long as possible. It means an autonomy to choose where and how one lives (especially when the individual is perfectly able to do so);
- *Fostering a sense of ownership and personal responsibility* towards community. A personal desire and conviction, which can also be encouraged through participation and volunteering in community;
- *Responsive, people-centric approach* (social surveys, needs assessments, housing audits) for better identifying individual needs, wants and interests. This is an essential in placemaking (Project for Public Spaces 2009);
- *Changing attitudes*. Understand that ageing is a gradual and natural process. Each generation of older adults should help set expectations of those around them;
- *Technology* supports the older person's continued independence. Urban environments have to be responsive to its increasing pervasiveness in society. In the discussions, technology was a means of continued mobility and an unobtrusive

[11]We note that in choosing the term 'independence', we do not intend to stigmatise the older person's dependency since a decrease in physical and cognitive function is a normal process in an ageing individual and with it the desire or need for assistance or company. Likewise, in reality, there is no absolute independence, which counter-intuitively means social isolation. Again, this does not represent our objective and the range of older people's desires. A preferred term may be 'interdependence' and to support that would mean enabling the older adult to live independently but be easily connected to social groups, assistance and support. See Luskin School of Public Affairs (2014).

means to keep older adults in touch, informed and have the freedom to choose to participation.

With these, an *age-friendly house* is one that allows:

- *Continued independence in mobility around the home and its vicinity*. Universal design currently allows new and retrofitted flats to be physically well equipped for the aging and decreasingly ambulant individual. The issues for consideration are several. First, BFA features like ramps need to abide by design guidelines to ensure they function as intended. Second, universal design has to continually evolve with changing trends, especially with the increasing use of motorized mobility aids; not only do interfaces (level change, doorway) need to be barrier-free but there should be sufficient space for mobility aid manoeuvre from corridor into rooms and within bathrooms. Third, all new homes should be 'future-proofed' to streamline later life additions such as grab-bars (e.g. through the reinforcing of toilet or corridor walls);
- *Continued maintenance, adaptation and affordability*. The general sentiment is for a 'life-long' home and to remain rather than move with age. In order to facilitate this, home maintenance and improvement services and subsidies should be widely accessible, effectively publicised to its target audience and streamlined for easy application. Quality control (in design, workmanship, finishes and materials etc.) in new and upgraded developments should be kept in check in order to prevent unnecessary repairs in the future, which would affect the quality of life of seniors. Some participants were unfazed by relocation while others were less in favour of it. They wanted a smaller place to lessen financial burden, housework or for independence from children. Others desired retirement homes for its personalised services, companionship and support. Hence, home-related financials such as payment plans, mortgage schemes, home improvement subsidies and rent need to be flexible, easily manageable and affordable to accommodate each older person's unique lifestyle choice;
- *Reintroduction of spatial conviviality*. The coveted *kampung spirit* can be attributed to the environmental design or spatial planning of past villages and a sharing of communal spaces such as house-front main streets (and less a concerted effort in community programming). Today, privacy has come at the expense of this neighbourliness. This has led to the perception of public housing shared areas as liabilities (e.g. untidiness and hogging in common corridors, football and sleeping in void decks) and their elimination in newer block designs (e.g. point-block, Punggol New Town). The social value of public housing shared spaces has to be reassessed. So does the degree of privacy and conviviality that people want. Reintroduction of shared spaces within public housing blocks could be an informal means of promoting neighbourly relations and could work in tandem with more formal and programmatic efforts.

By extension, an *age-friendly neighbourhood* would:

- *Be people-centric, responsive and adaptable*. Survey the needs and interests of residents in each geographical area to allow for appropriate, context-specific urban planning, design and upgrades that work for its community. This borrows from the concept and process of placemaking, which "*reinforces the relationship between places and the people that use the space, take advantage of the community's resources, aspirations and potentials, and it eventuates in the generation of great public spaces that contribute to people's health, happiness, and wellbeing and nourish communities*" (Project for Public Spaces 2009). Improve channels for resident feedback, streamline process for ground-up initiatives and publicise and educate seniors of the programmes and subsidies catered towards them;

- *Be envisioned as a self-sufficient, compact node*. Pragmatically, the neighbourhood would provide the older resident but also any individual with essential amenities, services, open spaces and places of social interaction and recreation without needing to travel outside this urban unit. Long distance travel is eliminated. Eradicated is also the encounter of large roads or highways, long and arduous crossings on overhead bridges (and the question of lift or ramp)—problems encountered when the older adults have to move out of residential surrounds into more car-oriented spaces. Hence, an age-friendly neighbourhood or town shares similar characteristics with a sustainable and happy one (Montgomery 2013), which relies less on automobiles and is well connected on a pedestrian scale. When older people can remain living in their neighbourhoods, they do not fear being separated from society and other age groups, simultaneously ensuring there is always a natural and healthy mix of demographic;
- *Have decentralised spread of essential amenities and services,*[12] *within a radial limit of one (walk) to four (bicycle, motorise aid, or bus) kilometres of residential areas*. In this compact model, independent proprietors (*mamak shops*, small coffee shops, barbers, etc.) operating at void decks also prove extremely useful, and participants have expressed favour for them over central hubs. Participants have repeatedly highlighted the need for 'more covered linkways'. However, this need might be reduced if they did not have to travel as far. Covered linkways need only be provided in the first/last mile, especially if there are ample and well-located greenery, which can provide just as much relief. To improve connectivity to less scalable amenities such as hospitals that cater to several public housing towns, shuttle/feeder bus services should be provided to and from bus interchanges and train stations;
- *Have ample walkable open spaces and informal gathering spaces*. Open spaces and neighbourhood parks provide relief from small apartment confines and dense block arrangements. International studies have shown that it improves physical and psycho-emotional wellbeing (Project for Public Spaces 2009). Social activity in these spaces can curb loneliness and depression in older adults (ibid). FGD participants conceded that development at the cost of public open spaces and greenery is a wrong move. Public seating is crucial in these spaces. However,

[12]FGD showed that wet market, hawker centre, convenience store, supermarket, polyclinic, banks and ATM are viewed as essential amenities and services by participants.

they have to be well designed and strategically located for rest, shade, thermal comfort, people-watching (and to be seen by others),[13] social interaction, board game players and their audiences. They would complement the void deck *mamak* shops and services as an informal gathering point. Auxiliary facilities such as toilets, 'cool' sheltered areas and safety features like emergency call buttons (in parks) could be provided to help seniors maintain a sense of security. As with the public housing shared spaces, when designed fittingly, these facilities would not be a liability but public space enhancing features. Finally, when informal social spaces are well utilised (with self-organised activities), reliance on formal community spaces and programmes could be reduced. This might relief pressure on manpower, spatial requirements and funding in formal set-ups;

- *Have formal community spaces and organised activities*. These can adopt a ground-up approach. For example, a degree of freedom for grassroots and community members to use spaces like Residents Corners and Senior Citizens Corners to run their own activities. Support, space and funding for proven popular activities. Second, needs and interests surveys conducted at zonal, town and regional level to implement suitable activities. Formal community spaces like Seniors Activity Centres to be envisioned and promoted as reliable, 'go-to' centres where the older person can expect to receive necessary social interaction and support. Have gender-based, mixed-gender, elderly as well as intergenerational activities;
- *Have accessible and well-connected transport*. Ensure different transport types complement one another to avoid gaps, which make transportation inconvenient. Feedback from users should be duly received, internalised and implemented. Bus stops and transit stops should be at regular walkable intervals. Decentralised spread of amenities can help reduce strain on transport. Feasibility studies into new transport forms (e.g. mobility aids, autonomous vehicles), dedicated lanes and accommodating their coexistence with existing transportation. Free or subsidised public travel for older people (current subsidies were highly appreciated); transport costs were a deterrent to income-less seniors, especially when compounded with entry fees into events and local attractions.

Acknowledgements We gratefully acknowledge the following for their generous support and assistance with the focus group discussions: our colleagues, Debbie Loo, Laura Jasmine and Bernadette Teo for helping us with the focus group organisation, facilitation and translation; and the participants who generously shared their ideas and views during the focus group discussions.

[13]Community Design Workshop Report, another qualitative data-collection instrument in our project, See Chap. 10.

References

Krueger, R. A., & Casey, M. A. (2015). *Focus groups: A practical guide for applied research*. Thousand Oaks, CA: Sage.

Liamputtong, P. (2011). *Focus group methodology: Principles and practice*. London: Sage.

Lindlof, T. R., & Taylor, B. C. (2002). *Qualitative communication research methods*. Thousand Oaks, CA: Sage.

Luskin School of Public Affairs, Lewis Center for Regional Policy Studies. (2014). Place making for an aging population. http://www.lewis.ucla.edu/wp-content/uploads/sites/2/2015/04/Seniors-and-Parks-8-28-Print_reduced.pdf/. Accessed December 10, 2015.

Montgomery, C. (2013). *Happy city: Transforming our lives through urban design*. Farrar, Straus and Giroux.

Phang, S. Y., & Kim, K. (2013). Singapore's housing policies: 1960–2013, *Frontiers in development policy: Innovative development case studies 123*. KDI School and World Bank Institute.

Project for Public Spaces. (2009). What is placemaking? https://www.pps.org/reference/what_is_placemaking/. Accessed October 20, 2015.

Yeoh, B., & Kong, L. (1995). (Eds). *Portraits of places: History, community and identity in Singapore*. Singapore: Times Edition.

Yuen, B., & Soh, E. (2017). *Housing for older population in Singapore: An annotated bibliography*. Berlin: Springer.

Chapter 3
Sharing Key Informant Views

Belinda Yuen and Sue Wei Cheong

Abstract This Chapter looks at the experts' views using the method of key informant interviews. The key informant interview is a rapid, relatively inexpensive data collection method that takes the form of a qualitative in-depth interview with a small select group of expert/industry/key individuals who work with the subject group, in this case, older people. The findings gathered their perceptions and insights on prevailing practices in housing and outdoor recreation spaces for older population, the challenges and opportunities for further development towards ageing in place and age-friendly communities.

3.1 Introduction

This Chapter looks at the experts' views through the tool of key informant interviews. The key informant interview is a rapid, relatively inexpensive data collection method that can generate information, insights and recommendations for project/programme design (USAID Center for Development Information and Evaluation 1996). It takes the form of a qualitative in-depth interview with a small select group of expert/industry/key individuals who work with older people. The purpose is to collect their perceptions and insights on existing provision and practices in elderly housing and outdoor recreation spaces, challenges and opportunities to help understand not only what designers do but why they do it, the underlying motivations and attitudes towards the provision and interventions in these areas.

The methodology and key findings are reported in this Chapter.

B. Yuen (✉) · S. W. Cheong
Singapore University of Technology and Design, Singapore, Singapore
e-mail: belinda_yuen@sutd.edu.sg

B. Yuen (ed.), *Ageing and the Built Environment in Singapore*,
https://doi.org/10.1007/978-3-319-92444-1_3

3.2 Methodology

A total of 33 key informant interviews were conducted between October 2014 and August 2015. The interviews were conducted on an individual basis or in small groups of two to four people, at the respondents' convenience. There were no incentives for participation. A uniform set of questions was used as a topic guide for all interviews; not all questions were asked during the interview. The interviews were conducted in English. All responses were voluntary and were completely anonymous. The interview proceedings were voice-recorded when all participants within that session consented to it; when consent was not given, hand-written or type-written notes were taken. Written notes and transcribed recordings were then collated and analysed. In the analysis, all responses were not identified by individual, but by pseudonym (Group A Participant, Group B Participant and so on).

A purposive sample of 43 experts and professionals responded during the 33 interviews conducted. Respondents were drawn from a wide range of sectors including agency representatives, community leaders, caregivers, developers, planners, service providers and professionals of elderly housing and outdoor recreation spaces. Broadly, they can be categorised under 4 groups:

A. Agency and public sector professionals (6 interviews; 10 experts)—government agencies and public sector workers;
B. Development industry experts and professionals (12 interviews; 12 experts)—planners, architects, engineers, developers, etc. in for-profit and non-profit real estate and infrastructure development;
C. Medical and healthcare experts and professionals (4 interviews; 4 experts)—doctors, geriatricians, healthcare and hospital workers, etc.;
D. Caregivers, non-profit organisations, charities and service providers (11 interviews; 17 experts)—caregivers, voluntary welfare organizations, social workers, family and welfare services, sheltered homes for the aged, etc.

Like the other research methods described in this Volume, key informant interviews have advantages and limitations. The main advantage is that these interviews provide unique perspective and much more detailed information from knowledgeable people (experts), e.g. the experts may tell of local happenings or conditions that explain implementation. But, because it is largely qualitative, these interviews provide very limited basis for quantification. Also, because key informants are not selected through random sampling, they are prone to interviewer bias. Notwithstanding, such research offers a useful exploratory method to uncover beliefs, values and motivations that are associated with a given outcome. There are methodology books on how to plan and conduct these interviews so as to reduce common sources of bias and improve reliability of key informant studies (see e.g. Miles and Huberman 1994; USAID Center for Development Information and Evaluation 1996; Powell 1999).

3.3 Findings

The aim was to learn more about age-friendliness from the experts' perspectives: their understanding, assessment and professional insight on age-friendly housing and outdoor recreation spaces, the concerns, innovations, barriers and opportunities in current housing and outdoor spaces, and their suggestions for future provision. Recurrent themes in participants' discussions have been content analysed and will be examined under the following headings:

- **Housing** Issues include the physical components of elderly housing, types of elderly housing, the physical, social and healthcare structures needed to facilitate ageing-in-place and a typological proposal for ageing in community;
- **Neighbourhood and City** Issues include features that make outdoor spaces more age-friendly, urban planning concepts such as car-lite development and less tangible attributes such as building the 'heartware' as a means to create a more age-friendly city;
- **Society and Individual** Issues include the role of the community and individual, values of society and empowerment of the individual for the making of a more compassionate and age-friendly society.

These themes will be illustrated with respondents' verbatim wherever appropriate. It should be noted that these verbal expressions might not always be grammatically correct.

3.3.1 Housing

On housing, respondents frequently began the age-friendly discussion with physical attributes that supported accessibility, safety and comfort for the older user within the home. About half of the respondents (19) identified the need for conversations of housing design and provision to be holistic and encompassing qualities beyond infrastructure—communal living, the need for social programmes and community support structures to be developed concomitantly and embedded within housing were deliberated. Beyond individual homes, housing typologies such as retirement, nursing and dementia care facilities were mentioned and discussed.

3.3.1.1 Being at Home

Adapting the Home Respondents across groups espoused that age-friendly conversation should begin at home since the older person spent the most time in this space. As one Group D participant said, 'home is where the heart is'. Other responses supported the notion of ageing in place and adapting the house to allow ageing people to remain in situ and feel independent in their homes. As some respondents opined,

> Home-age friendliness should start in the house and then outside…They should be able continue to live on their own homes as they age. (Group A Participant)

> Most elderly are content with staying at home. (Group C Participant)

> Old people cannot just move to smaller apartment just because it is more convenient. Most people do not prefer to move from their own house when they age due to the familiarity of the space. Hence, it is a better option to convert home to be age-friendly. (Group B Participant)

Respondents highlighted various key age-friendly hardware for the home: *to prevent falls* (e.g. provide hand-rails, non-slip flooring), *to facilitate wheelchair-user/those unconfident on their feet* and *to be readily connected to aid* (e.g. provide emergency cord, CCTV or other monitoring device). Toilets were another point of emphasis. Respondents stressed the importance of providing sufficient space in the toilet for easy manoeuvre (especially important for those with disabilities), keeping floors dry (to reduce fall risk and accidents), the installation of grab-bars and non-slip tiles (to reduce fall risk and give confidence of movement). Other physical considerations included the layout of rooms, height of switches and windows, relocation of distribution boxes outside of bomb shelters, width of corridors and steps before a house entrance. The design should consider the needs and capabilities of the older person. As various experts emphasised,

> Elderly need elderly friendly facilities to live; easy for them to move around in a place. Particularly in area of toiletry facilities – must be easy for them to access, enough space for them to go in, flooring must be suitable – having non-slip tiles, railings along the corridors, around the staircases just before they enter the rooms to hold on to prevent slip. (Group D Participant)

> [Elderly have a] tendency to fall. Emergency cord useful in the kitchen, bathroom, and bedroom too as they spend more time there. The relationship between bathroom, kitchen, living, bedroom in the layout is key. (Group B Participant)

> Layout of bathroom is a key area. Switches should be accessible. Height should be ideal range of 600 cm to 1.1 m from floor level. People require varied height. Reduce effort for elderly on wheelchair…Distribution box (DB) is usually in storeroom or bomb shelter. If power trips, it should be easy to access. If elderly need to call for help, DB should be near door. All aspects of living must be friendly. People should be able to live independently at home, unless they are bed ridden. (Another Group B Participant)

Thoughtfulness in future proofing of homes was brought up by a Group A respondent. When considering life-long homes, the respondent highlighted that it was unrealistic to expect or fix elder-specific installations in young people's homes. To get round the problem, provision for such fixtures could be made at the outset to enable installation later whenever the need arises, e.g. by building in reinforced partition walls for future handrail installation, adequate waterproofing coverage for future bathroom extension. This could save costs as well as make subsequent modifications easier. Another Group A respondent raised the importance of understanding the needs of older adults and this should start from planning stage to ensure that adequate facilities and amenities were provided.

Efforts in Private Housing and Cost of Age-Friendly Design Five respondents from Groups A and B talked about age-friendliness in private housing. A Group

B architect considered current public housing to be more age-friendly than private housing because of the mandatory need for universal design compliance in public housing. Private housing did not have such a requirement, which might be problematic for the future with population ageing.

> Public housing is very prepared...Universal design compliance is regulatory and is needed for land use approval...Currently no requirement for any house (public or private) to cater to elderly unless it is an elderly studio apartment. (Group B Participant)

Other experts shared that private housing like the condominium was largely designed for the able-bodied and a younger target market. Aesthetics were commonly selected to suit a younger audience and the provision of elderly studio units within condominium projects was generally unpopular. Several respondents variously illustrated this point,

> Many people do not build grab bars in condominiums as these condos are built according to age group demands – 20s to 50s where there is no need for such provision. We choose materials that are light and are aesthetically pleasing. (Group B Participant)

> In new housing, especially condominiums, toilets are so small that it may become a problem 20–30 years later for occupants. (Another Group B Participant)

> These days, granny flats and flat allocation is being done by the HDB.[1] Private sector, there was a small initiative like this but it was not too popular...In condos, too, 1% or 2% of units could be set aside as granny flats to integrate elderly into the newer developments. (A third Group B Participant)

In spite of this, these respondents (architects and developers) agreed that there was a growing awareness and increasing attention being paid to age-friendly design in recent years by individual clients preparing their homes for old age. A developer respondent acknowledged the need to look into building homes for an ageing population. The Hillford at Jurong Kechil was mentioned by three respondents as the first elderly condominium development in Singapore though some had reservation on the adequacy of its age-friendly features,

> In private housing, it is becoming more obvious that people are taking note of such needs. Clients request for lifts or to provide room on the first floor for needs of the aged. Began 5–10 years ago and trend is obvious. (Group B Participant)

> As developers, products [housing and built environment] need to be future-ready. Spaces we create focus on the outcome of inclusiveness. We are opening ourselves to take up demand. It calls for creating spaces where amenities cater to the elderly. (Another Group B Participant)

Two Group B experts opined that contrary to popular belief universal design was not necessarily more expensive than non-age-friendly design, especially when developed from the outset. The common view was that higher cost seemed to be a misconception amongst private clients or developers. As two participants explained,

[1]HDB is the abbreviation for Housing and Development Board, the public housing agency in Singapore. Public housing is popularly referred to as HDB housing/block/flat.

> It is not more expensive to provide senior facilities. Many things will change when seniors get more independent, when they have a bargaining power over their income and can be more demanding. (Group B Participant)

> Some clients see it as additional cost to the budget. However, if designed well and with it in mind from the start, I believe it does not necessarily cost more. Wider spaces or corridors may translate into cost, however, as the country moves toward being an inclusive society, this is a necessary step forward. (Another Group B Participant)

Interestingly, two other respondents from Group B shared that elder-related housing was not profitable. Even so, a couple of respondents predicted that age-friendly design in the private sector would become mainstream with the emerging cohort of older people who were more highly educated, wealthier and conscious of better design,

> The silver business is not profitable; not enough energy, need business model to cater to people whose income is falling. (Group B Participant)

> Opportunity cost for building [retirement home] in Singapore was too high and it was better building a condo on the same piece of land…There is also fear of not being able to resell a unit after the death of the occupant, a real concern in a relatively superstitious Asian society. (Another Group B Participant)

> There are also many opportunities for private industry. The silver industry is a growing one. For instance, in Japan, private operator services cater to different needs of the elderly, from homecare services, travel and leisure, grocery delivery to health and alert systems. (Group A Participant)

Rental Flats and Affordability Four respondents from Groups A, C and D were concerned with the older person's ability to afford housing. A respondent commented that the current cost of owning a flat was out of reach for many poorer older residents. Two respondents mentioned that the poorest elderly were often renters. The difficulty of getting a rental flat was highlighted,

> Key concern is that elderly cannot afford their own housing. The poorest of elderly are all renting. These 1-room rental flats lack many amenities. When they are discharged from the hospital and sent home, nurses visit and follow up. They [discharged patients] often do not leave their homes…lower income families, environment not very good, live alone, do not go out, do not feel safe, no one bring them to exercise…New rental flats are better; they are more integrated with the community. (Group C Participant)

> Money and affordability are major barriers. Each flat is money. Currently, HDB is very commercial and sell flats to the market. The queue [wait] is very long for rental flats and there are not enough rental flats. The current market rates are also high. So elderly cannot afford to buy. (Group D Participant)

Some respondents suggested setting aside some flats for older people at the outset and prioritising these for their purchase or rental. Other respondents brought up the need of housing subsidies and more crucially, increasing awareness and normalising the use of such financial aids. For instance,

> In future, we should allocate flats for elderly before selling out. (Group D Participant)

> Integrate and set aside sites for the elderly right from the start of new town development. Land should be set aside to build more retirement villages/homes. If the aged homes are priced appropriately, the aged will be able to afford it. (Group B Participant)

> Lack of money and finances is a big issue according to me. Even though we have subsidies for the elderly, the living expense rates have increased substantially… In these cases, they firstly need awareness to the source of money… In addition to this, most of them tend to have a negative perception of getting financial aid. (Group C Participant)

> Cost of living—need some subsidies, grants from social network. But, in our culture, subsidy is a taboo word; much is driven by money and profit. Australia—there are a lot of grants. (Another Group B Participant)

Respondents who had visited 1- and 2-room HDB rental flats lamented that these were typically in dismal conditions as the older residents were sometimes too weak or poverty-stricken to maintain a clean and hygienic living environment for themselves. They indicated,

> Rental flats are often infested with bed bugs. The elderly have physical limitations and therefore find it hard to keep their housing environment clean and conducive for healthy living. "Can die but cannot afford to fall sick". Money needs to be topped up constantly for utilities. There should be subsidies for the necessities of day-to-day living. (Group C Participant)

> …in single one-room flat, some of them are really pest-infested, with bed bugs, are old and smelly. Think it is about time that the authorities take a look at that and spend some money to refurbish that area. Usually only exterior touch-ups and clean ups. Would be good to start cleaning up the insides of the houses for the elderly. (Group D Participant)

Integrating Elderly Housing One point repeatedly emphasised by all groups was the need to integrate elderly housing with the larger community and mainstream elderly housing. This could occur at various scales or with different housing types, e.g. mixing elderly studio apartments with other units within a housing block, integrating elderly housing blocks with other blocks in an estate or ensuring rental flats (where many poorer older people lived) were not inferior in quality and appearance to the rest of the neighbourhood. Respondents from Groups A, B and C remarked that the segregation of elderly housing (whether geographically, organisationally or visually) from other housing could create aversion towards the differentiated group (the older individual) and a choice for avoidance (e.g. 'not in my backyard' NIMBY attitude). Housing blocks isolated from the larger precinct/neighbourhood area were also inconvenient for reaching amenities and would deter social interaction (discussed further in next section under Neighbourhood and City). They argued,

> Do not agree with building elderly housing on its own unless it is medically needed e.g. for acute care…A block of 2-bedrooms or 1-bedrooms just to cater for senior housing would encourage a lot of NIMBY [Not In My Backyard] attitude. Better to insert and integrate them together, have a mix of the room types within the block…Much needs to be done for existing housing and precincts as well, those which are not subjected to the SERs [selective

en bloc redevelopment scheme] programme, e.g. in Chai Chee and Bedok estates.[2] (Group B Participant)

Design can make old folks home more acceptable in HDB estates; need to blend in with surrounding environment…When design elderly housing, there should be certain portions for old and some for young. Dangerous for certain portions of the city to be built just for the older population, misses point of creating a more humane city…The design of the HDB is not of such a poor quality that it is difficult to fuse it into middle income household area. (Another Group B Participant)

It is the people. Young people refuse to stay around the older people. However, it is important for the intermixing of different generations in the same living environment. (Group C Participant)

Isolated blocks are the worst designs. The need to go out is not practical without other activities in precinct. These activities will promote social interaction. There could be a community hub to draw people together and to help people come together. (Group A Participant)

3.3.1.2 Enabling Ageing in Place

Ageing in place was another recurrent theme in respondents' conversation on elderly housing.

Our survey findings tell us that an overwhelming majority of our elderly residents desire to age-in-place. Overall, the majority of elderly (77.5%) [HDB] residents were content with the flat types that they were currently residing in. (Group A Participant)

Much effort was being directed to enable older residents to continue living within their own homes for as long as possible. Those efforts include physical enhancements like making older public housing flats age-friendly and safe for the older residents (e.g. the HDB Home Improvement Programme, Enhancement for Active Seniors Programme[3]), providing more age-friendly facilities to make it easier for older residents to go out and about (e.g. Lift Upgrading Programme, Neighbourhood Renewal Programme, 3-Generation playgrounds) and special design housing for older people (e.g. studio apartments, two-room flexi scheme).

More than physical change, an important consideration was to engage people. As a couple of the respondents shared,

We need to go in the shoes of old people, understand their needs, experiences and design for them. (Group B Participant)

In terms of accessibility, an engineer cannot decide the needs of the user. Only providing facilities is not enough. The user must be consulted to test for user friendliness of the amenities and provisions. (Another Group B Participant)

---

[2]Chai Chee is a public housing estate located in Bedok new town (204,300 population). Developed since the 1970s, this new town is located in the eastern part of Singapore.

[3]See Yuen and Soh (2017) for further details of these housing schemes.

As one Group D participant suggested, planners need to talk to people (old and young) to better understand their needs when making lifetime homes. She further put forward the concept of lifetime homes and neighbourhoods (from Joseph Rowntree Foundation and Habinteg Housing Association, UK) as an example when providing accessible, easily adaptable, age-friendly homes for all ages. Others advocated experiential learning as a way to better understand the older person's needs and wants.

> Planned camps for budding professionals, teaching the elderly's experiences through their disabilities, etc., will be experiential and useful while planning spaces for them…e.g. Singapore Polytechnic has 'dinner in the dark' where people experience what the visually challenged go through in their everyday lives. (Group C Participant)

Various other housing options for older population were mentioned.

Assisted Living Two Group D respondents bemoaned that there were currently little long-term care choices for the older person. Most older people required basic care, which could be facilitated at home. Assisted living seemed to present a viable option. As one expert opined,

> Innovative ideas need to be used to meet the living needs of the elderly. Most of them need only minimal care like someone to serve them medicines at night. Currently, there's no middle option, [there is] only home or nursing homes. We don't have any long-term care option. (Group D Participant)

The term 'assisted living' appeared as a concept familiar amongst healthcare professional participants; it was brought up by a majority of Group C and a few Group D respondents who were medical practitioners. According to the responses, the term could be understood as a form of decentralised or community-based elder support structure. It allowed older people some form of assistance with everyday living or activities (ranging from household chores, groceries, medical care or rehab, etc.), from a variety of people (such as befrienders, neighbours and other older or professionals like nurses and social workers, etc.), all while living in one's own residence. Responses suggested assisted living could exist in a variety of settings, in public and private housing, be provided by voluntary welfare organisations or hospitals, etc., on a range of levels including for the individual household or for an entire rental public housing block, etc. For instance,

> More assisted living - private or HDB. Safe and secure buildings that will allow for services and where seniors can feel safe and be assured to get help. Including daily living activities services like meals, groceries, etc. will be beneficial for even the future ageing. (Group C Participant)

> Maybe some levels [of HDB flats] could be renovated to enable assisted living facility. Instead of moving out of their community, the elderly could just move floors or to the next block within the same neighbourhood…Assisted living should be supported. There should be room for such facilities and to accommodate VWOs and such within the HDB estates… HDB should not sell all units in the beginning. Instead should reserve units to plan to house people for life. These reserved units could operate as service apartments or room rentals by assisted living care. (Group D Participant)

> Assisted living relates to elderly living together and taking care of each other. (Group A Participant)

Even though the term 'assisted living' might not have been specifically used, the concept and its constituent support elements were alluded to by several other Group C and D respondents,

> The befriending programmes target socially isolated elderly or the ones staying alone. We do home visits, once every fortnight depending on situation. Recommendations for homes to visit come from everywhere. Mostly people call us through word of mouth. We have about 30-40 befrienders. We cater to 250-300 elderly at a time. (Group D Participant)

> We don't want to hospitalise him. Long-term hospitalisation was not the solution to this as he was an active person and spent a lot of time outdoors socialising. So, we decided to extend home care facilities for him. We send him food and appointed a person to give him his medicines on time. This way we could monitor him closely within his own comfort space… Recently high-risk patients are being attended to by nurses in their own homes. This has proved to reduce the number of hospitalisation. Facilities like these will enable them to continue to stay in their homes without disturbing their daily routine. (Group C Participant)

> Giving access of homes to doctors and healthcare experts or telemedicine is another possibility to ensure that they [the older person] stay healthy. Mobile apps can be developed for the elderly. In future, people may not want to cook, as they are not used to it in their younger days [habit of eating out]. In this scenario, some provision for food to be made available in their apartment might be useful. (Another Group D Participant)

Group D respondents brought up a couple of examples of initiatives, which could support assisted living like the National Initiative for Care of Elderly (NICE) service by volunteer nurses in Singapore, Mobile Care for older people in Australia, etc.

> In terms of medical care, e.g. in Khoo Teck Puat [hospital], there is a 'Keep an Eye on Neighbour' programme. Neighbours knock on doors of elderly living alone or with one other…There are about 80 to 100 volunteer nurses from government hospitals who go round to the list of patients to keep an eye on them, they are outpatients and staying in their own flats, but they are under medication. So, nurses provide free uncharged work for us after the hospital hours. NICE service is one that elderly can register with at their local CC [community centre]. Covered eastern and western Singapore. Started in late 2012–2013. This cuts down on strain on hospital beds—'sick-in-place'. (Group D Participant)

> Mobile care for elders with insurance coverage till death of an individual. For example, there could be a nurse on call for every 10–15 flats maybe. Australia is a good model for a successful mobile care system. We could learn from them. (Another Group D Participant)

From the responses, the advantages of assisted living included reducing the number of hospitalised individuals, reducing strain on healthcare personnel and facilities, and supporting ageing in place.

Village Typology The notion of 'village' was sometimes suggested as a housing typology, which could facilitate assisted living and other satellite centres. More holistically, the village was envisioned as a model of housing development, which could be self-sufficient and embedded with social support structures, spaces and amenities for the entire community. Kampong Admiralty by WOHA Architects was mentioned as an example and where more studies could be conducted to assess its validity. As one respondent suggested,

> We suggest a village model in high-density development accompanied by nursing support, palliative care (on mobile). This could change the way we organise community care—homes can be easily adaptable to elderly care by sharing services and support structures. (Group B Participant)

In the context of land scarcity and high-rise vernacular of Singapore, the village was seen as a vertical, high-density model. One respondent imagined that a graduated eldercare support system could be expressed in a stacked urban form,

> Singapore will have problem with space but can look at doing it vertical. Several level of living, one level is independent living, another level is with some assistance from professionals and nurses. The next level could be in the style of a nursing home. This village would be comprehensive and helps many people come together, e.g. friends who decide to retire together and stay together. (Group D Participant)

Likewise, different types and forms of private and public spaces for social interaction and gathering could be expressed in the verticality. A design respondent observed that the equivalent of a front-lawn or neighbourhood street—traditional shared spaces that promoted human encounters—could be explored and articulated within the high-rise model,

> In Singapore, our vernacular will be high-rise apartment, so need to design to help promote this kind of social connection, e.g. open corridors to allow people to walk pass, similar to garden in American houses – semi-public/public space to make connection with neighbours, so one can bump into their neighbours, etc. That is, third spaces for purposeful meeting areas, gardens in the community. Designers neglected this in the pursuit of private spaces. (Group B Participant)

Other communal spaces such as cooperative supermarkets and farms (community gardening) within the village were suggested. These amenities within the housing estate could bring a variety of benefits to older residents, in terms of, for example, convenience, mobility, sociability and even employment and sustenance. As several participants suggested,

> Once met a farmer who invented rooftop gardens to create food – planting rice, vegetable, etc. Can look to creating orchid farm near to house for the elderly to work where elderly do not have to travel far from their home, within their capability and at same time, beautify the environment. (Group B Participant)

> Enable older people to have more bonding with other seniors…Useful to encourage some cooperative store, where more abled elderly can jointly start a cooperative, sell groceries. They can share their expertise, e.g. baking/cooking a meal and having a meal together. What old people need, when you are losing your friends, you need to make new friends so this is one way to grow old graciously. (Group D Participant)

> Need to facilitate the elderly and make them feel mobile. They should easily be able to move on their own, look for friends, services, market, etc. How do we cut down on the barriers of elderly mobility spatially? Suggests a community mall that is completely pedestrianised…Elderly can come down from their houses to an environmental deck where they gather and meet other friends. Space for clubs and activities etc. (Another Group B Participant)

Two Group B respondents suggested that support services for both the old and the young could come together through the bundling of elderly activity centre, nursing homes, kindergartens, childcare, etc. The proximity of such services could complement the needs of each group—the older person being around young people and play while the young have the older people as guardians and educators. As they discussed,

> We can think about how we can cluster homes such that they are safe for the elderly to age in, conducive for ageing and the young – their needs are similar. Elderly become guardian of their community – have job of looking after community…e.g. put elderly home next to school can contribute to the programmes, entertainment and work with old people – old can teach young, help with their homework - create an environment of cross-age groups by placing them close to each other. High-density environment can support this kind of cluster of activities. (Group B Participant)

> Hoped to find a solution where elderly care centre can be designed next to a childcare centre, sight, sound of children's laughter can help stir the emotions for the elderly, where they have activities that reinforce each other's needs. (Another Group B Participant)

The village typology could further be enriched by technology like remote and personalised monitoring systems, and accommodate 3-generation living options as espoused by some respondents.

> Family activity and get-togethers are eroding away due to space constraints and smaller homes. 3-generational families in a 3-tier space could be important, from an architectural and spiritual point of view. There is no need to live in the same accommodation, but if they lived on the same floor or in the same block that would allow calling on one another and get-togethers to occur easily. Preventing elderly from getting lonely is critical especially in urban environments. (Group B Participant)

Technology and Smart Homes Respondents had varying opinions on utilising technology to enhance age-friendliness in homes. Several respondents suggested using technology to enhance elderly support systems (e.g. in assisted living), with remote monitoring systems and live-chat programmes as a means of staying connected (e.g. Skype). As they explained,

> Establish a monitoring system for the elderly within the housing estates. The best option will be the granny flats sort…This way children will live nearby can keep an eye on older parents along with having privacy for themselves. (Group D Participant)

> Sometimes Skype is used from a nursing station at the void deck[4] to communicate with residents. Laptop is fixed up in homes for them or donated or lent to them. In Thye Hua Kwan, there is also advancement in technology in personalising care, away from the actual medical facilities. (Another Group D Participant)

> Master planning needs to consider these needs – old people and young people can walk from any blocks to these centres/hubs, to pain management system – all tied up with town level community hospital for continued health support system, supported by remote monitoring with human touch at certain intervals – link up health system with satellite centres. (Group B Participant)

---

[4]Void decks can be found on the ground floor of apartment blocks in public housing. They have been purposefully left vacant (open space) and typically used for communal activities.

With the onset of big data, one respondent recommended employing data to better understand the older residents' profile in housing estates while another suggested the creation of digital platforms that would connect needs of older adults with services. To illustrate,

> Need to leverage technology to create more transparent, real time feel of what's going on in ageing people dominated estates. Right now we are not there yet. More can be done. (Group B Participant)

> A portal can be formed that pulls together, coordinates volunteers and matches them with services, i.e. match people's wants and needs. (e.g. a recycling repository) Interesting idea to capitalise on modern technology to help match users and providers (volunteers) such as using Craig's List. Start bridging gaps between the ageing-in-place team, the post-discharge team and so on. (Group C Participant)

Two Group B respondents further focused on how technology could enrich quality of life. One emphasised that users should be accustomed to and be at ease with technology before their implementation while the other suggested that design and technology could cater to those older adults who selected to continue to work from home—this being particular to the larger numbers of working professionals who made up the emerging old.

> Fill the place with technology…The place should be filled with technology I am familiar with. Can have technology to help learn new things. As you get older, learning capability decline. (Group B Participant)

> We put in grab-bars, ramps thinking they are going to die but need to re-think and enable them with technology, keep them in touch with the world, be of service – can they run their own business at home. Senior accountants can help SMEs [small and medium-sized enterprises] to run their accounts. Today's context, homes do not allow for businesses like these. (Another Group B Participant)

Despite the advantages that technology could bring, a few Group A respondent cautioned that challenges remain and person-to-person encounters should not be totally eradicated.

> But, there are still challenges in installing these systems in all buildings. Community based personal checking (neighbours and community workers keep check on the elderly) is a more reliable method as of now. (Group A Participant)

> …technology (whilst a great enabler) would also be a major barrier to the outreach to the elderly depending on the tech-savviness of the elderly individual…The challenge will be to make new technologies more intuitive and elder-friendly. (Another Group A Participant)

3.3.1.3 Specialist Living

Retirement Villages Seven respondents from Groups A, B and D raised the issue of retirement village. Responses revealed a presupposition that this housing type was a luxurious option for the wealthy, taking the form of a self-sufficient housing village, which allowed older people to live amongst themselves in spacious, quiet environments. Some respondents noted that retirement villages were uncommon in Singapore and referenced overseas (Australia, USA, Europe and Japan) models as good examples. Four respondents appeared in favour of setting up more retirement villages in Singapore and for its pricing to be controlled for affordability. Some said,

> In Singapore, there is not much talk about retirement homes, mostly only about nursing homes. In Maryland, US, housing village called Asbury Homes has graduated living conditions for people of different stages of ageing. A self-contained village development with medical, grocery, sports and recreation, games, theatres. (Group D Participant)

> Land should be set aside to build more retirement villages/homes. If the aged homes are priced appropriately, the aged will be able to afford it. Important to keep price low. Have nurses, doctors stationed at the retirement homes to assist in the care of the elderly. Learn from Japan, Europe. (Group B Participant)

Others disagreed that retirement villages were an age-friendly solution, particularly in Asia. To them, retirement villages segregated the older person from the larger community, which was socially unhealthy and did not align with the Asian cultural tendency to live with/in proximity with family. As one respondent offered,

> In Asian culture, retirement village is not the answer for age-friendly living. We prefer to live close to family. Elders might prefer to live alone but close to kin. There should be a policy that enables people to live with/nearby elderly. Children can be a phone call away. (Group B Participant)

Nursing Homes Nursing homes were mentioned by respondents from Groups A, C and D. Three respondents pointed out that the interior design of nursing homes could be improved. Poor examples of nursing homes looked too 'hospital-like', which encouraged negative connotations of the facility. More could be done to create a warmer and more welcoming environment to help alleviate these sensitivities.

> In Singapore, nursing homes are an extension of hospital design (a big room with rows of beds). That itself is a mental barrier for elderly people who become hesitant to move to such facilities. Hence, design of these facilities has a huge role to play. (Group D Participant)

> …age-friendly elements in a public space look much medicalised, e.g. all railings and other amenities made of only stainless steel. This gives an old age nursing home feel to users who prefer to not use these as they do not want to feel old. There should be way to fine-tune these design details to make them more attractive to the users. (Another Group D Participant)

Another respondent shared an example of poor programming in nursing homes where a lack of activities could lead to occupants feeling purposeless.

> Residents on wheelchair are aimlessly, sitting around – can do more with activities. (Group B Participant)

As with the isolated housing blocks and retirement villages, the common opinion was that nursing homes should be better integrated with the community. This 'integration' could be achieved in several ways—in terms of embedding nursing home or elderly care in housing estate, serving residents of the estate, employing staff and volunteers from local talent pool and seamless blending of nursing home design into local urban fabric. One participant from Group C went so far as to argue that nursing homes should be given sites within housing estates and not at the void decks. In his view, void decks could be used for day activities but not with beds or nursing. Others opined that,

> In housing, elderly care should be part of society. Recently, people were against having nursing home on ground floor of building. Day centre and activity centre should be promoted. More like day care and activity centres instead of nursing homes within the HDB blocks. Nursing homes should be allotted within housing estates rather than in the same block…Void deck can be used for community day activities of elderly (not nursing). Like educational, play or therapy. (Group B Participant)

> Nursing homes can be integrated with local communities; volunteers can come from the same community, nursing homes without walls, nursing homes and childcares integrated into one place etc. …At the same time there should be rehabilitation in the nursing homes to actively aid individuals to go back home to live. Usually need two people to take care of elderly and when one cannot afford two carers, these elderly end up in nursing homes … Nursing homes can have outreach programmes or an operation bandwidth of a certain allowance to have carers out and about in neighbourhoods. (Group C Participant)

> Could we do this in Singapore? We need to camouflage such facilities to attract people to use them. We need to accept it as part of communal life first. It has to be gradual. Nursing home could be built on a mixed-use property. They can mix well as a community within community. (Group A Participant)

Dementia Care Facilities A majority of Group C respondents and a couple of Group D respondents talked about housing for older people living with dementia. The health-care experts referred to international and local examples. International examples included a specialised close-circuit facility assimilated into the community, which offered residents a sense of freedom and normalcy but with the safety of an enclosed space. For instance,

> In Netherlands, [there] is a purpose-built home caring for people with dementia. This facility is integrated into the community with only one barrier of a glass door. One could get in anytime, but can't get out without help. The presence of amenities like hair salon, café, minimart drew people from outside too. They pay with fake money. (Group A Participant)

> In Scotland, there is more funding for a smaller number of people. There are specially designed-villages ('prosthetic environment') that can support dementia patients – a typology that can be replicated everywhere in the world. (Group C Participant)

Local examples included Apex Harmony Lounge (a private dementia-specific nursing home), in-hospital dementia care ward at St. Andrew's Community Hospital and hospital-run services (Khoo Teck Puat Hospital). The latter offered training to members of the community (including shopkeepers) to recognise dementia and send the older person back home instead of to the hospital. As the participants shared,

> Apex Harmony Lounge (first dementia specific nursing lounge) at Pasir Ris, designed by David Lynn from Brisbane is a good example. This designer was specific that every patient should have at least a room to themselves. But the cost implication is high. (Group D Participant)

> Currently, for dementia patients, there is only one registered care provider in St. Andrew's Community Hospital. There are only 6 beds now, a common TV corner and a space to roam

> around. Cooking is not allowed. There is a wandering garden as they walk a lot. Walking therapy sessions are provided. Games (ball games, mah-jong sessions) are provided. The age range is 50–80. It is 100% lock down. Doors and walls are camouflaged. Mostly grey colours. Dementia patients are not sensitive to grey, white or black. Flooring is bright for them to see, WCs [water closets] have to be colourful. Dementia patients have a tendency to run, and are different from psychiatric patients. Language of communication is important. There is also a tendency to get physically violent. 60–70% in SACH [St Andrew's Community Hospital] are foreign nurses. The dementia ward is one of the most challenging wards (for nursing). (Group C Participant)

> Rehabilitation training is required for patients as well as the caregiver and maids. Khoo Teck Puat hospital works with the police to train them to return those with dementia back to the community instead of sending them to the A&E [accident and emergency unit] in hospitals. They have also trained people in the community such as coffee shop owners to recognise dementia. (Another Group C Participant)

The accounts indicated that there was room for improvement for dementia care in Singapore. Currently, such care was costly or reliant on the support of acute hospitals, which could place a strain on staff and beds. In another example, dementia care was perceived as a 'locked-down' facility that was purposefully isolated from the rest of community. International evidence showed that though there were challenges, people with dementia might be cared for in the community if appropriate services were available (Alzheimer's Australia Inc. 2014; Green and Lakey 2013).

A Matter of Preference The choice of ageing in place or moving into a retirement village, the choice of using technology or not in one's home and life were matters of individual preference and effort should be taken to understand individual needs, preferences and options. Several respondents emphasised that an age-friendly home was one that was individualised and tailored to suit the needs of the user. The idea of home was different for everyone. Thus, meaningful conversations should be had with the older adults themselves about where and how they would like to live. As some Group D Participants proposed,

> These aspects depend on what you grew up having, and your preference for having them. Someone who has grown up in a flat without a garden and lawn may not necessarily find such a thing as important as another who has. Those who have grown up in HDB flats may find void deck space important. (Group D Participant)

> Elderly have to be housed as per their preference, life and character. Most elders in Singapore do not want to go to day care as they think it's socially degrading for them. In Singapore, elders prefer to age-in-place. To live with children and within community. The newer cohort of ageing population might be more mentally prepared to move to old age homes in the future. (Another Group D Participant)

3.3.2 Neighbourhood and City

The respondents were asked to rate the age-friendliness of Singapore's housing and neighbourhood spaces on a scale of 1–5 with 1 being very age-friendly. A number of the respondents (39.5%) declined to provide an explicit score as they believed that spatial age-friendliness was dependent on multiple factors such as location (where you go, topography e.g. hilly landscapes are tough to be age-friendly), age of housing estate (when it was built), type of residence (public or private), age groups/needs of the users (different mobility/sensitivities), personal attitudes (awareness, NIMBY, do we all see a need to equip places to be age-friendly?) and so forth. They argued that it was hard to give a single score; some places were more age-friendly than others.

One respondent illustrated this by describing different areas in Singapore—northern Singapore would score a 3, southern Singapore 4 and eastern Singapore 2-1.5. He went on to give the example of Marine Parade, which he would give a general score of 1.5, encouraged by physical redevelopment and improvement in infrastructure but felt there could be more attention on people factor and social consciousness.

Most respondents acknowledged that an increasing amount of work had been done in recent years to make Singapore more age-friendly (e.g. installation of ramps, adoption of universal design guidelines, different government departments are more willing to look into concerns of old people). But, more remained to be done (not all places had age-friendly equipment/provision, fragmented in approach, connectivity still difficult in some places). Some respondents gave examples of design features that seemed to work against age-friendly spaces,

> Some examples of age-unfriendly areas: On a walk from Whampoa to McNair Road RC, the ramp for disabled is very long and far away. Ramps are far and access is convoluted. Furthermore, there's one step after the ramp, which still makes it difficult for disabled users without a caretaker to move around. Telok Blangah Heights – lift stops at every floor but there are 3 steps to enter each flat. Not enough space to put up a ramp. At Telok Blangah Crescent [public housing] areas, lift does not stop at all floors. In Whampoa, there is pole in the middle of the lift to prevent the kids from running in, this is not friendly for all ages. (Group D Participant)

> Shallow drains are found all around void deck. This leads to tripping hazards and is not easy for wheelchair users to manoeuvre. These days the drains are found to be covered with metal covering. This should have been executed from the beginning. (Group B Participant)

> There are some buildings with layouts that are difficult to navigate and with designer signage, some of which are difficult to comprehend. Edges of steps are not conspicuous. This makes staircases harder to navigate. Handrails are a must and must be extended till the last step. (Group A Participant)

There was a general sense that there was more awareness now on ageing issues, studies were being done (with industry and research institutions to develop new technologies to help the older population, for instance) and places becoming more age friendly, less homebound for older people. One Group C participant narrated the story of how the bus driver would stop, come down from the bus to help old people get up/down the bus.

Among those who gave a rating, most scores were in the less age-friendly range; none of them thought that the physical spaces were very age-friendly while 23% (6) rated the spaces a 2: age-friendly. Several mentioned that public housing had become more age-friendly over time; newer HDB towns had implemented age-friendly design (e.g. Punggol Town) and upgrading made to existing estates. As one respondent observed,

> For existing estates, HDB has implemented a lot upgrading programmes to make both the external environment as well as the homes to be more pleasant, accessible, inclusive and safer. After implementing the island wide Barrier-Free Accessibility programme through the Town Councils all our estates have at least one barrier-free route. (Group A Participant)

A large number of the responses about age-friendliness outside the home focused on infrastructure for accessibility, mobility, safety, comfort and connectivity, particularly in accessing shops and services. A majority of the respondents spoke about the social space organisation in the neighbourhood that facilitated community interaction and activity. The question of the quality of space and the need for a consultative approach in planning and design of outdoor spaces were highlighted. Yet, others stressed the need to address the issue on multiple fronts since the older population's needs are multi-faceted. For example,

> As the needs of the elderly are multi-faceted, we are addressing the needs on multiple fronts, e.g. housing, transport, healthcare, employment, retirement adequacy, etc. Singapore's population is ageing, but the full impact of the ageing population has not been felt, and that means we can still put in place the necessary measures to prepare ourselves as a nation. (Group A Participant)

3.3.2.1 Going Outdoors

Accessibility and Mobility On reviewing the outside home environment, the age-friendly physical provisions that respondents cited most frequently were related to protection from the elements and preventing falls. They included covered walkways (eight mentions), safe and comfortable walking surfaces (e.g. levelled and anti-slip flooring and covered drains) and traffic crossings (e.g. convenient overhead or under passes and longer green-man light).

> Spaces that are convenient for them to move about are most preferred by the elderly. Most of them do not like to go out when it's very sunny. So covered walkways might be good. (Group C Participant)

> Gathering spaces have to be sheltered (to provide relief from the sun). Even a courtyard or a space surrounded by buildings or trees would help. A sheltered drop-off point is useful for helping elderly or PWD [people with disabilities] alight from cars. (Group A Participant)

> In order to attract elderly to use a public space, there should be more toilets at regular intervals. Readable signboards. More sheltered walkways. The presence of these amenities must most importantly be easily visible as they cannot read location maps. There could be more friendly telling machine or more conspicuous kiosks. Trees and landscapes are important too. (Group D Participant)

> We don't have open drains in most places, which is good for uninterrupted movement. (Another Group D Participant)

Other features to improve the older person's outdoor experience included the provision of toilets and shaded seating at convenient intervals (five mentions), legible signage, appropriate lighting, visible amenities (e.g. ATM machine), sheltered gathering spaces and drop-off points and sufficient greenery (to provide shade and a pleasant journey). Groups A and B respondents highlighted that these components were crucial in parks as well, with the addition of conveniently located drinking fountains, public phones or emergency call points (especially if the grounds were large). A couple of respondents from Groups C and D suggested that outdoor spaces and shops could be pet-friendly and child-friendly as older people often had them in company.

> Spaces should be convenient for them to take pets or grandchildren for walks. (Group D Participant)

> Ikea has designated spots to anchor pets. Such a provision encourages the elderly to walk outdoors and even go shopping with the company of their pets. (Group C Participant)

Two respondents suggested that even though safety and conveniences of spaces were paramount, the physical challenge of 'inconvenient spaces' might be beneficial to health and fitness and need not be eliminated altogether.

> Some studies show that it is not necessary that every area needs to be accessed. 'Inconvenient design' might also be beneficial to the elderly at times. (Group D Participant)

> All of the physical environment needn't be made easy to use (inconvenient architecture). Slopes and small risers are still needed for the elderly to climb. They need to lift their legs and move it. It gives them a chance to exercise and to move limbs. Give room to accommodate their movement in design. If everything is near, it might be too harmful to their health in the long run. But, we should give them security and safety mentally and physically. My recommendation—let them walk freely, if they fall, let them fall safely. (Another Group D Participant)

Proximity and Connectivity Proximity to amenities was discussed by a few respondents from Group B and Group C. Respondents proposed a radius of not more than 400 m to 1 km to market centres, essential services and schools, etc. to promote walking. Distances could be longer if the need to cross roads was eliminated. For example,

> Elderly should not be isolated from these services. Not all have cars or access to MRT to go to mall to access all of this. We need to see what the population actually needs primarily. Ideal location for medical facilities - within precinct 800 m^{-1} km radius. So, maybe every 5 blocks. Good size can be 10–20 blocks within the defined area, without having to cross roads. (Group A Participant)

> 400 m radius to school, shop, etc. should in practice mean: do not encounter car, no threat of car. Put need in forefront, not as after-thought. (Group B Participant)

Connectivity between components was crucial. About a third of the respondents (11) from all groups highlighted the need for connectivity—terms such as 'first mile and last mile', 'integration', 'seamless', 'spaces in between', 'whole journey experience' were used. While some respondents (2) opined that connectivity was currently at a commendable standard, a majority of the respondents felt that more could be done. To many respondents, an absence of good connectivity could discourage older people from moving out of their homes into more social setting outdoors. As several argued,

> Last mile connectivity should be accounted for when planning facilities for mobility. There's no point designing modern transport facilities and not being able to connect the users to the last mile. The whole journey of user should be considered. (Group B Participant)

> Some gaps in planning include: In connectivity, decreasing the number of steps. Visibility of signboards can be improved, better lighting at road junctions. Adding new features that help to increase the safety and accessibility for seniors. (Group D Participant)

> The entire thought process about design needs to be integrated…Need to think about the people who are using these spaces…Lot of modifications need to be made to let people move around freely like closer drop off points, ramp, etc. (Group A Participant)

> Lots of emphasis on buildings – can put more emphasis on space between buildings. Start to fill the void deck with support infrastructure, consider design for social infrastructure to provide glue to society – started to do this. There is opportunity to do this in Singapore compared to other cities… Khoo Teck Puat Hospital has a nice park and across the road is an old folks home. But, the old people have to chart a bus to go this park opposite the road… Shared facilities matter – need to consider the hospital, school and elderly home, park and community centre by clustering them together to activate the space. (Another Group B Participant)

Two respondents from the building and planning industry suggested that the lack of good connectivity between buildings and spaces in the city could potentially signal a lack of integration of policies amongst agencies.

> There is a lack in integration/connectivity between buildings; this affects the accessibility in the commute and the precinct. More integration of policies amongst agencies within the built environment sector. (Group B Participant)

> The planning policies should give permission for developers to develop and design till the edge of the site and not just around the building. The covered walkway needs to be extended all the way from one boundary till the next one. (Another Group B Participant)

3.3.2.2 Vision for an Age-Friendly City

In discussing the ageing population and outdoor space, respondents from all groups (particularly Groups B and C) expounded their vision for a car-lite environment and for an alternative model involving public health in design. They felt that these visions could begin to redefine the quality of space that a society wants to achieve beyond the model of capitalism and efficiency. While these ideas were not all directly related to the ageing population, they were about people-centricity, which could lead to a more age-inclusive environment. As one Group A participant deftly summed up,

> One of the challenges that Singapore faces is the high density and the dependence on cars. With the trend towards other forms of transport like bicycles and motorised personal transport systems, even our pavement and footpaths are becoming less pedestrian-friendly. We need to address these competing needs for road space and footpaths. It is a challenge to make the environment more elder friendly. Wide roads also make it less friendly for elderly to cross. (Group A Participant)

A correlation could perhaps be drawn between *reducing cars and roads in the urban environment*, *embracing rehabilitation over care* and *creating a more age-friendly environment*.

Car-Lite Neighbourhoods Two respondents from the building and planning industry advocated reducing or eliminating cars (and car-related infrastructure) from housing estates. Other respondents that indirectly championed car-lite development wanted to reduce traffic crossings and make roads safer (e.g. better designed overhead bridges and longer green man). They indicated that car-lite development could improve safety, mobility, connectivity and convenience for older people, promoting a more age-friendly environment. For example,

> ...Lesser cars, lesser car parks, lesser driveways within housing estates. Simplistically, just reduction of number of car parks and increase in length of footpath will bring noticeable difference in walkability...Segregate car traffic from housing environment. This can help mobility. (Group B Participant)

More than reducing cars and traffic was to diminish the fear of being outdoors (for older and younger adults) and create a more walk-friendly environment that supported human encounters. As some argued,

> Is there a way we can design, so that it gives people a reason to walk and create opportunity for interactions? Walkability is a desired outcome. How to make spaces walkable? And how to minimise conflict between humans and cars? Comfort, shade and walkways are important along with creating destinations within walkable distance. (Group A Participant)

> Make space more defensible, design cars to disappear to periphery or underground...People feel safe, people will venture out – need to have no fear of infrastructure. (Group B Participant)

Healthy City, Healthy Parks Groups C and D respondents espoused the importance of building a public health dimension in design. They brought up the recent mindset shift in healthcare – there had been a push for choosing recovery over care (i.e. getting people back on their feet and out of hospital rather than to remain in treatment), encouraging preventive action through health promotion and education in self-therapy. Having this public health outcome in mind would instinctively influence the design of space. As some experts observed,

> We need to look at ageing in a much more holistic way. The government dreads over-medicalising ageing, but this is what happens when they are inadequate in addressing and managing the independence of individuals—there is no system in place to incentivize people to choose Recovery over Care. People should not over-consumer healthcare, they should over-consume rehabilitation instead... Send a therapist to make sure someone is back on their feet for free. The value of rehabilitation has to be demonstrated. If we have this conversation, then housing design will come intuitively. (Group C Participant)

> In a health perspective, the indoor and outdoor environment should support to promote physical and mental wellbeing of the elderly. A recommendation would be to have physiotherapists train a group of elderly in the neighbourhood parks who in turn can help to train the rest of the community. This will very beneficial as not all elderly can afford to visit the hospital regularly to do physiotherapy. (Another Group C Participant)

> Giving an incentive to stay healthy instead of giving incentive if they fall sick. Train people to get into active routine, which they can continue in later life. (Group D Participant)

There was potential in the aforementioned village development and satellite health-care structure as a housing and infrastructure typology to be developed in tandem with the recovery-over-care model.

Related to public health was the need for parks to preserve a quality of serenity and be a place for relaxation. Several respondents identified parks as places of organised activity and social gathering, and that many people, particularly older adults, also visited parks to relax, unwind and 'escape'. A balance, therefore, should be struck between these innate park qualities.

> The goal is how we can make the park even 'prettier', and a place for people to 'escape' to. This quality is lacking. The quality of space feels 'right' when you feel like you have escaped the density and stress. While people like parks, they do not want to get too far away from a city or their house. A park should have intrinsic quality that makes one feel relaxed and calm down. There is programming in the park… These are no doubt popular, but the organization of activities, especially larger ones, require much resources and preparation. These might not be sustainable in the long run. They also drive away people who want a nice tranquil park. (Group A Participant)

An emerging sense was that championing public health to promote health prevention (e.g. build upon the role that parks could play in the health of people) could create a more age-friendly urban environment. The ultimate goal was for such environments to foster health and wellbeing and the participation of people as they age.

3.3.2.3 Building 'Heartware'

Relating to the issue of a healthy and people-friendly city was the question of 'heart-ware'—building heart and humanity in the physical environment. Respondents from Groups A and C remarked that Singapore was lacking not in hardware but in the 'soft' interfaces that made a more compassionate environment. On these facets, responses highlighted the need for a personal touch, sense of ownership, building human bonds, rallying groups and the formulation of people services to accompany infrastructure development. Some experts observed,

> In Singapore, we are not desperate for hardware. We are lagging behind in adequate software…We are good at delivering services, not in formulation of services. We should have clear objectives and deliver on time. All these efforts are software in form of resources and capital management. It can make and break any effort. (Group A Participant)

> Infrastructure needs to be present to support activities and programmes. However, it needs to be supported by 'software'; it will not be successful unless people want to come out of their houses. There is a need for leaders, people, programmes, to encourage participation. Programmes need to be based on people's interest. (Another Group A Participant)

> Public spaces are quite sanitised, neat and clean, no man's land, do not belong to anybody. Need to shift this, need to colonise and take charge of space. Good at admin, need to allow people to use and take ownership of the space. (Group B Participant)

> Need to look at the softer issues of society and how that translates into the creation of space, design of living environment… (Another Group B Participant)

> It needs people to draw people out. No matter what programmes or spaces you create, there needs to be that personal touch. Grassroots leaders, NGOs, volunteers, etc. can help to achieve this. But, all of them need to be more coordinated and all efforts need to be pulled together effectively. (A third Group A Participant)

Drawing the Older Person Out The use of purpose-built or general outdoor spaces for older people had to be synonymous with 'people', 'participation', 'programme', 'activity', etc. This suggested that the social function of an outdoor space was equally, if not more important, than its physical attributes such as size, availability of seating and the fact that it was purposefully-designed. Various responses (from across all groups) recommended that outdoor spaces should be designed to accommodate activities already present (e.g. tai chi in the park) or be enriched with programmes that incentivise participation of older adults. For example,

> Senior activity centres are good spaces. In order to enable the elderly to use outdoor spaces, organised activity is important to bring them together. (Group D Participant)

> A sitting area is a natural place for congregation and interaction but there needs to be activities or an incentive to draw people down. (Group A Participant)

> Introducing more spaces to accommodate such activities in our community neighbourhoods will prove beneficial to improve the wellbeing of the elderly (e.g. tai chi, qi gong, community gardening). (Group C Participant)

> Space planning is more important than size. Arts and cultural events could act as a catalyst to get the elderly to participate. Can introduce free participation. Innovative ideas are required. (Group B Participant)

> In Pasir Ris Park, there is a group of elderly that gather to do tai chi every morning. There is no designated space for the activity and as the group grows, it spills out onto the jogging path. Hence, an area could be designed for this existing activity to support and encourage its growth. It brings the community together. (Another Group B Participant)

For programming to be sustainable, respondents stressed that community-driven initiatives and self-organisation were essential. Other responses suggested that social services were often needed to kick-start the initial engagement and encourage community leaders including from among the older population to emerge. As several respondents shared,

> In Tampines,[5] void deck was converted to Seniors Corners. Someone has to take initiative. One leads to the other. We also trained some elderly. They learnt to facilitate the session on their own. Then they conducted these sessions on their own and this is still continuing. So, the key is to identify a small group to start. (Group A Participant)

> Programmes are important; what holds back the elderly more is how to organise groups together. Do we have the people who organize the events? In Singapore, if you want to find space to walk or do some exercise, we do not have far to go. Question is more of whether they [older people] can organize themselves in groups on their own to take part in such activities. (Group D Participant)

---

[5]Tampines is a public housing new town in the eastern part of Singapore. Constructed since 1978, Tampines is the 3rd largest new town by area (2089 ha) and population (258,310).

> When it does not form naturally, we can provide social services to get things going in first instance, seed community action or engagement – do it in fun way, develop human connection. (Group B Participant)

A Consultative Approach to Design Apart from usage and programming, the design process and management of space could take on a more humane and gracious approach. Respondents stressed the need for public consultation in the creation of public spaces. People's voices, needs and experiences including the older person had to be heard, considered and articulated in their spaces. As various participants emphasised,

> More consultation with the public and users before designing spaces. (Group D Participant)

> More consultative in creating spaces – engage people to create the space. Need to listen to people – what are their needs and design in terms of those needs. (Group B Participant)

> More innovative efforts are needed that involve the users for the spaces to work. The success of spaces also depends on the culture and preference of the users… Letting the users be part of the process is a more engaging way of redeveloping and designing the estates. This process will also enable the user to engage with the designer and discuss the possibilities of what can or cannot be done will help them understand and think of what is doable and what is not. (Another Group D Participant)

> The thing that everyone values needs to be articulated in the design of social spaces. (Group C Participant)

A degree of spontaneity of use and the organic development of activities should be accommodated and encouraged. As some experts observed,

> There should be a bit of spontaneity, design chaos, organic design, less manicured gardens. (Group B Participant)

> Agency should allow and tolerate the organic development of elderly groups and activity. Users might not think pre-determined areas for elderly corners are suited for their needs. Therefore, communication with RCs [residents' committees] required. (Group D Participant)

A couple of respondents from Groups B and D observed that 'unfinished' or 'non-identified' spaces could be left (with a 'set of simple rules') for users to take charge/develop their interpretations of use.

> Leave unfinished spaces for elderly corners. Some MPs [Members of Parliament] are trying to create a social space for elderly to congregate, however, this is a top-down approach, regulated and directed, and is not a spontaneous gathering. Informal gathering requires responsible usage and should not obstruct others from using the space. (Group D Participant)

> Create some non-identified cluster spaces for people to affect and work on to become spaces of their own – different parties to take up different portions… e.g. eldercare is master organizer and let community rally around this space – get people creating space, set simple rules about how they should engage with each other to minimize conflict – over time, identity of space will emerge as well as culture. Different communities being known for different aspects or spaces, people will come and love the space. Do not give all spaces to be designed by professionals. (Group B Participant)

3.3.3 *Society and Individual*

3.3.3.1 Values and Mindset Towards Ageing

Several respondents across the groups suggested that current physical, social and healthcare structures were often underpinned by the perception that older people were vulnerable and dependent. To them, this mindset towards ageing was apparent at all levels—individual, familial, professional and governmental—and was identified to be a concern and a barrier to creating a more age-friendly city.

> Mindset and attitude are the biggest barriers… We tend to have a very morbid view of ageing; we can be more positive about the future of ageing. Need to rethink NIMBY [not in my backyard] syndrome. (Group B Participant)

> Ageing is not a disease in itself but a way of life. Key concerns are not necessarily to do with physical spaces, but attitude towards ageing. In families nowadays, if an elder finds difficulty to walk, they think of wheelchair first rather than trying to help them walk again. Awareness is there, but limited… Social barriers mainly [include] the perception of ageing of an individual versus that of his family. Even if an individual perceives himself as healthy, family won't trust him to go out and exercise. (Group D Participant)

Many of the current attitudes towards the older person ranged from aversion, apathy to protectionism, e.g. young people not wanting to live close to older people, developers designing for space optimisation instead of people-centricity, national subsidies for medical care instead of rehabilitation and so on.

> It is the people. Young people refuse to stay around the older people. (Group C Participant)

> We have to stop thinking of seniors as the 'silver tsunami'. There are nice ways to describe seniors other than words that connote fear, exclusion. Second thing is how we treat them. (Group D Participant)

Several respondents opined that such mindset often informed common understanding of age-friendly design. The result was design taking on an approach of risk mitigation, special add-ons or simply, exclusion.

> User-friendliness for ageing population is different from risk mitigation – [we are] looking at a graceful, gracious environment, where ageing population could feel relaxed, feel accepted and integrated to the rest of the community – [an] inclusive society…Today, there is silo approach to planning and design, which rules out certain types of people. (Group B Participant)

> Ramps and lifts are usually tucked in an unseen corner in the building (e.g. a shopping mall), while the main design is usually for able-bodied. UD [universal design] facilities are there but one does not feel a sense of being welcomed. Welcoming spaces are when everything just falls in place and one does not have to think twice about obstacles that may impede one's journey…This 'welcoming' aspect is not designed in totality. Retrofitting does not solve the issue. (Group A Participant)

A third of the respondents emphasised the need to reframe people's understanding of age-friendliness and universal design and to internalise the values of inclusivity and compassion in these concepts.

> Universal design is for all sectors and population, for everybody. Not specifically design for one group of people. (Group B Participant)

> We should design for the entire life course - young, old or disabled. It must suit everyone…How to make a city evolve to make it flexible for any percentage of any age group of people? (Group D Participant)

> We should not look at the elderly as helpless, and need to be rescued. We should shift our paradigm and approach – it is an opportunity to continue the development of the city, time to be all inclusive… Society becomes more inclusive when both young and old needs are addressed. By addressing needs issues, the attitude does not become one of fear and exclusion but inclusiveness…How to start thinking of 'kind space', 'loving space' – emotional/moral/social thinking needs to be cultivated alongside cognitive development. (Another Group B Participant)

> Development needs to be future-ready. Spaces should focus on the outcome of inclusiveness - across all ages. Developers are opening ourselves to such a demand. (A Third Group B Participant)

Education was identified as a concern and also an important change agent.

> Education is a concern. Inculcating the right values, respect for the elderly is important. Need to start young so one will be the first to get up and offer a seat. (Group B Participant)

> Education and continuous awareness is important. Creating a continuous awareness that ageing is not disease in itself but a way of life. Younger people should be aware that ageing is not equal to wheelchair in reality. It doesn't mean immobility in all cases…So the key concerns are not necessarily to do with physical spaces, but attitudes towards ageing. (Group D Participant)

> Education - reaching out to developers and architects. Awareness of design and giving guidelines. Hospitals take that into account while designing spaces but not commercial spaces…BCA [Building Control Authority] only develops guidelines, only the developer can ensure more space. Overcome barrier through education, then developers can make their choice. (Group C Participant)

Several other respondents brought up the need to institutionalise mindset change as a long-term goal—to change mindsets on a societal level and the normalising (mainstreaming) of inclusive design, step-down care and bottom-up initiatives—all of which was suggested to create a more age-friendly city and society.

> Change of mindset. If idea of universal or inclusive design (age-friendly design) can become the new norm like idea of 'Garden City', and the whole nation can be unified in this thinking, then more can be done faster. (Group A Participant)

> In Singapore, we work on making the home age-friendly first, institutionalisation should be the end result. People are moving gradually from dependency to independence. They need to be empowered with ability to rely and integrate into the population. The agencies should work together with the end users from beginning to end. Thirdly, recognize that growing old is tough. (Another Group A Participant)

> There is a pressing need for institutionalising the care giving network in Singapore…. Long-term hospitalisation was not the solution… Institutionalising of the care network can prove to be very beneficial in the wellbeing of the elderly in the long run especially in cases like this one where the elderly are socially active. (Group C Participant)

> Mindset of the leadership; if we don't change that and look differently from the elderly's perspective, none of the efforts will work. We are trying to change this by example in Whampoa. But it's difficult. We need to compliment top down approach with empowering the bottom levels. Channel these task oriented and skilled people to be open to elderly opinions. (Group D Participant)

3.3.3.2 Empowerment

When asked to discuss the factors that would empower older people to participate more and take on an active role in their community spaces, respondents stressed the need for *awareness, information* and *participation (communication)*.

> Empowerment is a function of awareness and information. At this current generation, the older people do not have enough information. (Group C Participant)

Participation in itself was a form of empowerment. As stated above, public consultation was considered vital in the creation of a more age-friendly and gracious city. Along with this was the parallel need to encourage the older person's *participation in the consultative process*. This could be facilitated through raising awareness and providing information. Equally important was the willingness for agencies and practitioners to listen and receive feedback in a reflexive, ongoing process; this was necessary to long-term sustainability. As some respondents observed,

> If we want Singapore to be city for all ages, we should start our planning process in a more participatory process. It is about empowering all the residents, starting with the elderly as they are the most vulnerable. The process of community building should be the first step to design. (Group D Participant)

> First, listen to them [elderly], facilitate more engaging activities and support them to do things for themselves…[Their] needs are reasonable. So, we need to find out what they really want instead of assuming what they want… Leadership style has to be facilitative style. This means that there should be allowance for vulnerability. Inspire them. Processes might not be so orderly, but you should allow for flexibility. Practitioners should use their skill to coordinate… After you improve facilities, you need to go back to them again for feedback. The communication should always be ongoing. The key is to consult people or ask them what they want. (Another Group D Participant)

At the community level, *participation in local activities* could be a means of empowerment—by showcasing skills, stepping up leadership, organising events, maintaining social ties.

> First, we need to engage [retired professionals] in more intellectually stimulating activities. They need to be able to showcase their own capabilities… This is how they can be empowered. (Group A Participant)

> We need to bring more awareness of the positive impacts of a more active social life on the health and wellbeing… The elderly should have a sense of purpose. Their lives need to be meaningful with a proactive approach. (Group C Participant)

> Seniors want a place where they can come and go freely, and take a proactive approach in setting up, managing and running the place. Besides the software is important. Need to work

> with HDB resident committees, which are nodes of contact with residents; RC [residents committee] needs to be given more empowerment, to reduce paperwork in order to engage the seniors. (Group D Participant)

> Organised activities will empower and enable seniors to participate in community. 'Activity begets activity'. Mental stimulation, psycho-emotional support is important. Developing a sense of independence, positive outlook and a value system, e.g. belief in God to provide a spiritual wellbeing. Belief in a system that will allow empowerment. (Another Group D Participant)

Another example of the empowerment of older adults was through *staying healthy, having health knowledge and management*. Respondents identified the importance of involving older people in health-related dialogues and equipping them with knowledge in health and sickness management. As they remarked,

> First thing is health. Health is one of the key empowering factors because without health, you don't feel like doing anything. (Group D Participant)

> In families nowadays, if an elder finds difficulty to walk, they think of wheelchair first rather than trying to help them walk again. Awareness is there, but limited. Healthcare has to be an ambassador to learning about ageing. (Another Group D Participant)

> There should be public education efforts to understand the problem solving as a result of a disease instead of just the disease. Health care providers have never had these conversations with patients so they do not know how to deal. (Group C Participant)

> Pre-retirement counselling, etc. can greatly empower [emerging old cohort]. There must be more mindset preparation. But, with the current oldest old cohort this is more difficult. These people need more moral encouragement to imbibe the process of being able to enjoy in whatever predicament they are in. (A Third Group D Participant)

Some respondents shared that *employment* could be a form of empowerment. Employment was related to continual engagement in workforce, keeping skills intact, income and opportunity for new skills training, etc.

> Healthcare needs to go hand in hand to make sure they are sustainable, to retain them in the workforce. Social engineering required? Need to design housing for old people better, enable them with technology, help them keep in touch with the world, run service/work from home. (Group B Participant)

> Employment is another way to empower them. If they need a job, there should be ways to re-train them to do other jobs more suitable do their energy and fitness level. Empowering means also the employers, all of us who employ people. To make it easier for people to employ us, we must lower our compensation expectations. (Group D Participant)

> Some steps that could be taken immediately could include improving employment for the elderly, Life-long learning…retirement adequacy, more research on aging etc. (Group A Participant)

3.3.4 Different Groups, Similar Concerns

By industry/sector classification, respondents might have belonged to different groups. But, the findings showed that their concerns for ageing and age-friendly spaces were often similar and crosscutting raised by all groups. This could be due to the salience of the topic in recent times, background training of the respondent or personal dealings with older people (e.g. ageing parents, patients/clients) that might have shaped their perspective regardless of occupation.

Many respondents (including a wheelchair user) who were also caregivers (with ageing parents) stressed the physical components of age-friendliness the most. This could be attributed to their hands-on dealings with older people in their everyday environment (e.g. flats and void decks) and thereby encountering issues of accessibility, safety and comfort on a higher frequency than other groups. Respondents from this group were also the only ones who highlighted the rental flats of the poor older adults, the dire condition they were in and the urgent need to look into this matter.

Group D respondents and caregivers from other groups also emphasised the social aspects of elderly life. These included tackling loneliness, engaging in community activity, counselling to cultivate better mindset towards ageing (for the older adults themselves and their family) and the need for more after-care services and workers/volunteers—points associated with day-to-day social needs of older adults in their homes and neighbourhood spaces.

Group B respondents and some Group A respondents (of building and planning background in government agencies) were inclined towards defining age-friendly design or universal design as designing for all people and for inclusiveness. This could be due to their familiarity with the Building and Construction Authority (BCA) Universal Design guidelines, which defined it as such.[6] These respondents also spoke about the more abstracted concepts of 'loving', 'caring' and 'inclusive' spaces. More could be done to realise such spaces. Importantly, to understand what such spaces might be like. They further stressed the need for inter-agency collaborations in inclusive spaces.

> There's room to improve laterally across government agencies and vertically across communities and community centres. The link must be improved. (Group A Participant)

The discussions of Group C respondents and healthcare experts from other groups regarding an age-friendly environment were often framed by a public health perspective. They opined that the physical environment should facilitate the older person's physical and mental wellbeing, decentralised elderly support (step-down care, assisted living) and embrace the concept of recovery-over-care. These also had the purpose of managing healthcare resources (e.g. acute care hospital beds, nurses, etc.). They emphasised the importance of going beyond economic model considerations to include public health and more people-friendly characteristics in the design of built environment. As with Groups B and A respondents, they highlighted the importance of collaboration in design thinking. More communication and collaboration were needed between those that care for, treat and accompany older adults, the older adults themselves and those who design the environments for them.

---

[6]BCA Universal Design guidelines were introduced in 2007, institutionally recognising the growing number of people with different degree of abilities including the ageing population in Singapore and the need to cater to their needs.

3.4 Conclusion

The key informant interview is a qualitative tool that allows the collection of insights from a selected range of experts. The findings highlight the role of an age-friendly built environment as a facilitator for physical and mental wellbeing for the older individual and community. Discussions brought up how the physical environment could be shaped so that it was safe, accessible, comfortable, well-connected, conducive for walking or taking public transport, supports community life, etc. While there was room for physical improvement to create more age-friendly housing and neighbourhoods in Singapore, many acknowledged the improvements made and stressed that what was more crucial was understanding the 'why'—why age-friendly housing and city needs to be done.

As one respondent put it, we would just be 'tinkering around the edges' if we were unclear of our motivations (be it barrier-free access add-ons or risk mitigation, among others). This awareness needs to be developed in tandem with a shift of mindset towards a positive view of ageing and the aged, cultivating societal compassion and graciousness, 'heartware' along with hardware. The key learning points from the findings are:

Housing

- Age-friendly hardware for the home was related to issues of safety, accessibility and comfort.
- Key age-friendly hardware included handrails, non-slip floors, emergency call points, appropriate height of furniture, windows and switches, room layout and sufficient manoeuvre space. Toilets were a place of emphasis.
- Include future-proof adaptations in house at the outset so as to allow older residents to remain in situ and independent in their homes as they age.
- Increase awareness about age-friendly design so as to mainstream age-friendly housing.
- Age-friendly or universal design was not necessarily costlier than non-age-friendly design.
- Need to look into the poor older adults' ability to afford housing or access financial aid. The poorest of the older adults were the most vulnerable group; usually renters and current condition of their rental flats need to be looked into.
- Elderly housing should be integrated with the larger community.
- Not one but a range of housing including, for example, specialist housing, village housing was needed to meet the differing needs and preferences of older people. Since the notion of home differed, meaningful conversations should be had with the older population about their preferences for housing.
- There was room for improvement for dementia-friendly facilities.
- Technology could be more used to enhance age-friendliness in homes and quality of life but personal touch should not be diminished.

Neighbourhood and City

- Age-friendly hardware for the outdoors was related to issues of safety, mobility, accessibility, comfort and connectivity.
- Key age-friendly provisions for the outdoors included covered walkways, toilets and shaded seating at convenient intervals, safe and comfortable walking surfaces and traffic crossings.
- Other age-friendly provisions for the outdoors included legible signage, appropriate lighting, sheltered gathering spaces and drop-off points, sufficient parks and greenery.
- 'Inconvenient spaces' might be beneficial to health and fitness and need not be eliminated completely. A degree of spontaneity in the use of outdoor spaces could be accommodated.
- Connectivity and proximity to spaces/amenities was crucial in encouraging the older person to step outside of the house.
- Going car-lite could improve safety, mobility, connectivity and convenience in neighbourhoods and reduce the older person's fear of being outdoors; it is a feature of a more age-friendly outdoor environment.
- Recovery-over-care public health model could inform the design of more age-friendly neighbourhood spaces and amenities.
- Championing public health in design (e.g. through the reduction of car-related infrastructure, design of better parks, study of progressive healthcare models) could create a more age-friendly city.
- 'Heartware' needs to be developed—humanising infrastructural development with a personal touch, people services, social bonds, sense of ownership, etc.
- Social and psychological functions of a recreational space were equally, if not more important, than its physical attributes.
- Social services could help kick-start the initial engagement in community activity.
- Community-driven initiatives and self-organisation were essential for sustainable community activity.
- Public consultation was necessary in creating user/age-friendly public spaces.

Society and Individual

- Ageist mindset was identified to be a concern and barrier to creating a more age-friendly city.
- Current attitudes towards the older adults ranged from aversion, apathy to protectionism.
- Ageist design takes an approach of risk mitigation, special add-ons or simply, exclusion.
- Understanding of age-friendliness has to be reframed to internalise the values of inclusivity and compassion.
- Establish inclusive design, step-down care and bottom-up initiatives as long-term goal.
- Older people can be empowered with information, awareness and participation.

- Empowerment can be in the form of participation and gaining awareness in: place-making consultation, community activity, employment, health and sickness management dialogues, etc.

Acknowledgements We gratefully acknowledge the following for their generous support and assistance with the Key Informant Interviews: our colleagues, Laura Jasmine and Parvathi Nair for taking notes and summarising several of the interviews; interns Lim Aunn Ning, Madison Landon and student helpers for assisting us with transcription of interview recordings; and all interviewees who spent their valuable time to meet and generously shared their ideas and perspectives.

References

Alzheimer's Australia Inc. (2014). *Living with Dementia in the Community: Challenges and Opportunities*. Alzheimer's Australia Inc.

Green, G., & Lakey, L. (2013). *Building Dementia-friendly Communities: A Priority for Everyone*. UK: Alzheimer's Society.

Miles, M. B., & Huberman, A. M. (1994). *Qualitative Data Analysis*. Thousand Oaks, CA: Sage.

Powell, L. (Ed.). (1999). *Conducting Key Informant and Focus Group Interviews*. Boise, ID: Mountain States Group.

USAID Center for Development Information and Evaluation. (1996). *Conducting Key Informant Interviews, Performance Monitoring and Evaluation Tips No. 2*.

Yuen, B., & Soh, E. Y. X. (2017). *Housing for Older People in Singapore: An Annotated Bibliography. Springer Briefs in Ageing*. Cham: Springer.

Chapter 4
Housing Audit

Belinda Yuen and Parvathi Nair

Abstract Housing conditions strongly influence people's quality of life and contribute to their health. Housing features such as entry stairs, narrow doorways could potentially help or hinder the older person's ability to remain active, independent and socially included in the community. In thinking about the influence the home environment has on the older person's quality of life, we have developed a Housing Audit Checklist to analyse the age-ready quality of individual homes. This Chapter introduces the Housing Audit Checklist, the results and recommendations from its pilot testing and validation on a sample of older residents in Singapore.

4.1 Introduction

Housing conditions strongly influence people's quality of life and contribute to their health: 'a complete state of physical, mental and social wellbeing' (World Health Organisation 1946, p. 100). International evidence suggests that the home is where many older people spend the majority, over 80% of their time (Help the Aged 2006; US Department of Labor 2014). Ageing impacts many activities of daily living within the home such as moving around the home, home upkeep, among others (World Health Organisation 1998).

Housing features such as entry stairs, narrow doorways could become a challenge as people age in their homes. Small changes to housing design (e.g. having level access and flush thresholds at the entrance, bathrooms and toilets that are capable of conversion to disabled standards) can potentially improve the older person's

B. Yuen (✉) · P. Nair
Singapore University of Technology and Design, Singapore, Singapore
e-mail: belinda_yuen@sutd.edu.sg

P. Nair
e-mail: parvathi.snjv@gmail.com

B. Yuen (ed.), *Ageing and the Built Environment in Singapore*,
https://doi.org/10.1007/978-3-319-92444-1_4

wellbeing and quality of life (World Health Organisation 2007). They can help or hinder the older person's ability to remain active, independent and socially included in the community. There is a compelling argument for housing design to be inclusive. If homes were better fitted to the needs of people at all stages of their lives, predictably there would be less need for specialist residential housing for the old. Experts predict that with the development of tele-enhanced medical and care interventions, the home will become the centre of health and social care in the coming decades (Mountain et al. 2015).

In thinking about the influence the home environment has on the older person's quality of life, we have developed a Housing Audit Checklist (HAC) to analyse the age-ready quality of individual homes.

This Chapter introduces the Housing Audit Checklist, the results and recommendations from its pilot testing and validation on a sample of older residents in Singapore.

4.2 Methodology

A quick scan of the literature shows that accessibility to housing and building is an important concern. Access audit checklists have been developed to help housing providers and landlords to identify accessibility barriers. An example is the Access Audit Checklist developed by the Royal Borough of Kensington and Chelsea, UK (https://www.rbkc.gov.uk/pdf/Access%20audit%20checklist%20communal%20areas.pdf). Still in the UK, the CABE (2004) has published a housing audit to assess the design quality of 100 recent housing developments in London, the South-east and East of England. The focus is on urban design (e.g. layout, place-making) and development process of housing. Using the national Building for Life Standard, housing is assessed along four key criteria of character; roads, parking and pedestrianisation; design and construction; environment and community.

Elsewhere, e.g. USA has developed housing inspection checklist to determine the state of infrastructure relating to water supply, toilet facilities, sewage and waste disposal, fire protection and safety of the living areas in the house. An example is the Labor Housing Inspection Checklist by the Department of Consumer and Business Services, Oregon Occupational Safety and Health Division (http://osha.oregon.gov/OSHAPubs/checklists/1876.pdf). The World Health Organisation (2007) under its age-friendly cities programme has developed an Age-friendly Cities Guide using a set of outcome and impact core indicators like neighbourhood walkability, accessibility to public transport, to measure the age-friendliness of cities.

Housing is one of eight domains reviewed in the World Health Organisation Age-friendly Cities Guide as essential to the older person's safety and wellbeing though concern is largely with macro provision issues like the availability of affordable housing, home maintenance and support services. While housing audits prevail, the great majority of them are prepared from the perspective of the developer and for their use, rather than from and for the older resident living in the housing unit.

The Housing Audit Checklist presented here is designed as a quick and easy-to-use instrument to help residents assess their current home's age-friendly aspects and identify potential areas that require reconsideration based on levels of usability and ease of the housing unit to support their daily needs. The aim is to:

- Identify current and potential barriers and enablers in housing design for ageing; and
- Understand the features that are important to older people in age-friendly housing.

The Checklist considers five evaluative components using the Likert scale (very difficult-very easy continuum of response categories). Likert scale is a psychometric scale, often used in survey research to measure respondents' attitude, belief or behaviour about a particular topic (Likert 1932; Dukes 2005; Burns and Burns 2008). Drawing from key determinants in the World Health Organisation (2007) Age-friendly Cities Guide and Universal Design (Mace et al. 1991), the five components are:

(a) Accessibility and Inclusivity—considers entry and exit to home, rooms and movement between spaces within the home, reach for things and visibility within the home;
(b) Adaptability and Flexibility—considers the ease of the individual to manage basic household chores during temporary disability, the ease by which the housing unit (and its components) can be modified to accommodate one's ageing needs as they change over time, the housing unit's potential to be used in a variety of ways as one's needs change over time;
(c) Maintenance, Services and Affordability—considers the availability of reliable home maintenance and home-help services when needed, the availability of healthcare facilities and services near to one's home, the availability of information about the services to help people age in place, the availability of financial assistance and the ease of affording home modifications, services and ageing-related facilities for ageing in place;
(d) Privacy, Comfort and Security—considers the degree of freedom from unwanted intrusion into one's personal physical or emotional space, the ability to feel at ease and relaxed at home, the degree of safety and security in one's home in terms of prevention from falls and hazards, feeling lonely or vulnerable to harm or abuse; and

(e) Sustainability and Liveability—considers the overall dwelling desirability and the integrated relationship of the housing unit within the overall living environment of the neighbourhood to support continued living in the home and/or neighbourhood over time as one grows older.

The Checklist is not designed or intended for use as an audit for accessibility code compliance because it does not cover the level of detail necessary.

During the present study, the survey was administered through a personal interview without requiring entry into the house. The Checklist questionnaire can also be self-administered, if desired. There are three sections to this Checklist questionnaire:

Section 1 – covering respondent's demographic information;

Section 2 – covering respondent's assessment (using five-point Likert scales) of his/her current dwelling in terms of the abovementioned five evaluative components; and

Section 3 – covering a summary of the evaluation scores.

A personal interview was conducted in English on a randomly selected non-probability sample of 159 older persons (above 55 years old) Singaporeans of both genders across public and private housing. Each interview took approximately 15–20 min to complete. The intent was to test the Checklist as a tool for age-friendly housing assessment.

4.3 Findings

In the present analysis, the respondents were split into the following age groups:

- Emerging old, 55–64 years (48.4%);
- Young old, 65–74 years (39.6%);
- Middle old, 75–84 years (10.1%); and
- Oldest old, 85 years and older (1.9%).

Among the sample of 159 respondents, almost an equal sample of female (47%) and male (53%) older persons was interviewed. As we were interested to study both public and private housing, 58.5% of respondents were from public housing and 41.5% from private housing (apartments/condominiums and landed houses). All responses were analysed as a group.

The analysis of results includes: (1) computation of the average score of each component as indicated by 159 respondents, which is plotted on the HAC age-friendliness matrix; (2) detailed analysis of each component as per age group, gender, housing type and length of stay; and (3) analysis of priority ratings.

4.3.1 HAC Age-Friendliness Matrix

At an aggregate level, the audit demonstrates that there are many areas where respondents find their homes to be age-friendly. The average scores of components indicate that the respondents perceive it is easy for them to attain privacy, comfort and safety in their homes, the item that receives the highest scores (Fig. 4.1). They seem to have more difficulty with the quality of maintenance, services and affordability and the adaptability and flexibility of their housing units. Both of these items registered the lowest average scores (both are 13 out of a possible score of 20), suggesting a potential area for recalibration.

At the individual dwelling level, the matrix can highlight potential gaps and barriers in general housing, especially when rating appears between 0–9 (under average). Figure 4.2 illustrates one respondent's individual ratings. The 'under average' scores suggest the components that need review and even recalibration when considering the home for ageing in place.

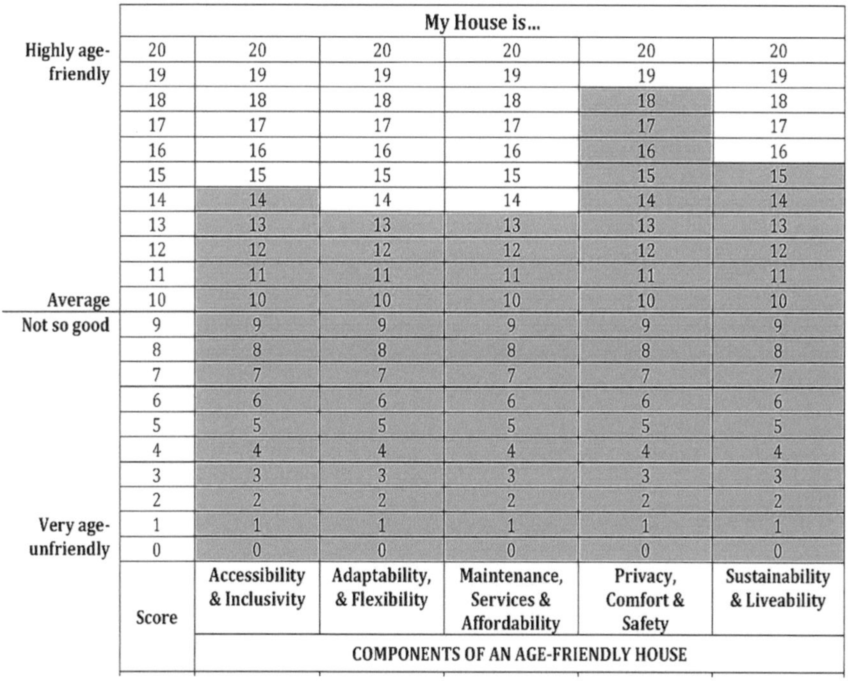

Fig. 4.1 Average score of components on HAC age-friendliness matrix

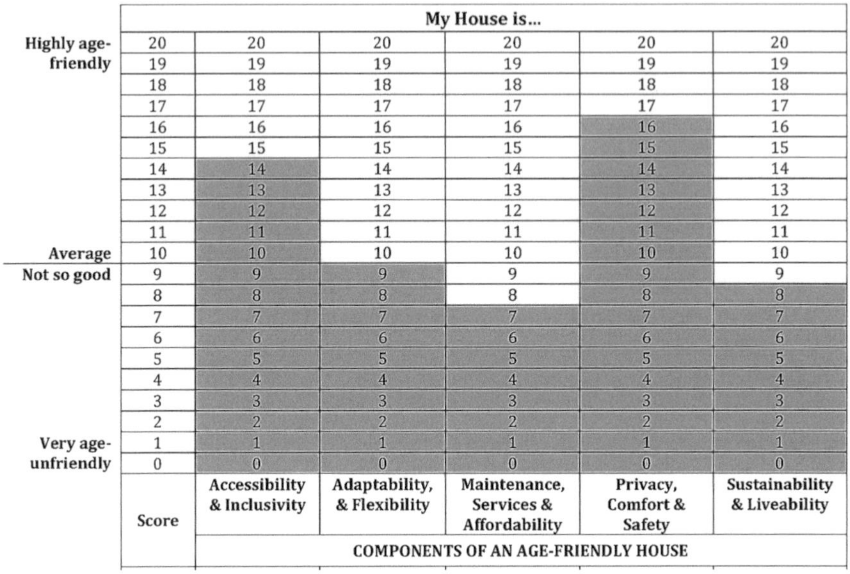

Fig. 4.2 Example of an individual rating of dwelling unit

4.3.2 Component-Wise Analysis

A majority (88.6%) of the respondents rate Privacy, Comfort and Safety as the component that is easy to have in their current dwelling units. This is followed by Accessibility and Inclusivity (69.8%), Sustainability and Liveability (58.7%) and Adaptability and Flexibility (53.7%). Only 48.0% of respondents rate Maintenance, Services and Affordability as easy/very easy within their homes. The highest mention of difficulty is reported for Adaptability and Flexibility (23.6%) and Maintenance, Services and Affordability (23.5%).

In order to understand each of these components in more detail, the data are further analysed based on age group, gender, type of housing and length of stay of respondents.

4.3.2.1 Accessibility and Inclusivity

Most respondents seem to feel a sense of easy accessibility and inclusivity within their homes. Almost 90% of respondents reported that it is easy for them to enter into their homes and to negotiate their way within their house. About half (58.5%) of the respondents indicated ease of moving across different levels and climbing the staircases within their house. Ease of accessibility is not limited to the respondents themselves; 84.2% reported that it is easy for their household members to use the spaces in their home. Almost three-quarters (72.9%) of respondents mentioned that it is easy to reach for things they need within their dwelling, there are no hard to reach corners while 71.0% found it easy to get a firm hold on lever handles, taps and switches in their house even when their grip is weak or limited. Visibility in the house is generally good; 83.0% of respondents found it easy to see around the house as there is sufficient lighting.

The majority (87.4%), however, reported a lack of provision of any alert systems within their homes that can be activated in the event of a fall or need to call for help. This suggests an immediate area for review, particularly when the older person wants to remain living in his/her own house—which most old people would prefer to do for as long as possible.[1]

Although respondents across housing types generally perceived a level of ease to access various aspects of their housing, private housing respondents are more inclined to express slightly higher ease of accessibility within their dwellings than public housing respondents. Their highest rated aspects include: ease in entering their house (92.5% among private housing respondents and 89.2% among public housing respondents), ease in negotiating their way within the house (90.9% among private housing respondents and 90.2% among public housing respondents).

Lower scored items include: ease of moving across different levels (68.2% among private housing respondents and 52.2% among public housing respondents), ease to reach for things that they need around their house (75.8% among private housing respondents and 71.6% among public housing respondents). The widest disparity in ratings among private and public housing respondents is in the ease for household members to use the spaces (92.4% among private housing respondents and 79.4% among public housing respondents) and the ease to get a firm hold on lever handles, etc. when their grip is limited or weak (78.7% among private housing respondents and 66.2% among public housing respondents).

For respondents living in public housing, the responses are further analysed based on the size of dwelling unit (one- or two-room, three-room, four-room, five-room apartments and others including elderly studio apartments, maisonettes and executive apartments).[2] The latter group of housing (others) is found to be the easiest to access across all aspects. The results point towards higher accessibility and inclusivity in larger apartments. The smaller flats have their areas of easy accessibility too. In terms of the ease to reach for things that I need around my house (e.g. hanging laundry, shelves are within reach) and there are no hard to reach corners, the one- to two-room public housing respondents tend to report more ease among respondents: 80.0% as compared to 60.9% for three-room, 75.8% for four-room, 55.0% for five-room and 90.9% for others public housing respondents.

The one- to two-room public housing respondents also reported that it is easy for them to get a firm hold on lever handles, etc. around the house when their grip is limited or weak: 60.0% as compared to 47.8% for three-room public housing respondents. For this as well as many of the other items, a high degree of ease is also observed among the larger flat respondents: 66.7% for four-room, 70.0% for five-room and 81.8% among others public housing respondents. Respondents living

[1] See Chap. 1.

[2] See Yuen and Soh (2017) for further description of these housing typologies.

in one- to two-room units often registered the highest difficulty level for household members to use the rooms and spaces within their home (60.0%) and with visibility around the house due to insufficient lighting (20.0%) among the different public housing types. These respondents also reported the lowest ease and highest difficulty in negotiating their way within the house. These findings suggest areas for further study including potential recalibration for future-proofing to meet ageing needs.

Across most aspects of accessibility and inclusivity within one's home, the ease of use is generally observed to decline with age. The highest ratings of ease are reported among the emerging old and young old respondents though there are instances where more among the oldest old said it is easy. These include: to overcome steps, thresholds or kerbs to enter into the house, to get a firm hold on lever handles to doors and cupboards, taps and light switches when my grip is limited or weak, to see around the house because there is sufficient lighting, for me to reach for the emergency pull-cord for the alert system in my bathrooms and bedrooms when I fall or need help, perhaps suggesting the availability of these provisions in their homes and the lack of them in the younger age cohorts' homes.

The gender differences in accessibility and inclusivity are in the range of <13% though male respondents seem to find greater ease in reaching for things and getting a firm hold on lever handles, etc. around the house than their female counterparts. Considering the effect of the length of residency, the results suggest that the ease of accessibility is most often mentioned among those with 11–20 years of residency, except for two items—ease to overcome steps to enter house and ease to reach out for alarm system where those with less than five years residency register higher degrees of ease. The former may be a result of home modifications carried out to units after 10 years while the latter is perhaps a product of self-selection for age-friendly housing.

4.3.2.2 Adaptability and Flexibility

Respondents' evaluation of ease of adaptability and flexibility within their dwelling is in the range of 35.8–69.2%. The highest rated aspects are the ease to manage basic household chores and to maintain house even during a temporary disability (69.2%) and ease to rearrange the furniture and objects in house (67.3%). About one in two respondents reported that it is easy for themselves and their family members to adapt and modify the house as their physical needs change (57.9%), easy to convert or modify a part of the house to accommodate live-in helper or caretaker (55.3%), easy for guests to visit, especially those who have different disabilities (52.2%), easy to adapt house design—with even surfaces, passages wide enough for wheelchair access and appropriately designed bathrooms, toilets and kitchens (49.7%).

Amongst the different age groups, the young old respondents registered the highest level of ease with all aspects of adaptability and flexibility within their homes. The most often cited 'it is easy' adaptability issue for all age groups seems to be the ease to manage basic household chores and to maintain house, even during a temporary disability (59.7% for emerging old, 79.4% for young old, 75.0% for middle old and

66.7% for oldest old). Perhaps to be expected, each age cohort has its own worry about adaptability and flexibility.

Among the emerging old respondents, the most often mentioned difficult items are with height-adjustability of kitchen counters, table tops, etc. (39%) and the adaptability of house design (36.4%). Among the young old and middle old respondents, height-adjustability of kitchen counters, table-tops, are their most difficult item (42.8 and 43.8% respectively) followed by ease to improvise space to cater to any new difficulties (28.6 and 37.6% respectively). The oldest old respondents appeared to have more concerns; generally registering over 33% (sometimes over 60%) of difficult responses for all but one item. That item is: It is easy for my family and me to adapt and modify the house and its rooms to suit my specific needs as and when my physical needs change.

Among the genders, male respondents generally considered most aspects of their homes easy to adapt and adjust when compared to female respondents. Female respondents tend to report more ease than male participants along two adaptability dimensions—it is easy/very easy for me to rearrange the furniture and objects to suit changing needs (64.0% when compared to 58.3% among male respondents) and ease to convert or modify part of the house to accommodate a live-in helper or caregiver when necessary (49.3% when compared to 45.2% among male respondents). More women (about 10–25%) tend to report levels of difficulty for all aspects of adaptability and flexibility of their homes when compared to their male counterparts.

Considering the effect of respondents' duration of stay in their current dwelling, ease of adaptability is most frequent among those with one to five years residency. On the item: this house design is easily adaptable, 53.6% of these respondents report that it is easy/very easy to adapt (compared to 33.3% for <1 year, 52.9% for 6–10 years, 40.0% for 11–20 years and 32.7% for >20 years of stay) while 46.4% of them register ease to improvise existing physical space to cater for any new difficulties (compared to 33.3% for <1 year, 41.2% for 6–10 years, 35.0% for 11–20 years and 30.9% for >20 years of stay). The majority of them (75.0%) further say that it is easy for friends and relatives with different disabilities to visit (as compared to 33.3% for <1 year, 44.1% for 6–10 years, 47.5% for 11–20 years and 30.9% for >20 years of stay). There are exceptions.

Ease to manage basic household chores and to maintain house, even during a temporary disability is most frequently mentioned among respondents with 6–10 years of residency (79.4% as compared to 66.7% for <1 year, 64.3% for 1–5 years, 65.0% for 11–20 years and 41.8% for >20 years of stay). More respondents with 11–20 years residency tend to score the following as easy/very easy: ease to rearrange the furniture and objects to suit my changing needs (70.0% as compared to 66.7% for <1 year, 60.7% for one to five years, 67.7% for 6–10 years and 49.1% for >20 years of stay), ease for family members and self to adapt and modify the house to suit my specific needs (67.5% as compared to 33.3% for <1 year, 60.7% for 1–5 years, 41.2% for 6–10 years and 43.6% for >20 years of stay) and ease to convert or modify part of the house to accommodate a live-in helper or caretaker when necessary (62.5% as compared to 33.3% for <1 year, 57.1% for 1–5 years, 50.0% for 6–10 years and 43.6% for >20 years of stay).

Even though several expressed ease over many of the assessment items, about a quarter of the respondents with one to five years length of stay (17.9–25.0%) are ambivalent, scoring 'neither difficult nor easy' in their evaluations. This sentiment seems most marked among those with longer residency (>20 years); 23.6–38.2% of them rate 'neither difficult nor easy' in their assessments. More among them (>20 years) and those with <1-year length of stay tend to indicate difficulty with the various adaptability and flexibility dimensions. These findings are interesting from the adaptable housing design perspective and further research is needed to clarify the causality and the feasibility (and effectiveness) of whether it is better to renovate or build a new, custom-designed home for one's ageing needs.

4.3.2.3 Maintenance, Services and Affordability

Fewer respondents seem to rate this aspect as easy to achieve in their current dwelling. The highest rated aspect is ease to locate healthcare facilities and services near home (72.3%). This is followed by ease of maintaining the immediate common spaces around house without inconvenience to oneself (71.7%) and ease to get reliable home maintenance services when needed (59.7%). Less than half of the respondents considered the following issues as easy: ease to get help should I need any assistance around house (45.2%), ease to afford house modifications, services and facilities I need to help me age in place (39.0%), ease to get good information of the services available to help me age in place (36.5%), ease to arrange for home-help services during temporary disability (36.5%) and ease to get financial help for ageing-related facilities and services (23.3%). The proportion of respondents indicating difficulty to get financial assistance for ageing-related facilities and services is more than those expressing ease (44.1% difficult/very difficult as compared to 23.3% easy/very easy).

Respondents who live in private housing tend to express ease in maintaining and affording services for their homes. Three-quarters of these respondents reported that it is easy/very easy to get reliable home maintenance services when needed (71.2% as compared to 52.1% in public housing) and to maintain the immediate common spaces around house without inconveniencing them (74.2% as compared to 70.7% in public housing). More public housing respondents (75.0%) reported ease to locate healthcare facilities and services when compared to those living in private housing (69.7%).

Among public housing respondents, those living in 'others' public housing category are likely to express ease of maintenance of their homes than the rest of public housing respondents. The possibility to get financial assistance for ageing-related facilities and services is scored by all public housing respondents as difficult; it registers the most scores—40.0% of one- or two-room, 65.2% of three-room, 60.6% of four-room, 45.0% of five-room and 27.3% others public housing respondents mark 'difficult/very difficult' on this item.

Among the age groups, the oldest old respondents are more likely to score levels of ease on all aspects of maintenance, services and affordability (except for ease to get financial assistance for age-related facilities) discussed in the checklist. Perhaps this is because by that age, they are no longer directly involved in issues of maintenance and services. All age groups unanimously (one third to half of the respondents) report that they experience difficulty in getting financial assistance for any age-related facilities and services. The possibility of easily and quickly maintaining the immediate common spaces around the house without inconveniencing me gets the highest 'easy/very easy' scores across age groups—the emerging old (67.6%), young old (74.6%), middle old (75.0%) and oldest old (100.0%).

Considering the responses based on gender, the various aspects are rated as easy by more men than female respondents, by more than 10%. Female respondents tend to express higher difficulty for all aspects compared to male. However, female respondents express more ease to locate healthcare facilities and services near their homes (74.7% when compared to 58.3% by male respondents).

Ease of maintenance and services of the dwelling is found to be consistently low for respondents with long residency (>20 years), except for ease to get information of services to enable ageing in place (25.5% as compared to 33.3% for <1 year, 46.4% for one to five years, 17.6% for 6–10 years, 37.5% for 11–20 years). The results signal the importance of designing easy to maintain housing at the outset if the ultimate objective is ageing in place. Another area to review is the availability of financial assistance for ageing-related facilities and services. The ease to get financial assistance for ageing-related facilities and services is low, irrespective of how long respondents may have lived in the house.

4.3.2.4 Privacy, Comfort and Safety

Privacy, comfort and safety is tested in the HAC by two sets of questions—one set assessing the level of ease of various aspects relating to the privacy, comfort and safety within the home and another set assessing the frequency of vulnerability to falls and hazards within the home. In general, most respondents (over 85%) have rated this component as easy. The most rated aspects are: ease to feel safe and secure at home, without feeling vulnerable or in danger (92.5%), ease to find personal space, and enjoy peace and quiet (89.9%), ease to relax and enjoy any activities at home (88%), and ease to live and move around independently in my house, without fear of accidents (85.6%).

Most respondents (92.5%) reported that it is seldom/very seldom for them to trip over furniture or loose items around floor of the house. Many others said that they seldom lose balance and slip on the kitchen or bathroom floor (88.1%), fall in the house in the last 3 months (86.8%), or injure or knock into objects or edges around the house (85.5%). It is possible that despite assurances of confidentiality, some respondents may not fully share their risks and vulnerabilities during the survey. Notwithstanding, these results give a glimpse of the respondents' perception and self-assessment of the safety and comfort of their homes.

Private housing respondents seem to report more ease and lower frequency of vulnerabilities to the various aspects of privacy, comfort and safety when compared to public housing respondents. More private housing respondents rate levels of ease to relax and enjoy any activities at home (96.9% as compared to 82.6% in public housing) and that it is easy to feel safe and secure at home (95.5% as compared to 91.3% in public housing). Marginally more among private housing respondents also expressed that they seldom trip over furniture or loose items around floor of the house (94.0% as compared to 92.3% public housing respondents).

Among public housing respondents, there is a trend for more among those living in the 'others' housing category and those in 4-room flats to rate ease with privacy, comfort and safety and lower levels of vulnerabilities within their homes when compared to others. Interestingly, all respondents of one- or two-room and 'others' public housing rate the lowest possibility to lose balance and slip on the kitchen or bathroom floor (both 100% as compared to 73.9% for those living in three-room, 84.8% in four-room and 80.0% in five-room), trip over furniture or loose items around floor (both 100% as compared to 91.3% by those living in 3-room, 93.9% in four-room and 75.0% in five-room) and to injure or knock into objects or edges around the house (both 100% as compared to 73.9% by those living in three-room, 90.9% in four-room and 65.0% in five-room).

When comparing the age groups, more among the middle old respondents reported levels of ease to find personal space and enjoy peace and quiet (100.0% as compared to 83.1% among emerging old, 96.8% among young old and 66.7% among oldest old respondents) and to relax and enjoy any activities at home without being disturbed or restricted (93.8% as compared to 83.1% among emerging old, 93.7% among young old and 66.7% among oldest old respondents). The ease to live and move around independently without fear of accidents is reported as easy by the oldest old (100.0%) while the ease to feel safe and secure at home, without feeling vulnerable or in danger is reported easy by most young old (98.4%). Perhaps to be expected, the young old reported the lowest frequency to falls in the house in the last three months (90.5%) while the emerging old recorded the lowest levels of vulnerability to injury or knocking into objects or edges around the house (87.0%).

A salient observation is that the oldest old respondents reported the lowest levels of vulnerability; 100.0% of them say they seldom lose balance and slip on the kitchen or bathroom floor or trip over furniture or loose items around floor of the house. This seems to suggest an age-friendly living environment. Even so, 33.3% of the oldest old respondents informed that they often fall to the ground in the house over the last 3 months (even though 66.7% say they seldom fall during this period) while a third of them return 'neither often nor seldom' to the statement: It is very common for me to injure myself or knock into objects or edges around the house (66.6% respond seldom/very seldom). Given that falls and injuries related to falls are common among older people, these results call for further research on safety within homes.

Looking at the gender data, more female respondents (20–50% more than men) assign levels of ease to privacy, comfort and safety and lower levels of vulnerabilities within their homes. Respondents who have lived in their current dwelling for less than one year expressed the greatest levels (100.0%) of ease to find personal space, and enjoy peace and quiet, feel safe and secure, without feeling vulnerable or in danger and to relax and enjoy any activities that without being disturbed or restricted. They also scored the lowest levels of ease to live and move around independently in house without fear of accidents (33.3% as compared to 85.7% for one to five years, 91.2% for 6–10 years, 82.5% for 11–20 years and 70.9% for >20 years of stay).

Respondents who have lived in their current dwelling for 11–20 years reported the lowest levels of vulnerability to lose balance and slip on the kitchen or bathroom floor (92.5%), trip over furniture or loose items around floor of the house (100.0%) and lowest frequency of falls in the past three months (97.5%). The lowest frequency to injury by knocking into objects or edges around the house is reported by those who have lived in the present house for 6–10 years (94.1% as compared to 66.7% for <1 year, 75.0% for one to five years, 87.5% for 11–20 years and 70.9% for >20 years of stay). These results prompt further investigation to better understand the relationship between length of stay and vulnerability to falls.

4.3.2.5 Sustainability and Liveability

Apart from 3 items, most respondents (over 52%) have rated this component as easy. The three items are: ease to use sustainable building materials and construction methods to build/modify the house as I age (32.1%), ease to find alternative housing options in the local area when one is no longer able to live in present house (34.6%), connect and interact with other young and old people in my community (49.1%). In contrast, eight in ten respondents say it is easy for them to continue living in one's house as they age (83.0%) and to enjoy clean air, water and environment from home (82.4%). These two items also register the least number of neither difficult nor easy responses (12.7% and 14.5% respectively). The other positive aspects where two in three respondents say that it is easy are: to feel that they are a part of their local neighbourhood (68.5%) and to maintain their home in good and comfortable condition with minimal repairs and no major renovations as they continue to live in it (67.3%).

Along a couple of the dimensions, there are more responses of difficulty than ease for: finding alternative housing options in the local area when they are no longer able to live in present house (37.7% report difficult as compared to 34.6% easy) and using sustainable building materials and construction methods to build/modify house as age increases (34.6% report difficult as compared to 32.1% easy). About a third of the respondents are undecided and score these two dimensions as neither difficult nor easy (27.7 and 33.3% respectively). The highest neither difficult nor easy responses are recorded for the statement: It is easy to connect and interact with other young and old people in my community as I grow older (37.1%) while 49.1% consider this as easy to do and 13.8% difficult.

More private housing respondents reported levels of ease with sustainability and liveability of their homes, except for two items: ease to continue living in the house as they age (80.2% in private housing compared to 85.9% in public housing) and to build trust and develop a sense of community (50.0% in private housing compared to 55.4% in public housing). Among public housing respondents, those living in others housing category tend to score higher levels of ease with sustainability and liveability. The ease to find alternative housing options in the local area when they are no longer able to live in the present house is rated lowest by respondents of most house sizes—54.5% by others, 15.0% by 5-room, 33.3% 4-room, 13.0% by 3-room and 0.0% by 1–2 room respondents. Fewer 3-room public housing respondents scored the ease to use sustainable building materials and construction methods to build/modify house as they age (8.7% as compared to 40.0% by 1–2 room, 27.3% by 4-room, 30.0% by 5-room and 63.6% by others public housing respondents).

Most age groups reported 'it is easy' to continue living in their house as they age (76.7% among emerging old, 88.9% for young old, 87.5% middle old and 100.0% oldest old). Looking at the easy ratings of various age groups, the most commonly voted item among the emerging old respondents is: It is easy for me to enjoy clean air, water and environment from where I live (80.5%) while the ease to continue living in their house as they age appears more frequently among the young old and middle old (88.9 and 87.5% respectively). For the oldest old, the ease to continue living in their own house and to maintain their home in good and comfortable condition with minimal repairs and no major renovations are both scored 100.0% easy, no difficulty or ambivalence are expressed on these items.

More male respondents expressed levels of ease with most aspects of sustainability and liveability when compared to female. More female respondents tend to find it is easy to continue living in their own house (86.7% when compared to 60.7% among male respondents), ease to maintain home in good condition with minimal repairs (70.7% when compared to 66.7% for men) and ease to enjoy clean air, water and environment (82.7% when compared to 52.4% for men).

Where length of stay is concerned, most groups seem to express levels of difficulty with finding alternative housing options in their local area (28.6–66.7%), to connect and interact with other young and old people in their community as they grow old (<25%), to build trust and develop a sense of community (<30%), to use sustainable building materials and construction methods to build/modify house as they age (29.4–66.7%). In particular, two groups of respondents—those with less than one year and those with more than 20 years of residency—are more likely to report difficulty with several of the dimensions.

Where responses of ease are concerned, more among those with 1–5 years residency tend to express levels of ease in house maintenance—to maintain it in good and comfortable condition with minimal repairs and no major renovations (71.4% as compared to 33.3% for <1 year, 67.6% for 6–10 years, 65.0% for 11–20 years, 49.1% for >20 years), to find alternative housing options in local area (46.4% as compared to 0.0% for <1 year, 41.2% for 6–10 years, 30.0% for 11–20 years, 23.6% for >20 years), to build trust and develop a sense of community (64.3% as compared to 33.3% for <1 year, 32.4% for 6–10 years, 55.0% for 11–20 years, 47.3% for >20 years) and to use sustainable building materials and construction methods to build/modify house (35.7% as compared to 0.0% for <1 year, 20.6% for 6–10 years, 32.5% for 11–20 years, 30.9% for >20 years).

More respondents with 11–20 years residency seem to indicate levels of ease: to connect and interact with other young and old people in their community as they grow older (65.0% as compared to 0.0% for <1 year, 39.3% for 1–5 years, 26.5% for 6–10 years, 41.8% for >20 years), to enjoy clean air, water and environment (85.0% as compared to 66.7% for <1 year, 82.1% for 1–5 years, 76.5% for 6–10 years, 70.9% for >20 years), and to feel a part of their local neighbourhood (75.0% as compared to 66.7% for <1 year, 57.1% for 1–5 years, 55.9% for 6–10 years, 58.2% for >20 years). For the statement: ease to continue living in my house as I age, more among those with 6–10 years of stay tend to indicate levels of ease (88.2% as compared to 33.3% for <1 year, 78.6% for 1–5 years, 77.5% for 11–20 years, 69.1% for >20 years).

4.3.2.6 Component Comparison

In this part, we examine whether respondents' evaluation of housing quality (i.e. the above mentioned five components) differ significantly between the population/housing groups. A one-way analysis of variance tests whether the average scores assigned by respondents of different groups are different from one another. Table 4.1 summarises the ANOVA test results between public and private housing respondents.

The F statistics in the ANOVA tests for accessibility and maintenance scores are significant at 1% level while the statistics in the ANOVA tests for privacy (and vulnerability) and sustainability (and liveability) scores are significant at 5% level. These results indicate that respondents' scores for these four housing quality measures are significantly affected by whether they live in a public or private housing unit. Respondents' evaluation for adaptability is only marginally affected by the type of housing they reside (F statistic significant at 10% level).

Given that the majority (82%) of Singapore's resident population resides in public housing, we are interested to compare respondents' scores among different public housing types. Table 4.2 summarises the ANOVA test results.

For those respondents living in public housing, their evaluations of all five housing qualities are significantly affected by the types of public housing they live in (F statistic significant at 5% level).

Table 4.1 Respondents scores by housing type

Housing type		Accessibility & inclusivity	Adaptability & flexibility	Maintenance, services & affordability	Privacy, comfort & safety	Sustainability & liveability
Private	Mean	30.1	28.4	28.2	36.2	30.0
	Standard deviation	4.9	6.4	6.3	3.6	4.5
	N	66	66	66	66	66
Public	Mean	27.5	26.6	25.4	34.9	28.3
	Standard Deviation	6.1	6.3	5.6	4.3	5.1
	N	93	93	93	93	93
	F	8.36	2.99	8.86	4.19	4.68
	Prob > F	0.004	0.086	0.003	0.042	0.032

Table 4.2 Respondents scores within public housing

Public housing		Accessibility & inclusivity	Adaptability & flexibility	Maintenance, Services & Affordability	Privacy, Comfort & Safety	Sustainability & Liveability
1 or 2 room	Mean	23.4	20.6	24.4	34.6	26.8
	Standard deviation	7.4	5.9	5.4	4.2	6.9
	N	5	5	5	5	5
3 room	Mean	25.4	24.3	24.0	33.8	26.5
	Standard deviation	6.4	5.8	5.2	5.0	5.5
	N	24	24	24	24	24
4 room	Mean	28.3	27.1	24.6	35.9	29.1
	Standard deviation	5.9	5.2	4.9	3.3	3.4
	N	33	33	33	33	33
5 room	Mean	26.9	25.8	24.8	32.9	27.0
	Standard deviation	4.7	5.9	4.6	4.4	4.6
	N	20	20	20	20	20
Others	Mean	32.3	34.5	32.0	38.0	33.0
	Standard deviation	5.3	4.6	6.3	3.2	6.3
	N	11	11	11	11	11
	F	3.46	8.36	5.26	3.88	4.22
	Prob > F	0.011	0	0.001	0.006	0.004

Table 4.3 Respondents scores within private housing

Private housing		Accessibility & inclusivity	Adaptability & flexibility	Maintenance, services & affordability	Privacy, comfort & safety	Sustainability & liveability
Apartment/Condo	Mean	29.5	28.9	28.3	36.3	29.7
	Standard deviation	5.0	5.5	5.3	3.3	3.5
	N	34	34	34	34	34
Landed	Mean	30.7	27.8	28.0	36.1	30.4
	Standard deviation	4.8	7.4	7.2	4.0	5.4
	N	32	32	32	32	32
	F	1.01	0.51	0.04	0.09	0.39
	Prob > F	0.318	0.479	0.851	0.771	0.533

Table 4.3 summarises only the private housing respondents' scores. For those respondents living in private housing, their evaluations of the five housing quality measures do not appear to be significantly affected by whether they choose to live in an apartment or a landed house.

Table 4.4 presents the respondents' scores by age groups: emerging old, young old, middle old and oldest old. Similarly, the F statistics in the five ANOVA tests are insignificant, implying that on average, respondents' scores of any of the five housing quality do not seem to differ significantly across age groups.

Table 4.4 Respondents scores by age group

Age		Accessibility & inclusivity	Adaptability & flexibility	Maintenance, services & affordability	Privacy, comfort & safety	Sustainability & liveability
55–64	Mean	29.3	26.7	26.1	34.9	28.5
	Standard deviation	5.9	6.6	5.9	4.4	5.3
	N	77	77	77	77	77
65–74	Mean	28.1	28.3	26.5	36.2	29.4
	Standard deviation	5.5	6.4	6.2	3.6	4.6
	N	63	63	63	63	63
75–84	Mean	26.7	26.9	27.8	35.2	29.6
	Standard deviation	5.4	5.2	5.9	3.4	4.2
	N	16	16	16	16	16
85 and older	Mean	30.0	26.0	32.0	33.7	32.0
	Standard deviation	8.7	8.2	3.0	7.6	8.0
	N	3	3	3	3	3
	F	1.14	0.74	1.23	1.43	0.86
	Prob > F	0.336	0.527	0.300	0.237	0.465

Table 4.5 Respondents scores by gender

Gender		Accessibility & inclusivity	Adaptability & flexibility	Maintenance, services & affordability	Privacy, comfort & safety	Sustainability & liveability
Male	Mean	29.2	27.6	27.2	35.5	29.0
	Standard deviation	5.6	6.5	5.9	4.2	5.1
	N	84	84	84	84	84
Female	Mean	27.8	27.0	25.8	35.3	29.1
	Standard deviation	5.9	6.3	6.1	4.0	4.9
	N	75	75	75	75	75
	F	2.35	0.34	2.37	0.05	0.01
	Prob > F	0.127	0.561	0.126	0.828	0.934

Next, we examine the effect of gender (Table 4.5). The corresponding F statistics in the five ANOVA tests are not significant at 10% level. Thus, respondents' average scores of any of the five housing qualities do not appear to be significantly affected by their gender division.

4.3.2.7 Priority Rating

In order to understand user priority, the respondents are asked to rank the 5 components in order of their importance to them. The results indicate that more respondents

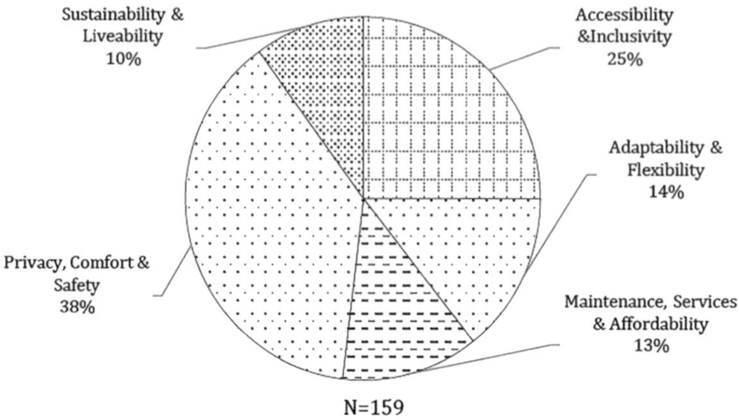

Fig. 4.3 Respondents' no. 1 priority ranking

(38.4%) ranked Privacy, Comfort and Safety as their number 1 priority in housing consideration (Fig. 4.3). There is strong evidence at the international level to indicate that privacy and feeling safe in the home environment are regarded as important for older people (World Health Organisation 2007). Respondents also ranked the following items as the number 1 priority: Accessibility and Inclusivity (25.2%), Adaptability and Flexibility (14.5%), Maintenance, Services and Affordability (12.6%) and Sustainability and Liveability (10.1%). Sustainability and Liveability as well as Adaptability and Flexibility are placed among the least important components (ranked 5th) by most respondents (33.3 and 20.1% respectively).

When considered together, the top three priorities as indicated by the respondents using weighted sum analysis (Table 4.6). They are in descending order: Privacy, Comfort and Safety, Accessibility and Inclusivity, and Maintenance, Services and Affordability.

Table 4.6 Respondents' component ranking

Components	Respondent ranking (%)					Average weighted scores	Overall ranking
	Rank 1	Rank 2	Rank 3	Rank 4	Rank 5		
Accessibility and inclusivity	25.2	12.6	21.4	23.9	15.7	2.89	2
Adaptability and flexibility	14.5	21.4	21.4	21.4	20.1	3.08	4
Maintenance, services and affordability	12.6	24.5	18.9	25.2	17.6	3.07	3
Privacy, comfort and safety	38.4	23.9	15.7	11.3	9.4	2.26	1
Sustainability and liveability	10.1	12.6	22.6	19.5	33.3	3.48	5

4.4 Conclusion

This is a small sample study and the findings may not be generalizable. Even so, the results seem to indicate that most components have been rated above average on the HAC age-friendliness matrix. While these scores may indicate that respondents' housing is currently age-friendly, there is opportunity for all components to move towards the top of the matrix—highly age-friendly, considering the rate at which the population is ageing. Singapore is projected to become the world's 4th oldest country over the next three or four decades. The number of older population (65+) is expected to triple to 25% of Singapore's population (900,000) by 2020. Average lifespan is projected to extend to 97 years old for women and 92 years old for men.

Respondents' top four priorities in age-friendly housing are: Privacy, Comfort and Safety, Accessibility and Inclusivity, and Maintenance, Services and Affordability. Areas of difficulty include: Maintenance, Services and Affordability, and Adaptability and Flexibility. The degrees of ease with these components are consistently rated lower than others. Further research on these issues would be useful as if not addressed, they can potentially affect people's ability to stay independent and remain in their homes as they age.

It is important to be mindful of the changing needs of older people and to direct attention towards inclusive design, preventing architectural discrimination or disability in the built environment. Many argue that rethinking architectural disability, i.e. how the physical design, layout and construction of buildings lead to hazards and barriers, making the built environment inconvenient, uncomfortable or unsafe to use or even prevent some people from using it at all, can be a creative starting point for inclusive design (Goldsmith 1997).

Experts estimate that as many as 90% of people in USA may be architecturally disabled in some way or other in using the built environment at some time in their lives (Wylde et al. 1994). Some such as Hanson (2001) suggest that at times, people are more vulnerable to architectural disability than medical or physical disability. Hanson (2001) further posits that the roots of most architectural disability can be traced to one of two design deficits—difficult changes of level or insufficient space. The findings from the housing audit remind the importance of inclusive design of homes to de-barrier architectural disability.

Another key finding is that home maintenance and modification services are important to help older people maintain their independence within the comfort of their own home. Several factors will weigh in on their decision to live at home like location and accessibility, home accessibility and maintenance, support available, isolation, medical conditions and finances. Keeping the home well maintained takes a lot of work and age-related declines in capabilities may erode the older adults' ability to maintain their homes. Our results on HAC have shown that participants are concerned about this issue. Maintenance, services and affordability are among their top 3 priorities in housing consideration. This is a component that they seem to have more difficulty with (another is adaptability and flexibility of their housing units). Both items get the lowest average scores, suggesting a potential barrier and area for recalibration. The finding lends support to Age UK research that the cost of home maintenance is a major worry for the 60+ with 1 in 5 older adults saying it is their biggest concern when thinking about their home as they age (Age UK Enterprises 2015). UK spending on home maintenance is estimated to cost £6.9 billion in 2014.

A common way to help the older population overcome home maintenance challenges is to provide advisory and information about affordable and reliable home maintenance services. There are many examples. In the UK, non-profit organisations such as the Elderly Accommodation Counsel First Stop Advice maintains a national home services database under its Care and Repair Services programme to help the older person find services in local areas. In USA, senior community services help mobilise skilled workers and volunteers to bring affordable home maintenance services, from housekeeping to handyman tasks and painting, to older residents (60+). These services may be ordered online.[3] In Singapore, the Enhancement for Active Seniors programme offers eligible residents the opportunity for subsidised home modifications in public housing. The availability of such advisory and assistance gives the older person the confidence to age in place. Studies show that relevant information at appropriate times contributes to personal empowerment as well as to healthy behaviours among the older adults (Lam and Lee 2009; Beales 2012).

More than accessible information and assistance, there is opportunity to leverage emerging technology to develop living environments that meet unique needs (e.g. the older person's memory, mobility, etc.). Provision like the installation of emergency alert alarm system in bathrooms and other living spaces, grab bars, height-adjustable tables and shelves, non-slip flooring, zero-step entrance, wider doors and hallways, though basic and simple, have proved to offer an added measure of confidence as people age. As alluded to earlier, the ability to age in place is determined by several factors including the physical design and accessibility of the home. Against the fast growth of technology, access to technology innovation (e.g. robotics, assistive technology, sensor-based networks, etc.) is no longer an optional extra but fast becoming recognised as a basic enabler to create a safe environment for the older person to age independently.

The European Commission has adopted an Action Plan on ICT for Ageing Well in the Information Society to help older residents continue living independently at home, in the workplace and in society. The UK Department of Health has published its vision for health and social care a decade ago, identifying the potential of tele-care to support individuals to live at home and complement traditional care. In addition, it has initiated a Preventative Technology Grant (2004–2008) to encourage local authorities to initiate step change in the design and delivery of health, social care and housing services and prevention strategies to enhance and maintain individual wellbeing and independence. Supported by this and other funding (e.g. Department of Health £31 m Long Term Conditions Whole System Demonstrator, Technology Strategy Board Assisted Living Innovation Platform), UK industry and research

[3] See, for example, the Household and Outside Maintenance for Elderly (HOME) Program provided by Senior Community Services in Hennepin County, Minnesota, the Housing Opportunities and Maintenance for the Elderly (HOME) services for low-income older people in Chicago, USA.

communities are active in advancing technology to meet the demand for independent living, especially for those suffering from chronic conditions and wanting to live independently.

Despite the potential benefit from new technologies, concerns remain. It is important to consider technology acceptance when designing with technology, issues such as the older adults' concerns about useability (e.g. fear that technology may be hard or impractical to use), appearance (e.g. the technology is too noticeable or obtrusive within their homes), control (e.g. no control over the activation and de-activation of the technology), consequences of using technology (e.g. the burden it might put on their children in their role as caregivers, possible negative effects on their personal health) (Yuen and Withanage 2015).

Regular housing audits are a crucial process to understanding and identifying whether the home is supporting the independence of older residents. It is a way of getting more evidence for the planning and design of age-friendly spaces, with a view to improve the interaction of the user and physical environment. The home (inside and outside the dwelling) is a fundamental physical space in the older resident's everyday life that can facilitate or hinder their way of living and quality of life. Housing audits offer an important tool for developments targeting the older person and should be conducted at regular intervals to monitor the performance of spaces.

We have piloted the HAC in this study, providing a simple to use tool where the older resident can describe how a checklist of certain spatial conditions of their home allow individual accessibility, adaptability, maintenance, privacy and liveability, without actually entering the home. It presents the older person a starting point to think about their experience of housing with age and the designer greater knowledge with which to develop housing that takes account of the older person's experiences and needs. When collected from a large enough sample, the varied responses to the HAC suggest that some features are more relevant to different individuals. For example, among housing types, private housing respondents are more likely to report ease of accessibility and maintenance than public housing respondents. Female respondents are more likely to have difficulty with adaptability issues and more ease and lower vulnerability to aspects of privacy, comfort and safety than men.

It is likely that completing the HAC will highlight a number of actions needed. Positive, age-friendly features as well as barriers and areas that require review and improvement may be identified. As shown in the present study, the areas to review (and thus, the opportunity to enhance age-friendliness of homes) include: the provision of alert alarm system (88.7% of respondents reported the absence of such provision to alert for help during any falls or emergency), height-adjustable kitchen counters and table-tops, financial assistance for ageing relating facilities. Other areas where respondents seem to experience difficulty include to: afford housing modifications to be able to age in place, arrange home-help during disability and to find alternative housing options in local area.

Adopting the Checklist is but a first step. Further assessment must be conducted to develop priorities and actions, especially for the lowly scored (under average) conditions, with emphasis on user 'enablement rather than disablement' (Plouffe and Kalache 2010). Even though the focus of this study is housing, there is a need to also consider and promote the concept of age-friendliness across the neighbourhood.

International evidence suggests that beyond home, the immediate neighbourhoods are where older people tend to spend a lot of time (World Health Organisation 2007). A neighbourhood that is accessible, safe and welcoming can only enhance the older person's active ageing process while poorly designed neighbourhoods where the physical environment and social infrastructure is not fit for purpose can isolate and trap people in their homes (World Health Organisation 2002; Lui et al. 2009; Buffel et al. 2014).

Acknowledgements We gratefully acknowledge the following for their generous support and assistance with the Housing Audit Survey: our colleague Cheong Sue Wei, student intern Lim Aunn Ning and interviewers for helping us conduct the interview; our colleague Dr. Hou Yuting for helping us with the statistical analysis of data; residents who participated in our interview and generously shared with us their responses and experiences of their in-home environments.

References

Age UK Enterprises. (2015). *Cost of home maintenance key concern for over 60s*. http://www.ageuk.org.uk/latest-press/archive/home-maintenance-concerns/. Accessed March 8, 2016.

Beales, S. (2012). Empowerment and older people: Enhancing capabilities in an ageing world. In *Expert Group Meeting on Promoting People's Empowerment in Achieving Poverty Eradication, Social Integration and Productive and Decent Work for All*, September 10–12 2012, HelpAge International, UN Headquarters, New York, Accessed http://www.un.org/esa/socdev/egms/docs/2012/SylviaBeales.pdf. Accessed March 10, 2016.

Buffel, T., Mcgarry, P., Phillipson, C., De Donder, L., Dury, S., De Witte, N., et al. (2014). Developing age-friendly cities: Case studies from Brussels and Manchester and implications for policy and practice. *Journal of Aging and Social Policy, 26,* 52–72.

Bureau of Labor Statistics, US Department of Labor. (2014). *The American time use survey 2013*. Washington DC: Bureau of Labor Statistics.

Burns, A., & Burns, R. (2008). *Basic marketing research*. Pearson Education.

CABE. (2004). *Housing audit: Assessing the design quality of new homes*. London: CABE.

Dukes, K. A. (2005). *Likert scale, encyclopedia of biostatistics*. Wiley.

Goldsmith, S. (1997). *Designing for the disabled: The new paradigm*. London: Architectural Press.

Hanson, J. (2001). *From sheltered housing to lifetime homes: An inclusive approach to housing*. London: University College London.

Help the Aged. (2006). *Older people, decent homes and fuel poverty*. London: Help the Aged.

Lam, K. L., & Lee, S. L. T. (2009). *Elderly participation and empowerment: Experiences of SAGE*. APIAS Monograph Paper Series No. 21. Retrieved from Lingnan University website: http://commons.ln.edu.hk/apiasmp/21.

Likert, R. (1932). A technique for the measurement of attitudes. *Archives of Psychology, 140,* 1–55.

Lui, C. W., Everingham, J. A., Warburton, J., Cuthill, M., & Bartlett, H. (2009). What makes a community age-friendly: A review of international literature. *Australasian Journal on Ageing, 28,* 116–121.

Mace, R. L., Hardie, G. J., & Place, J. P. (1991). Accessible environments: Toward universal design. In W. E. Preiser, J. C. Vischer & E. T. White (Eds.), *Design intervention toward a more humane architecture*. New York: Van Nostrand Reinhold.

Mountain, G., Gomersall, T., & Taylor, J. (2015). *Developing medical, fitness and wellbeing environments to maintain health and wellbeing over the life course*. Future of an Ageing Population: Evidence Review, Foresight, Government Office for Science, UK.

Plouffe, L., & Kalache, A. (2010). Towards global age-friendly cities: Determining urban features that promote active aging. *Journal of Urban Health: Bulletin of the New York Academy of Medicine, 87*(5), 733–739.

Oregon, US Department of Consumer and Business Services, Oregon Occupational Safety and Health Division. (n.d.). *Labor housing inspection checklist*. http://osha.oregon.gov/OSHAPubs/checklists/1876.pdf. Accessed August 13, 2017.

Royal Borough of Kensington and Chelsea, UK. (n.d.) *Access audit checklist: Residential communal areas*. https://www.rbkc.gov.uk/pdf/Access%20audit%20checklist%20communal%20areas.pdf. Accessed August 13, 2017.

World Health Organisation. (1946). Preamble to the Constitution of the World Health Organization as adopted by the International Health Conference, New York, June 19–22 1946, Official Records of the World Health Organization, No. 2.

World Health Organisation. (1998). *Growing older—Staying well*. Geneva: World Health Organisation.

World Health Organisation. (2002). *Active aging: A policy framework*. Geneva: World Health Organisation.

World Health Organisation. (2007). *Global age-friendly cities: A guide*. Geneva: World Health Organisation.

Wylde, M., Baron-Robbins, A., & Clark, S. (1994). *Building for a lifetime: The design and construction of fully accessible homes*. USA: Taunton Press.

Yuen, B., & Soh, E. (2017). *Housing for older people in Singapore: An annotated bibliography*. Springer.

Yuen, B., & Withanage, C. (2015, November 17). Smart designs for an ageing population. *The Business Times*.

Chapter 5
Walking with Older Adults

Belinda Yuen and Penny Kong

Abstract As part of the methodology to better understand the lived experiences of older adults as they move through outdoor urban environments, an Urban Audit was developed comprising three tools—onsite observation, walking survey and photograph interview. This Chapter explicates one of these tools—the walking survey where older respondents were invited to walk along and share their experiences and feelings about their everyday neighbourhood spaces. The findings showed that social and recreational activities were strong pull factors attracting older people to public spaces.

5.1 Introduction

There is a growing body of literature on the importance of the neighbourhood environment to older people's quality of life, including support for physical activity, leisure and social needs (Brownson et al. 2009; Everson et al. 2009; Arango et al. 2013; Poortinga et al. 2017). Yet, as Hofland et al. (2017) observed, resident participation in neighbourhood auditing remains largely limited. In the present study, an Urban Audit Toolbox was developed to assess neighbourhood environments to better understand the salient features and qualities that are important to the older person. A secondary goal is to identify potential weaknesses and key strengths in the design of outdoor urban environments for ageing. The toolbox comprises 3 components: onsite observation, walking survey and photograph interview.

The walking survey will be presented in this Chapter while the other two tools will be reported in subsequent chapters.

B. Yuen (✉) · P. Kong
Singapore University of Technology and Design, Singapore, Singapore
e-mail: belinda_yuen@sutd.edu.sg

P. Kong
e-mail: penny.kong@tum-create.edu.sg

B. Yuen (ed.), *Ageing and the Built Environment in Singapore*,
https://doi.org/10.1007/978-3-319-92444-1_5

The walking survey or 'Walk and Talk' is a survey where older respondents are invited to share their experiences and feelings about their neighbourhood spaces, both positive and negative traits/design of space and the obstacles that hinder their enjoyment and use of that space. It aims to assess the existing conditions and opportunities for improving outdoor recreation spaces through the eyes of the older person as they move through these spaces on their daily routine.

Four public spaces with different characteristics were surveyed:

A. A successful (well-used) senior activity centre site: a precinct square;
B. An unsuccessful (not well-used) senior activity centre/senior care centre: a senior residents' corner in a void deck[1]
C. A site which presently does not have any elderly service facility: a void deck; and
D. A neighbourhood park, which is the site for community design workshops (another research instrument, see Chap. 10).

5.2 Methodology

The Walk and Talk Survey was conducted as a mobile interview—a walk with small groups of two to three older participants (aged 55 and above) through the selected public space. During the walk, participants were invited by the facilitator-researcher to talk about the space using a uniform discussion guide. The facilitator-researcher was supported by a note-taker who took notes of the discussion and observations made during the walk such as what the site looked like, its facilities, atmosphere, strengths and problems. The aim was for the participants to give a guided tour of how they used and perceived the space. Unlike offsite sedentary interviews (e.g. asking people to sit in a room and answer questions about the space), Walk and Talk respondents could set the route of the walk and have their memories provoked by what they see or sense on site.

Photographs of the site (not of the participants) were taken to complement the notes. At the end of the walk, the researcher worked with the participants to mark on a map of the study space the areas which they liked or disliked and where they wanted to see changes and improvements. Conducted largely in English, each Walk and Talk session took between 40–60 min to complete. The sessions were conducted between June and October 2015.

[1]Void decks can be found on the ground floor of apartment blocks in public housing. They have been purposefully left vacant (open space) and are typically used for communal activities.

Even though the Walk and Talk sessions are useful for collecting qualitative data relating to people's use and understanding of place, they can be unpredictable. As with sedentary interviews, some Walk and Talk sessions will be rewarding, providing rich and more-than-anticipated information while some might be brief and uninformative. For the analysis, responses are ascribed to groups of respondents rather than individuals due to the nature of the data collection.

The sites of the Walk and Talk survey were all located in Bukit Panjang Town.

A. **256 Bangkit Road** (Fig. 5.1)—*site with successful senior activity centre*. This precinct square is situated within a cluster of public housing blocks and a busy local retail development. The Bukit Panjang neighbourhood centre is less than 5 min' walk away. The site features a children's playground in the centre of the space, an elderly fitness corner on its northern edge and seating along the edges. The site is surrounded by amenities such as coffee shops, grocery stores, shops and banks. Barrier-free ramps connect the site to its surroundings.

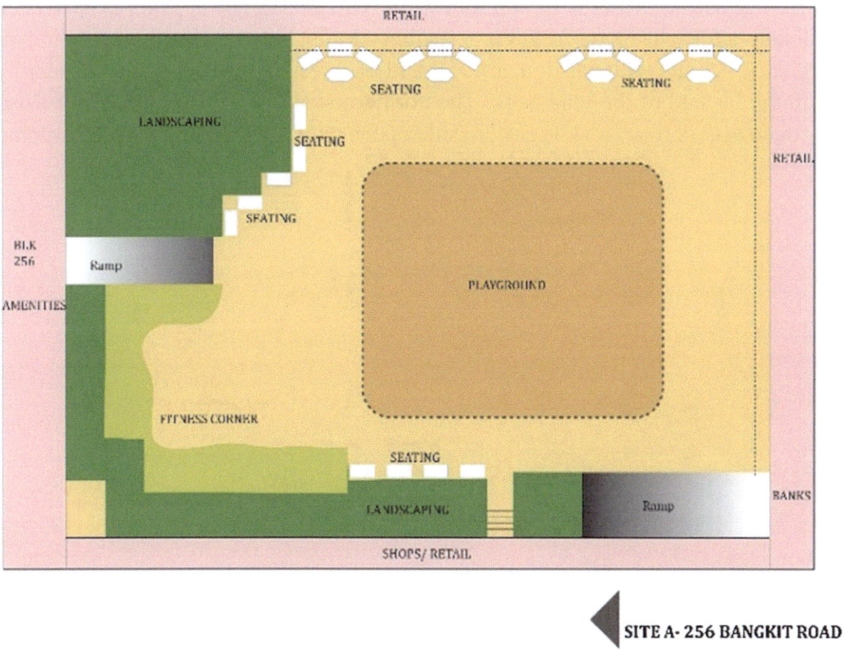

Fig. 5.1 Site A: 256 Bangkit Road

B. **434 Fajar Road** (Fig. 5.2)—*site with unsuccessful senior activity centre/senior care centre.* This site comprises a senior residents' corner within a void deck and an outdoor exercise area at an adjoining hard court. Facilities provided in the residents' corner included stone seating and tables, a toilet, vending machine, wash area and related furniture (some cupboards in the wash area have missing/broken doors, TV console is permanently locked). This site is part of a much larger void deck space, bookended by a community centre office and a bird singing corner, which functions only on weekends. The Fajar Light Rail Transit (LRT) station is located across the road. It is understood that the site has been earmarked for renewal.

C. **415 Fajar Road** (Fig. 5.3)—*site, which presently does not have any elderly service facilities.* This site is a common void deck space with no specific provision other than some seating. There are two sets of permanent seating—one a cast iron bench and another comprising two sets of stone table and seats, one located to the far end of the void deck near the cast iron bench and another set of stone table and seats closer to the lift lobby. The cast iron bench appears to have been a later addition. Aside from these seating, residents have contributed their own chairs—chairs with cushion, armrest and backrest around the cluster of seating in the far end of the void deck. The northern side of this site faces Fajar Road (main access road) and its eastern side opens onto a surface car park. Adjoining

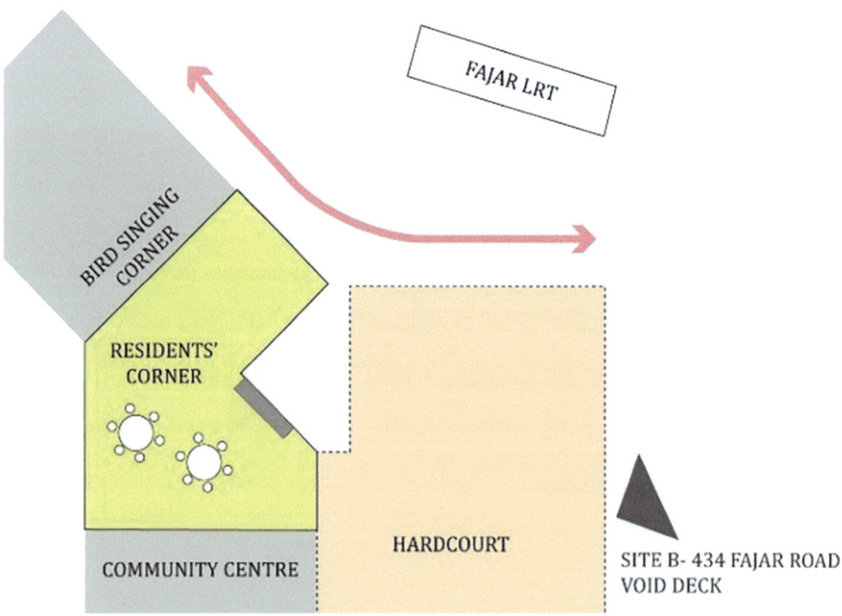

Fig. 5.2 Site B: 434 Fajar Road

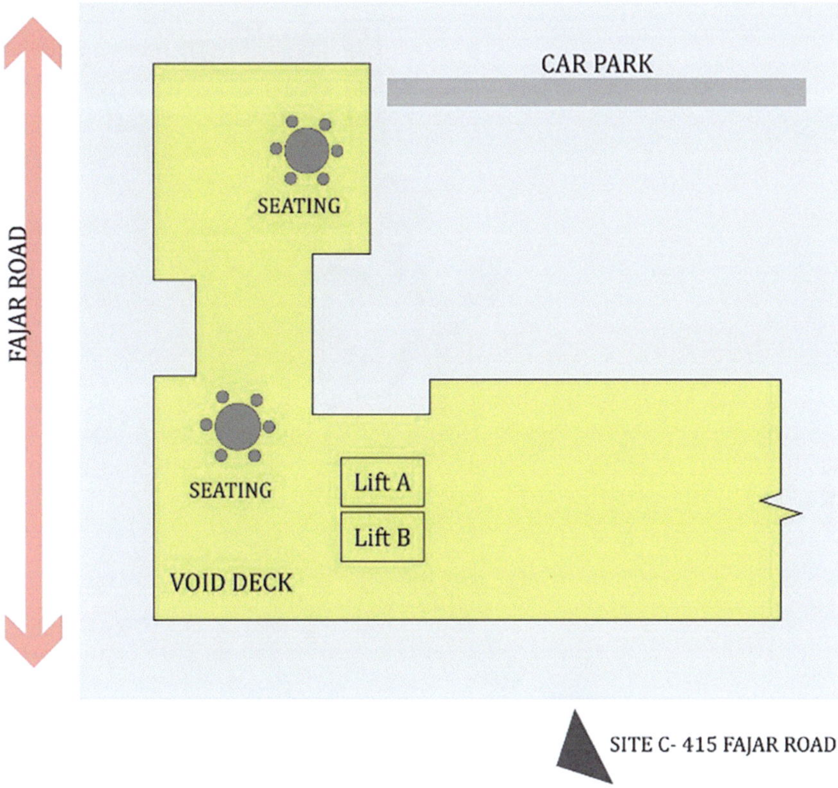

Fig. 5.3 Site C: 415 Fajar Road

uses include a Chinese temple on one side and public housing blocks on the other.

201 Petir Road, Bukit Panjang N2 Park (Fig. 5.4)—*site for community design workshops*. This neighbourhood park was used as a reference site for the project's community design workshops.[2] Of undulating terrain located atop a hillock, the site features two community gardens, a hard court, a portable toilet and such recreational facilities as individual physical proficiency test (IPPT)[3] and elderly exercise equipment at the top of the park, nestled among meandering park trails and greenery. Public housing blocks and new condominium development surround and overlook the park.

A total of 57 participants (20 groups) took part in the Walk and Talk survey. Where possible, on-site recruitment of subjects adhered to the demographics of current Singaporean society. The overall participant ratios of 47% male and 53% female

[2]See Chap. 10.

[3]This is a standard physical fitness test used by the Singapore Armed Forces, Police Force and Civil Defence Force to test the physical fitness and motor skills of its members, ranging from under 25–49 years old.

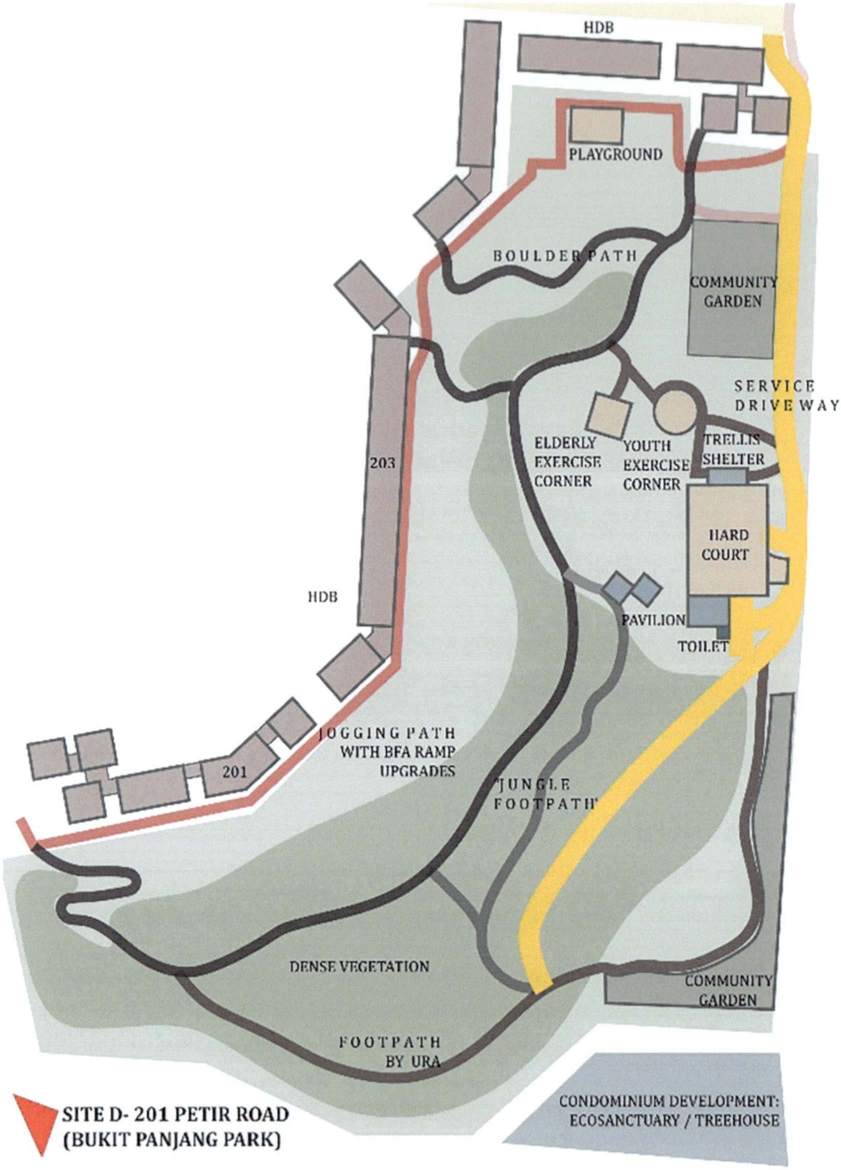

Fig. 5.4 Site D: 201 Petir Road, Bukit Panjang N2 Park

reflected gender proportions in the demographic group aged 55 and older (Department of Statistics 2015). The 'Others' ethnic group was, however, not represented. This was due to disinterest in participating (e.g. people declined to participate when approached) or not finding the particular ethnic group during the time of the Walk and Talk survey.

About 80% of the respondents were in the young old age group of 55–70 years while the remaining 10 respondents (17.5%) were in the older old group, above 70 years old. There were fewer older old respondents because many more young old people were found at the spaces. The older old person may lack mobility or good health to occupy these public spaces, have no interest or be unable to participate in English (Mandarin/Chinese dialects were used in some cases to facilitate participation). Sites A and C had more older old participants (25–30% of respondents) while Site B had no respondents in the older old age group. This is a study limitation as the age-70+individuals may offer different responses from the 55+.

5.3 Findings

5.3.1 Respondent Usage of Space

Reasons for Visiting the Sites Participants were asked to share the reasons why they visit the site and what activities they do while there. They were not limited to singular answer. Common pull factors included: meeting friends, visiting nearby coffee shops for meals, buying groceries at the market and participating in recreational activities. Some respondents would pass through the site on their way to public transportation (bus stop, LRT station) or amenities like the bank automated teller machines (ATM). One respondent would visit the area whenever there were religious events at the Chinese temple adjoining the site (Site C).

The activities named were grouped into four main categories:

1. Social, e.g. meeting friends to chat;
2. Recreational, e.g. exercise, tai chi, qigong; gardening; bird singing;
3. Relaxation, e.g. sitting; bringing grandchildren to play, watching other people;
4. Errands, e.g. accessing the letterboxes or ATM; going to the bank, shops, temple or LRT/bus stop; buying groceries; visiting the coffee shop for meals.

The first three categories contain activities that are conducted at the site while the last category embraces activities that provide a reason for passing through the site but are not necessarily conducted there (transition zone).

The most common activities across all sites were social and recreational in nature, with about half of all respondent groups mentioning either of these activity types. Running errands was a common reason for passing through all sites, except Site D (understandably because Site D is on the top of a hillock and can only be accessed through a long flight of stairs or barrier-free ramps). Site A, possibly due to the surrounding coffee shops, grocery stores and neighbourhood centre on its immediate fringe, had half of its participants going through the site for meals or grocery shopping. None of the Site A respondents mentioned recreational activities despite the presence of a senior fitness corner (perhaps because those who stopped to participate in this survey were there for a different purpose) though a few talked about relaxing in the space.

The void deck sites of B and C showed similar distribution of activities among their respondents: a majority mentioned meeting friends though activities in the other three categories were also mentioned a few times. The senior residents corner in Site B seemed to attract residents for recreational activities such as bird singing and tai chi. Aside from meeting to socialise in the void deck space, Site C participants visited the site to check the mailbox (located at the void deck lift lobby area) and as a shortcut to the nearby LRT station. The unanimous mentions of recreational activities from Site D respondents correspond with its park setting.

With regard to the aggregated participation in activities among age groups, the older old group tended to display a strong preference for social activities—respondents mentioned meeting friends as a primary reason for visiting. This seems to suggest the importance of social activities to older people. Activities performed by the older old group that fall under Errands relate largely to food: buying meals from coffee shops or grocery shopping. The young old groups had many more different errands, e.g. checking the mailbox, using automated teller machines (ATM), besides buying food and groceries.

The young old respondents appeared to have a wider range of activities than their older counterparts. They tended to participate in a variety of recreational activities, displaying an enthusiasm for staying active through physical exercise and traditional Chinese breathing exercise (e.g. tai chi, qigong) and participating in hobby-related leisure activities like gardening and bird singing. A few groups (from Sites A, B and C) also mentioned the passive activities of sitting and relaxing in the spaces surveyed though their reasons and activities for going to the sites were wide-ranging. One would pass through the site *en-route* to buy meals. Another was drawn to the site to chat with friends while a third would sit and rest at the site. The Site B respondents who talked about bringing grandchildren to play in the space shared that prior to this—taking care of grandchildren—they had no reason to visit the site. As Singaporeans delay marriage and parenthood, the age range of grandparents who engage in this activity may gradually shift towards the upper end of the age spectrum.

Frequency and Duration of Visits Most respondents visited their respective sites almost daily. Among Site A respondents who visited the site daily, the common activity was meeting friends. Frequent visitors at Site B either met friends or brought their grandchildren to the playground. The respondent groups who visited on a weekly basis mentioned bird singing as a reason for visiting the site. The most frequent visitors to Site C did so for social activities while the respondent group who visited on weekly or monthly frequencies went there for religious or brisk-walking events. Site D respondents all reported visiting the site up to 7 times a week for recreational activities like exercising and gardening.

Both age groups—young old and older old—visited the sites frequently. Half of the young old frequent visitors went to meet friends and at least three groups mentioned recreational activities like exercise and gardening. The young old respondents who visited weekly did so for recreational activities such as bird singing, gardening and physical exercise. The older old respondents tended to visit more frequently, perhaps because they were past retirement age and had more time for leisure compared to the young old group. A common motivation to visit the sites was to meet friends.

Site A respondents tended to spend more time at the site than respondents in other sites—perhaps there was more to do and see at this site. At Sites A, B and C, respondents who stayed between one and three hours mentioned meeting friends there. Site C respondents also mentioned participation in gardening and brisk-walking or temple events, while Site B respondents talked about engagement in recreational activities like bird singing or tai chi. Site A groups also spoke about buying groceries or meals and relaxing at the site. Site B respondents who spent the least amount of time there went to the site when bringing their grandchildren to the playground. Site D respondents who visited up to two hours went there to do gardening or exercise.

The overall trend for all respondents indicated that older adults tended to spend less than half a day at public spaces and while there, they mainly engaged in social and recreational activities. Across all sites, respondents had expressed dissatisfaction about outdoor heat and shared that the sites tended to be more vibrant during mornings and evenings (when it is cooler). This suggests that people would avoid public spaces when the weather is too hot and uncomfortable. Some respondents requested for the provision of water coolers, vending machines or public toilets in the sites studied, perhaps indicating a lack of these facilities in public spaces to accommodate their physiological needs.

Respondents from the older old group generally spent about one to three hours in the public space. This indicates an interest to spend time at public spaces. The duration of no more than three hours could be due to a number of factors such as weaker health, less mobility, disinterest in activities available or personal need to use facilities such as toilets. As healthcare and mobility aids for older people advance, future generations of seniors may become more mobile and spend more time outside the home, especially when there are appropriate on-site interests and provisions. Almost three in four groups of young old respondents spend less than two hours at the sites. The six groups who mentioned staying more than two hours generally participated in social, relaxation and recreational activities.

We turn next to look at each study site in detail through the eyes of the respondents.

5.3.2 Site A: 256 Bangkit Road

Five Words Describing the Site. Respondents from diverse demographics described this site as being in a convenient location, sitting at the foot of several HDB[4] blocks and featuring a variety of amenities (Table 5.1). They appreciated the range of shops and goods available in the nearby market and neighbourhood centre where they can purchase affordable fresh vegetables and food. Respondents felt that it was a convenient location to meet friends and socialise. As one young old male respondent

[4]HDB is the abbreviation for the Housing and Development Board, the public housing agency in Singapore. Thus, public housing is popularly referred to as HDB housing/block/flats.

Table 5.1 Respondents' description of Site A

Positive	Negative
1. Convenient	1. Hot during lunchtime
2. Market and shops	2. Overcrowded
3. A place to meet friends	3. Dirty and poor management
4. Bustling, brimming with life, lively	4. Incomplete facilities
5. Easy access, especially to the market	5. Insufficient coffee shops/car parks
6. No falls	
7. Friendly (for both young and old)	
8. Familiar; sense of belonging	

shared, they would meet their friends at the site to chat instead of at each other's homes; this was to avoid inconveniencing their families.

An older old group of respondents noted the easy access to amenities, especially to the market. They felt that it was a friendly place for both the young and old; the site gave them a familiar feeling and sense of belonging. One elderly couple stated "no falls", possibly referring to the site's universal design features that aid mobility, such as non-slip tiles and handrails along the barrier-free access ramp. Three young old groups of male respondents (seven men) described the site as bustling, brimming with life and lively.

One negative aspect shared by an older old couple was that the site could get too hot during midday. A young old couple reported that the site was poorly managed—it could get overcrowded and dirty.

Spatial Quality and Infrastructure. Several aspects are examined:

Lighting. The space was sufficiently lit both in the day and at night; activities could continue to take place through the night.

Wayfinding. There was a general agreement that wayfinding in the area is easy though some respondents said that signposts in Mandarin would help the older residents navigate and locate banks and clinics more easily. Multi-lingual signage could expand beyond English to cover the three main local languages (Mandarin, Malay, Tamil) or use pictorial representation in places frequented by older people.

Accessibility. Respondents felt that getting to the site was very convenient and older adults could easily access it. This is probably because this public space is located at the foot of residential blocks. An older old couple liked the public transport options afforded by the nearby light rail transit and bus services. One young old pair of respondents noted that outside the boundaries of the survey site (on the main road), there is a steep slope near the light rail transit station, which is the scene of frequent accidents.

Maintenance. The site lacked cleanliness, perhaps due to the space being overcrowded and sited next to grocery and food stores (goods often spill over onto the surrounding areas). A group of young old male respondents felt that the coffee shop toilets lacked proper hygiene.

Design. Respondents reported that the estate was a little old and due for an upgrade. Although the trees provided some shade, some respondents suggested installing shelter to provide more shade and protection from rain. Providing a separate smoking area and improving ventilation were other issues raised.

Urban Furniture. Respondents felt the site was a good place for the older residents to rest before heading home after shopping for food and groceries, though a young old male respondent said that there was insufficient seating due to the crowds. Two respondents from the older old group felt that the exercise facilities were not age-friendly to those above 80 years old (Fig. 5.5).

Programme and Social Activity. The main pull factor to the site, for all respondents surveyed, was its convenient access to amenities such as the grocery stores and coffee shops. Photographs of the site in the morning at 11 am compared to 2 pm showed the stark contrast between a large crowd and a virtually empty square. This could be because of people's lifestyles (e.g. going to the shops and markets to purchase groceries in the morning or to lunch) and hot weather (lack of shade, better to stay indoors). Most respondents would visit the site several times a week or daily.

The site was most vibrant in the mornings as people went to the grocery stores, shops and food stalls in the area. The evenings also saw a crowd when older people would sit around the place to relax (when not so hot) or bring their grandchildren to the playground at the site. Respondents shared that they liked to sit at the coffee shops

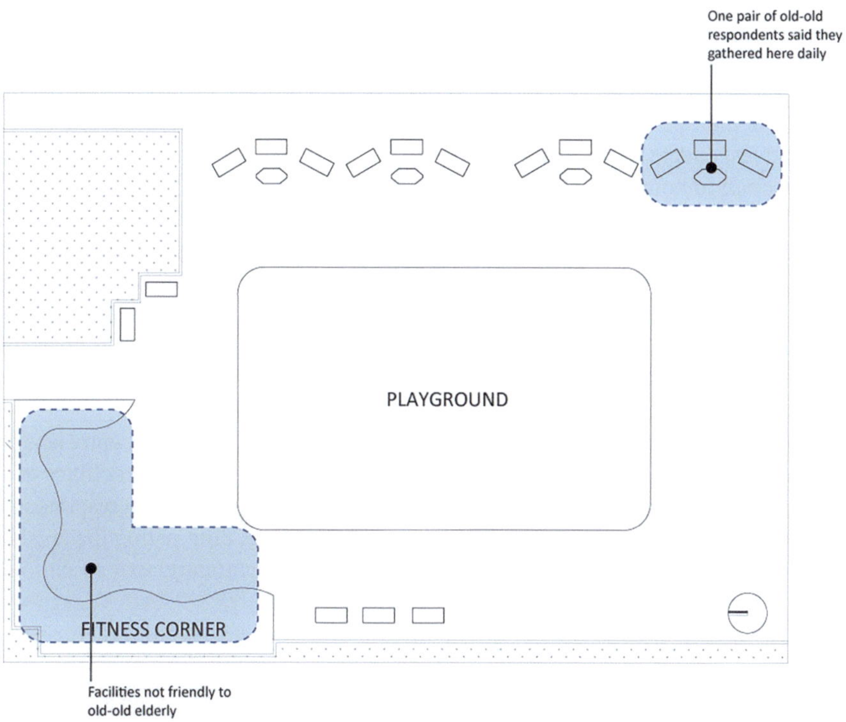

Fig. 5.5 Site A layout annotated with respondent comments

or in the shade and seldom stood or walked around when at the site, emphasising the importance of providing adequate seating. For example, one pair of respondents had a favourite spot to sit daily (Fig. 5.5).

The data suggest the importance of fringe activities as additional pull factors to the site. Opportunities for casual social interaction occur while people are going about daily errands like grocery shopping. Respondents attest to the conduciveness of the site for forming and maintaining human connections through their comments,

> Friendly meeting/chatting place, not only for elderly but also for young generation… This place is very convenient for elderly to meet friends. (Older old Chinese couple)

> Usually listen to stories by the elderly around the neighbourhood. (Young old Chinese couple)

> I always see my friends here. It is a good place to contact with them. (Male, young old Chinese respondent)

> Many people in the neighbourhood moved here from kampongs, so the human connection is definitely there. (Male, young old Chinese respondent)

Facilities such as playgrounds and exercise equipment have important draw power. Playgrounds would generally be used by children outside school-going hours, making them relevant in the late afternoons or evenings. The provision of different programmes in and around Site A like shops, food stalls, playground and an exercise area forms a node where people of all ages can converge, creating a vibrant people space with plenty of goings-on for older people to watch and enjoy. Respondents also enjoyed the presence of children in the area as they enlivened and energised the space with their play, chatter and laughter.

Some respondents shared that public spaces near their residential blocks give them a space close to home where they could meet friends without inconveniencing their families, suggesting a need for "third places", social spaces outside the workplace and home where people are free to gather casually (Oldenburg and Brissett 1982). A pair of young old male respondents recommended putting a reading corner complete with newspapers and books in the residential block void deck because most current programmes like tuition centres, childcare centre were largely targeted at children. A near-home older people-centric facility would be more accessible to older people as they are unable to walk long distances.

The presence of a 24-hour supermarket, a temple and kidney dialysis centre nearby were positive, people-attractor features. As the population ages, health facilities near residential neighbourhoods will become increasingly necessary. Indian respondents at this site appreciated shops selling Indian goods and food, eliminating the need to travel to Tekka Market[5] in downtown Singapore. It is important to extend cultural diversity to commercial provisions in the neighbourhood.

[5]Located in Little India conservation area, Tekka Market is a multi-use complex, comprising a wet market, food centre and shops selling primarily Indian goods.

Table 5.2 Respondents' description of Site B

Positive	Negative
1. Convenient and accessible 2. Friendly 3. Nice/good place to meet friends and spend time 4. Reminisce about kampong life 5. Great amenities 6. Relaxing/peaceful 7. Cooling 8. Helpful 9. Good air quality	1. Crowded 2. Dirty 3. Noisy 4. Dark 5. More space/seating needed

5.3.3 Site B: 434 Fajar Road

Five Words Describing the Site. All respondents from this site were from the young old group. They found the site convenient and accessible (Table 5.2).

Spatial Quality and Infrastructure. Several aspects are examined:

Lighting. The space suffered from poor lighting at night due to damaged lights that had not been repaired.

Wayfinding. Signage could be bigger; small block signage might pose a difficulty for older people to find their way as blocks were not always numbered in order.

Accessibility. Sheltered walkways provided good connectivity within the estate. Transportation options—both feeder routes and buses to the town/city centre—were appreciated.

Maintenance. The site suffered from poor cleanliness and damaged lights. One group of respondents felt that the previously employed cleaners did a better job than those now servicing the blocks.

Design. Respondents liked the air quality and ventilation in the space though some felt that more fans could be installed.

Urban Furniture. Respondents opined that there was insufficient seating and tables and pointed out the areas where they deemed this lacking (Fig. 5.6). The bird singing hangers could be improved with higher placement as the cages currently hang too low.

Programme and Social Activity. Unlike Site A, the void deck in Site B featured no other proximate amenity other than the bird singing corner. Its proximity to the light rail transit station and shops saw many passers-by but the space itself was underutilised due to a lack of programming. The provision of seating though important will not by itself encourage people to use it as a social space. Frequency and duration of visits were largely spread and random, reflecting a lack of strong spatial identity, which would pull people to the site.

The bird singing corner could become a potential source of conflict for residents who may find it unhygienic and noisy. A pair of enthusiasts suggested relocating the bird singing area away from the residential blocks. Regardless of preference,

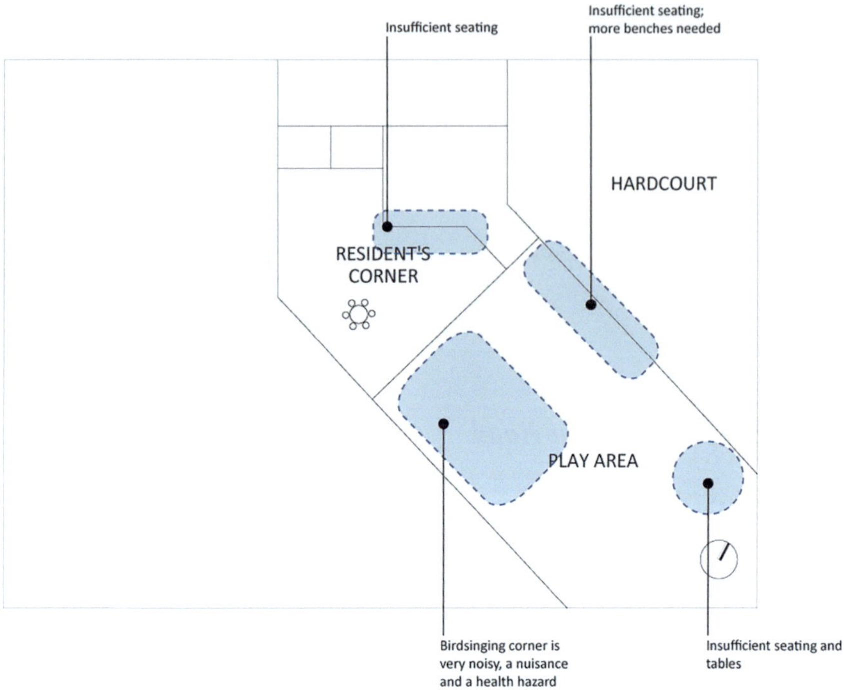

Fig. 5.6 Site B layout annotated with respondent comments

all respondents felt that the weekend bird singing events add to the vibrancy of the space.

Another common issue raised was about the demand and supply of facilities such as seating. The provision of more seating, spread across the space, may mitigate this problem. Some talked about insufficient space in the community gardens (located at the next block) to accommodate everyone who was interested; residents had to make an application and join a waiting list—queue to do gardening.

5.3.4 Site C: 415 Fajar Road

Five Words Describing the Site. Respondents across the demographics opined that the space was bright and cooling, desirable qualities making for a good outdoor environment (Table 5.3).

A young old couple stated that the site was on their way to the light rail transit station and referred to the site as their "home's void deck", perhaps suggesting good accessibility. They also had the impression that it was a place for older people to chat. A male young old group of Chinese respondents opined that it was a convenient

Table 5.3 Respondents' description of Site C

Positive	Negative
1. Good environment, bright, cooling 2. Good, friendly for elderly 3. Quiet 4. Easy to access 5. Safe 6. Peaceful 7. Good transport connection 8. Pathway to LRT 9. "Home's void deck" 10. Place for elderly to talk 11. Convenient to reach coffee shop 12. Friendly	1. Inadequate facilities 2. A bit hot during lunchtime 3. No one uses the space

space on their way to other destinations like the coffee shop. The same group thought that there were inadequate facilities at the site. A young old couple felt that it could get a bit hot during midday and noted that no one used the space then.

Spatial Quality and Infrastructure. Several aspects are examined:

Lighting. The space was well lit by daylight and electric lights at night. Not all respondents know who to contact when the lights malfunction.

Wayfinding. The space was easy to locate and navigate with the adjoining Chinese temple and light rail transit station nearby acting as landmarks.

Accessibility. A pair of older old male respondents pointed out a "nice path" for the older adults to walk (see Fig. 5.7) though they did not elaborate why. The convenience of a nearby traffic crossing was appreciated. But, one group of respondents had suggested improving its traffic light timings. Two respondents suggested installing a roof over the taxi stand for shelter when it rained. They felt that there were no taxi stands in good (easy to access) locations for the older person.

Maintenance. One pair of respondents mentioned the presence of rats and cockroaches in the area, though another group had said they liked the site's cleanliness. The lights in the space sometimes malfunctioned and would take a while to be repaired (Fig. 5.7).

Design. There was a general positivity about the space. Respondents liked the coolness in the shade provided by the trees (near the block) and the space being located on the ground floor of their apartment block. However, it could get a little too warm around midday and the installation of fans would improve ventilation. The site's proximity to the main road and traffic crossing was mentioned by three respondents, indicating its convenience as a throughway. The separation of the recreational and corridor space is an important consideration as there are many passers-by moving through. Two Malay respondents felt that there were occasional problems such as

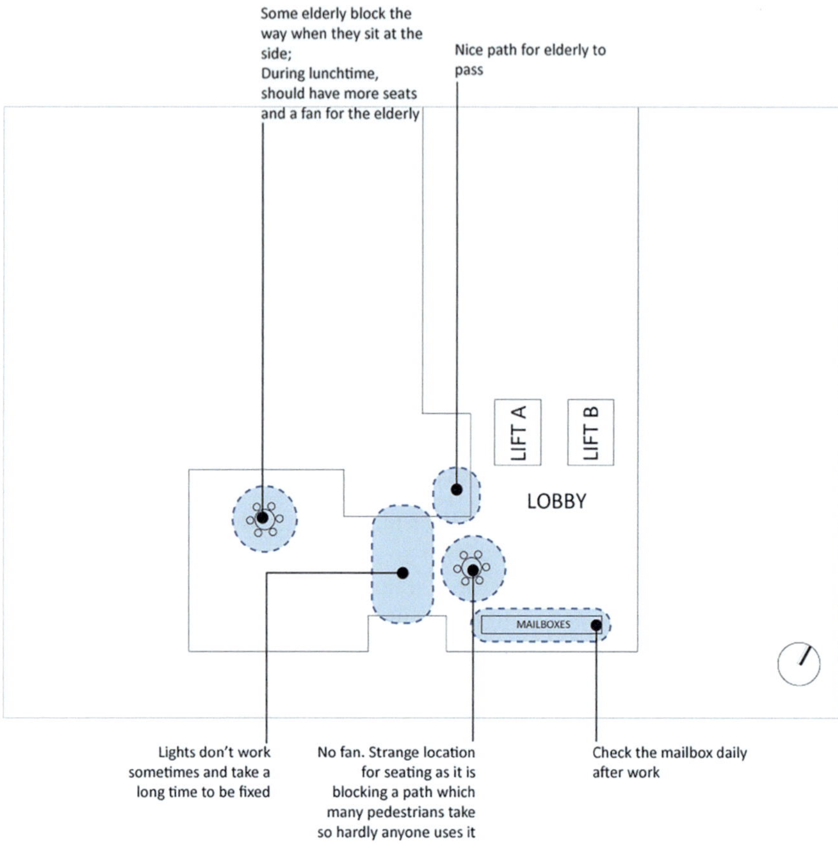

Fig. 5.7 Site C layout annotated with respondent comments

flying ashes generated from open burning of incense offerings during the Chinese Hungry Ghost Festival[6] and noise from Malay weddings in the void deck.

Urban Furniture. All respondents, including those who felt that there was already sufficient seating, requested more seating and tables. Some respondents pointed out the odd placement of a stone table and stools (Fig. 5.7), noting that no one uses them because they were in the way of pedestrian flows.

Programme and Social Activity. Respondents who used the space frequently met their friends in the void deck space for a few hours at a time. They suggested that the space was suitable for older residents to relax and socialise, as it was conveniently located directly under the residential block. In both age groups, except

[6]Observed during the 7th month in the lunar calendar every year, Chinese people of Taoist faith would pray to ghosts/spirits of their ancestors with offerings outside their homes. The offerings include the preparation of ritualistic food and burning of incense, joss paper and a plethora of *paper mache* material items, e.g. clothes, gold and other fine goods. See Fu (2004), Yee and Tan (1999) for more information on this festival.

for two participants from the young-old group, respondents cited socialising as a pull factor to the site. The older old group also enjoyed recreational activities in the space. However, a number of respondents reported that there was insufficient space for more active recreational activities.

The space saw more footfalls when people moved towards Zheng Hua Community Centre in the early morning for brisk-walking and in the evenings for enrichment classes or interest group activities. However, these were passers-by who did not actively use the space. Also, the proximity of the community centre to the residential block may negate the need for recreational space in the void deck itself. For example, two respondents mentioned that older people could go to the public swimming pools at Senja-Cashew Community Club though they did not state if they used the pools or club facilities themselves. The questions raised then are whether the community centre is sufficiently accessible to older residents and how to ensure a smooth, seamless journey from their flat and block to the community/activity centre.

5.3.5 Site D: 201 Petir Road, Bukit Panjang N2 Park

Five Words Describing the Site. Respondents described the site as a beautiful, natural environment (Table 5.4). Groups with both male and female respondents shared some negative points such as unkemptness of the park. One group with mixed genders felt it could become too dark at night.

Spatial Quality and Infrastructure. Several aspects are examined:

Lighting. Some felt that there was insufficient electric lighting and even during the day, there were some dark spots within the park (Fig. 5.8). Poor lighting coupled with the tall grass (reducing visibility and providing hiding places) led to many respondents feeling unsafe. Only the community garden was well lit at night.

Wayfinding. While the park could be conveniently accessed by people living nearby (they know how to get to the park), visitors from elsewhere may not know of its existence or be able to find it. Respondents felt that there could be a signboard with

Table 5.4 Respondents' description of Site D

Positive	Negative
1. Peaceful	1. Falling branches [from trees]
2. Beautiful	2. Dull
3. Scenic	3. Dark at night
4. Quiet	4. Untidy
5. Natural environment	5. More attractions needed
6. Spacious	
7. Good, clean, fresh air	
8. Rubber and coffee trees left behind from the kampong	
9. Variety of colours, sights and sounds	
10. Near home	

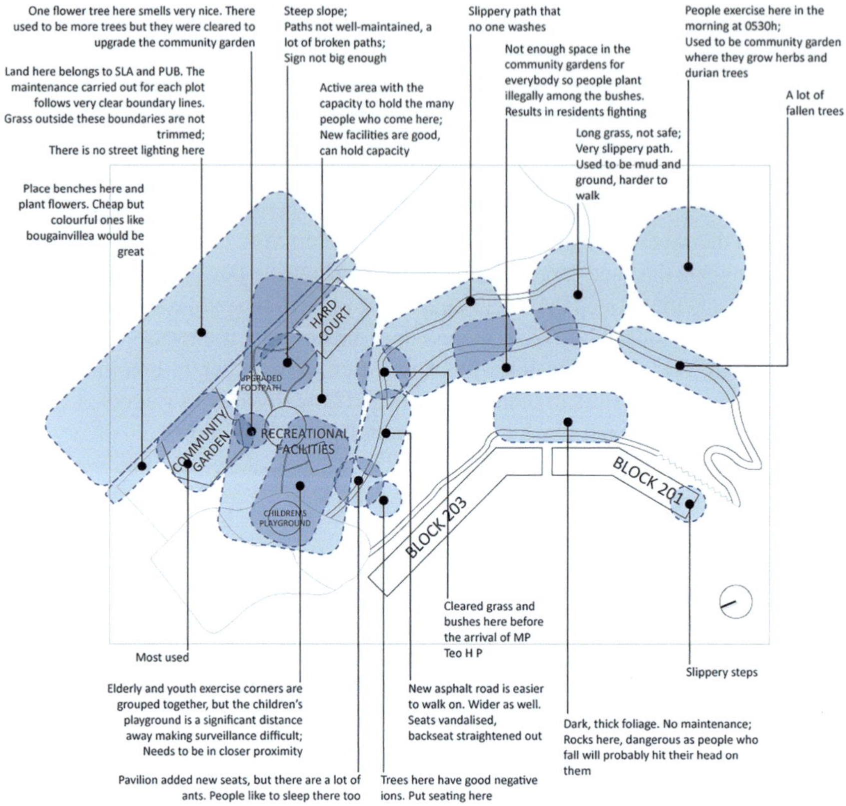

Fig. 5.8 Site D layout annotated with respondent comments

the park's name at its entrance; presently this was not obvious. Having the signboard as a place-marker could strengthen place identity.

Accessibility. Most respondents noted that older adults would have problems moving around the site. The natural topography of the park was appreciated, but some slopes were considered too steep to climb for wheelchair-bound patrons (sometimes difficult for helpers to push them up the slope) or the weaker older individual. Instead, they tended to gather at the flatter paths or foothill of the park and would not venture too far out/up the park grounds, preferring to remain close to the residential blocks. Wheelchair-bound residents were generally unable to access the community garden. One respondent group felt that motorised wheelchairs could make it easier for older people to move around the park.

The slopes could become slippery at times (when moss grew on them) and people had fallen while walking there. Stairs provided in the area were also quite slippery in places, rendering them difficult to access for seniors, especially those with mobility difficulty. Such accessibility difficulties might explain why there were few respondents from the older old age group during the survey period. Respondents suggested that more handrails could be provided as walking aids.

Maintenance. Poor maintenance was a severe problem at this site (Fig. 5.8). Major gripes included the tall, uncut grass (leading to poor visibility of what lies beyond the tall grass), piling up of weeds and dry leaves on the ground (leaves collected were not cleared away) and fallen trees and branches (these, like the leaves, were not cleared away). Rubbish was sometimes left in the central forested area of the park where there was a lot of overgrown vegetation (out of common sight). A group of female respondents liked walking on the path leading to Dairy Farm Park as it was wider and did not have long wild grass. Another respondent pointed out that in contrast, Fajar Park always had trimmed, tidy grassy areas and looked beautiful; it also had a good running track and was near the swimming pool.

Design. Some respondents said that the park was large enough but could use better space planning. They insisted that there should not be too much vegetation planted, as it was not a vegetable garden, but a park (should be green yet open). Respondents enjoyed the fresh air and nature. But, some opined that the landscaping could be improved with a greater variety of plants and flowers (more colour). Some lamented the cutting down of existing fruit trees (some planted from long ago before the park development) to make way for the new condominium development in the park's southern region (Fig. 5.8). They felt that the development removed direct pedestrian access to Bukit Timah Nature Park[7].

Some respondents requested for a children's playground to attract more children to the park. Many felt that the current play area (at the foothill of the park) was located too far away from the exercise corner (at the top of the park), making it difficult for older adult caregivers to keep an eye on the children while engaging in exercise. A group of female respondents observed that the area around the community garden had a nice atmosphere as the park landscape design could appear chaotic at times (especially around the forested area) and it could be too quiet, giving a "scary ambience". They noted that the shady areas provided a nice respite from the hot sun but the park atmosphere had changed somewhat since a murder incident occurred in the area a few years ago (felt unsafe).

Urban Furniture. The pavilions were sometimes used for small group functions and parties but they lacked washing areas so participants could not do much or stay long. Respondents wished to have vending machines or water coolers provided in the park. There was a lack of rubbish bins at the site. Some of the wooden benches along the quieter footpaths had been modified/vandalised by some people—deliberate flattening of the backs of the benches so they could sleep there. This coupled with the general poor maintenance in the area (e.g. fallen tree branches and leaves swept, collected but not removed) radiated a sense of insecurity for residents. This lends evidence to the "broken windows" theory where pervasive physical disorder and disrepair (e.g. vandalism or broken equipment) can lead to fear and a reduction in social control and ultimately, an increase in crime (Wilson and Kelling 1982).

[7]This is a nature reserve and Singapore's largest surviving primary rainforest.

Programme and Social Activity. All respondents visited the site to use the park for recreational activities like exercise, qigong or gardening at the community gardens. Of the three groups who shared their visiting frequency, two visited the park daily for exercise while the third group visited up to five times a week. This site had fewer respondents in the older old age group; two of them were regular exercise goers to the park (they enjoyed walking in the park). Another group of respondents stated that the older person had nothing much to do at the park, especially when they had alternative facilities at the foothill of the park closer to the housing blocks (no need to climb/walk up the park), and that working adults were too busy to bring the older adults to the park. Respondents remarked that there were fewer children playing in the park now, observing a slow but steady change in the area's demographics (more old people now).

One group of respondents shared that some residents had lived in the area for more than 25 years, even before the boulder landscaping in the park was put in (these boulders provided good alternative seating to the benches). They felt that in old age, going to the park to sit and relax was good for health (close to nature). Respondents enjoyed the variety of wild animals seen in the park such as monitor lizards, chameleons and monkeys.

The park could get crowded during the early morning exercise period and in the evenings, when people used the recreational facilities (e.g. exercise equipment) and hard court (Fig. 5.8). Some respondents shared that exercise would need accompanying music. They noticed that residents would move to the park to exercise—instead of using the void deck—so as to avoid disturbing residents with the noise generated from the music in the early morning. They noted that the park currently attracted some visitors, namely, couples (largely foreigners), domestic helpers or students who would hang out in the park (e.g. at the pavilion, away from common view) and engage in unsavoury activities. They observed that there were people (usually youths) who would sometimes stay overnight in the pavilions, creating too much noise. They were not sure whether those people were residents from the neighbourhood. A majority of the people using the park were residents in the area though they noted that there were some external visitors.

The community garden was a favourite park feature for five respondents who suggested that the nearby unkempt grass area of the park should be cut to make more space for gardening. Respondents shared that they developed friendships through exercise groups or community gardening. Programmes that run on a regular basis could help create opportunities for residents to socialize and bond.

5.4 Conclusion

The Walk and Talk survey revealed that social and recreational activities were strong pull factors attracting people to public spaces across all sites. The young old respondents used the sites as a go-through space when running errands (all with the exception of Site D because of its location) in addition to active engagement such as relaxing, socialising or participating in recreational activities like exercising and gardening (community gardening was only at Site D). The older old respondents mainly participated in social activities; errands performed were usually associated with everyday life activities like the purchase of food or groceries.

Most respondents visited the sites frequently, with many using the public spaces daily. This could be traced to the location of these spaces; they are found at or close to their housing. The proportion of older old respondents visiting frequently and for longer durations was higher than the young old group, perhaps suggesting that the older old had more time for leisure. Overall, the older old respondents would spend less than half a day in public spaces though improvements in mobility, healthcare or public facilities (like toilets) might encourage them to spend more time in public spaces and do not become isolated. A majority of the respondents felt that the sites were most vibrant in the mornings and evenings or when there were events happening.

In general, respondents had more positive than negative impressions of their visited public space as evidenced from their five-word descriptions of the spaces. Sites A, B and C were seen as convenient places to meet friends and socialise away from the home due to their location on the ground level of respondents' flats. Site A demonstrated that vibrancy could be generated by locating seniors facilities (fitness equipment and seating) next to amenities, forming nodes, which attracted various age groups to the site. The variety of on-site programming—playground for children, grocery shops and coffee shops for adults, seating for the older adults to watch the world go by—and easy accessibility (barrier-free ramps) drew crowds to the space.

Site B highlighted the challenges in accommodating different needs and demands in capacity when planning programmes for public use. The lack of seating discouraged some groups from using the space while other deterrents included noise generated from hobbies like bird singing. Some Site B respondents took issue with the location of the bird singing corner, which they felt could be moved away from the residential area to mitigate the noise generated from the weekly sessions. Some suggested that the bird singing hangers should be raised higher. Others suggested that more seating could be provided across the void deck at Site B so different groups could be in the same space. Interestingly, respondents at all sites requested more seating and tables, even if they already thought there were enough. This hints at the importance of seating to older people and perhaps, an unfulfilled need or hidden demand.

Site C provided examples of infrastructural design elements that are necessary in HDB void deck spaces. The high pedestrian traffic through the area generated a need to separate functional space from passageways. There was a lack of recreational space at the site, which may be circumvented by the nearby community centre. Sheltered sites like Sites B and C received better reviews for comfort in terms of shade, ventilation and air quality than the other open spaces.

Finally, the park at Site D emphasised the importance of universal design and facility maintenance in the provision of public space infrastructure. While the natural topography of the park enhanced its natural aesthetic, the hilly slopes and undulating terrain proved to be a challenge for many senior residents, especially the wheelchair-bound. The park was untidy and unkempt in places at times (e.g. fallen trees and collected fallen leaves were not cleared). This along with the dark spots and pockets of tall grass reduced visibility and respondents' sense of security. The maintenance of public spaces is an important issue. Respondents took notice when cleanliness and hygiene were lacking or lights did not work at their respective sites.

References

Arango, C. M., Paez, D. C., Reis, R. S., Brownson, R. C. & Parra, D. C. (2013). Association between the perceived environment and physical activity among adults in Latin America: A systematic review. *International Journal of Behavioral Nutrition and Physical Activity, 10*, 122. http://www.ijbnpa.org/content/10/1/122 Accessed September 3, 2017.

Brownson, R., Hoehner, C., Day, K., Forsyth, A., & Sallis, J. (2009). Measuring the built environment for physical activity: State of the science. *American Journal of Preventive Medicine, 36*(4S), S99–S123. https://doi.org/10.1016/j.amepre.2009.01.005.

Department of Statistics (2015). *Population Trends 2015*. Singapore.

Everson, K. R., Sotres-Alvarez, D., Herring, A. H., Messer, L., Laraia, B. A., & Rodriquez, D. A. (2009). Assessing urban and rural neighbourhood characteristics using audit and GIS data: Derivation and reliability of constructs. *International Journal of Behavioral Nutrition and Physical Activity, 6,* 44. https://doi.org/10.1186/1479-5868-6-44.

Fu, C. (2004). *Origins of Chinese Festivals*. Singapore: Asiapac Books.

Hofland, A. C. L., Devilee, J., van Kempen, E. & den Broeder, L. (2017). Resident participation in neighbourhood audit tools: A scoping review. *European Journal of Public Health* 1–7.

Oldenburg, R., & Brissett, D. (1982). The Third Place. *Qualitative Sociology, 5*(4), 265–284.

Poortinga, W., Calve, T., Jones, N., Lannon, S., Rees, T., Rodgers, S. E., et al. (2017). Neighbourhood quality and attachment: Validation of the revised residential environment assessment tool. *Environment and Behavior, 49*(3), 255–282.

Wilson, J. Q., & Kelling, G. L. (1982). Broken windows: The police and neighbourhood safety. *The Atlantic Monthly, 249,* 29–38.

Yee, P., & Tan, V. (1999). *Feasts and Festivals of Singapore*. Singapore: The Educational Publishing House.

Chapter 6
Seeing Through the Eyes of Older Adults

Belinda Yuen and Emily Soh

Abstract The See and Snap study uses a user-based qualitative research approach involving older people to take photographs of the built environment using their smart phones. It offers a way to see through their eyes how older adults perceive, negotiate and experience the outdoor built environment in their everyday journeys. It is one of three tools developed in the Urban Audit to understand the experiential perspective of older users and the barriers they face when using the outdoor built environment. The findings revealed that the respondents visited a range of places, structured and incidental, purposeful and social spaces, in their everyday lives within their neighbourhoods and outside of their town. They particularly liked the convenience and proximity of amenities like markets, local shopping and green spaces to their homes.

6.1 Introduction

The See and Snap (photo-taking) study used a user-based qualitative research approach to see through the eyes of older participants the way they perceive, negotiate and experience the outdoor built environment. This may be in the immediate neighbourhood or further afield on their travel journeys on foot or by other modes of transportation to reach destinations beyond their neighbourhood. The daily routes and places that the participants visit were not prescribed; they were self-selected and form part of their everyday routines. The aim of such an approach is to unearth, through the lenses of the participants, otherwise unspoken and innate travel patterns and preferences in older people's use of space within and outside the neighbourhood. Hurworth (2003) has recommended this method as "a component of multi-methods triangulation to improve rigour".

The photo-taking activity is one of the tools developed in the Urban Audit to understand the experiential perspective of older users and the barriers they face

B. Yuen (✉) · E. Soh
Singapore University of Technology and Design, Singapore, Singapore
e-mail: belinda_yuen@sutd.edu.sg

E. Soh
e-mail: emily.y.soh@gmail.com

B. Yuen (ed.), *Ageing and the Built Environment in Singapore*,
https://doi.org/10.1007/978-3-319-92444-1_6

within the outdoor recreational spaces around their homes. To recap, the main aims of the Urban Audit are to:

- Identify potential weaknesses and key strengths in the design of current outdoor urban environments for ageing; and
- Understand the factors and qualities that are important to people in age-friendly outdoor recreational spaces.

This Chapter presents the environmental and mobility issues shared by the participants through their photo-taking activity.

6.2 Methodology

The See and Snap survey was carried out over the period, July 2015–May 2016. It was conducted using photographs taken by the participants (aged 55 and older) to reveal the way they perceive and experience the urban environment. It was administered in two parts. Part 1 comprises participants' photo-taking activity and Part 2 is a follow-up discussion session where participants gather in a small-group setting to discuss the photographs taken.

In the photo-taking activity, participants were asked to use their camera-enabled smartphones for one week wherever they go and to take pictures of their neighbourhood or beyond as they go about their daily activities. Participants did not have to do anything special or go anywhere outside of their daily routines and activities. They were encouraged to take note of the time and place where the photographs were taken and to provide short captions when they submitted their photographs to the research team at the end of each day. The photographs could be of:

- Their favourite places;
- The things they liked or disliked about the places they visited;
- Their concerns about the built environment;
- What they see (in the built environment) to be important for their ageing process.

It was emphasised that they should take photographs of places and the environment rather than of people. After all photographs were submitted, the respondents were invited to a sharing session where they gathered in groups of two to nine persons to describe and explain their photographs and share their opinions, insights and perspectives in non-scientific, everyday language about the environment. Although a common discussion guide was used, the emphasis was on using the photographs to thematically guide and motivate the participants to speak freely about their personal views and values. Participants were able to choose to discontinue at any point of the process and their data would be accordingly omitted from the study.

This participatory photography method has been previously employed. In Singapore, town councils, in collaboration with various government agencies, have carried out photograph-based participatory town audits to identify physical hazards in residential neighbourhoods. These have been found to be useful to discover the "needs and gaps in the neighbourhood" and to suggest improvements (Ministry of Health 2014, p. 19).

In the wider literature, the seminal work of Collier (1967) was foundational in propagating the use of photographs (whether supplied by researchers or participants themselves) to garner information, opinions and insights (Prosser 1998; Harper 2002; Moore et al. 2008). Few, however, have used photograph-based neighbourhood or community audits among older participants (Mitra et al. 2015; Ronzi et al. 2015; Novek et al. 2012). This method of working with older participants to capture their concerns of the built environment has several advantages.

Since the built environment is inherently experiential and visual, photographs coupled with descriptions and details from the participants could provide rich insights into behaviours, perspectives and opinions than solely by verbal description alone. The taking of photographs is a relatively straightforward and user-friendly way to capture the older user's perception of their living environment, and to let photographs convey their opinions. These respondents preserve 'narrative-autonomy' and become research participants of their own homes and neighbourhoods as they capture images for the photo-sharing session and speak about their photographs in their own words. The process can help to uncover tacit knowledge about place, perceptions, and relationships, etc. that would not have been known through purely outsider, independent and objective observations.

Even though photograph taking is a useful method to explore what older people find salient about their neighbourhood from the user perspective, it is not without limitations. Some people may find this activity time-consuming to have to follow through one whole week of continuous photograph taking and thus, decline participation. Some gave up the activity after a couple of days. This potentially excludes people who are busier, engaged in full-time employment, less mobile or without cameras. This method requires the researcher and participant to have access to cameras and a process for checking and transferring daily photographs from participant to researcher. There is also the logistical challenge in scheduling the discussion session after photograph taking.

The other limitations of our study pertain mainly to the difficulty in recruiting survey participants. As the recruitment of participants was done in public areas in Bukit Panjang, we found that participants of this study tended to be generally able-bodied, mobile and active individuals. This could mean that older people with limited mobility (e.g. those on wheelchairs or require walking aid) are inadvertently omitted and a further study may be required to consider their perspectives.

Consequently, the study has only 17 participants and is over-represented by females. From the initial 22 persons (aged 55+ to 70+) who expressed interest to participate, three-quarters of them (17 participants) followed through and completed the photo taking and sharing session. There are 16 females and 1 male; 12 Chinese, 3 Malay and 2 Indians. They are all residents of Bukit Panjang Town, study site of this and other urban audit surveys.

Our discussions with the respondents revealed that some of them are long-time (over 20 years) residents of Bukit Panjang. At least two respondents are in full-time employment and the rest are either housewives or retirees. All participants are Singaporeans or Permanent Residents and are physically mobile. All participation and responses are voluntary and completely anonymous. Participants sent an average of about 38 photographs each, over a period of seven days. Participants' comments were included as illustration wherever relevant. It should be noted that these comments as written/spoken by the participants might not always be grammatically correct.

6.3 Findings

Respondents visited a range of places in their everyday lives, both within Bukit Panjang Town and their neighbourhoods, and outside of their town. The visual and textual data collected through the dual activities of photograph submissions and follow up discussion session were grouped and analysed around seven categories of outdoor spaces with the purpose to identify common themes, issues and concerns:

- Amenities and facilities;
- Structured social spaces;
- Informal spaces;
- Outdoor/green spaces;
- Outdoor infrastructure/street furniture;
- Activity nodes outside of Bukit Panjang neighbourhood;
- Transportation and travel journeys.

The first five categories are largely related to Bukit Panjang Town and its neighbourhoods. Category 6 (activity nodes) are places participants frequented outside their immediate neighbourhoods, while Category 7 (transportation) relates to issues on travel journeys within and out of Bukit Panjang Town by way of public transport or on foot (none of the respondents drive or cycle). The results indicate that the various ethnicities generally share spaces very well within Bukit Panjang though they do travel to their respective traditional ethnic/cultural or religious places beyond the neighbourhood during certain special days/festivals to purchase ethnic goods or meet with their ethnic communities.

6.3.1 Amenities and Facilities

The first category of photographs is the amenities and facilities, which would include the marketplaces, eateries, shopping areas and library, etc. in the neighbourhood (Fig. 6.1). Some 117 (18%) of the collected photographs show images from this category. Most of these places can be found within amenity areas like at Bukit Panjang

New Bukit Panjang Market and Hawker Centre — *Library* — *Kopitiam* — *Pasar (market place)* — *Neighbourhood retail*

Fig. 6.1 Neighbourhood amenities frequented by respondents (all photographs from respondents)

Central (comprising areas around Bukit Panjang mass rapid transit (MRT)/light rail transit (LRT) and Bukit Panjang Plaza), the area at Bangkit-Bukit Panjang neighbourhood centre (comprising areas around Bangkit LRT, Bukit Panjang Town Council office and Block 256 market places), around Fajar LRT Station-Fajar Shopping Centre and Greenridge Shopping Centre.

The two main marketplaces are at Bangkit-Bukit Panjang neighbourhood centre and Fajar precinct centre. As with many local marketplaces (*pasar* in local parlance), they consist of a local wet market that sells fresh vegetables, meat and seafood, and a hawker centre, which comprises a range of stalls selling different local cooked food (Malay, Chinese, Indian, Western). In both marketplaces, supermarkets are co-located nearby together with local retail shops such as hairdressers, bakeries, stationery shops, hardware shops, clothing shops, eyewear shops and a number of *kopitiams* (coffee shops in local Chinese dialect). These add to the variety of goods and services that can be accessed by the participants on any single trip.

Three-quarters of the respondents remarked that the marketplaces and local food establishments (the two key places being *kopitiams* and hawker centres[1]) are places that they would frequent on a regular basis—daily, several times a week, or weekly.

[1] *Kopitiams* and hawker centres are two types of food establishments that are ubiquitous in Singapore. They have a range of food stalls, often reflecting the multi-ethnicity of Singapore. *Kopitiams* are of a smaller scale and frequently found at the void deck of public residential blocks while hawker centres are purpose-built establishments, comprising hundreds of food stalls. See Lai et al. (2012); Kong (2007) for more information.

Aside from visiting these spaces for grocery (hawker centres are often co-located with wet markets) and for meals, respondents also see these places as social spaces to meet neighbours and people from the community. It is a place where they can stop, linger and sometimes watch the world go by. Respondents talked about how housewives and retirees (including themselves) often visit these places, passing time chit-chatting, playing chess, etc., enabled by the provision of seating, television sets (in some *kopitiams*) and availability of food and drinks. Men are more likely to be seen passing their time at *kopitiams*, drinking coffee and chatting with friends.

These neighbourhood marketplaces are valued for their ease of accessibility and convenience. Most participants informed that they could reach these places by walking. They are able to get a wide variety of daily necessities without the need to travel elsewhere. As one respondent who lives in Fajar explains why Fajar Central is her favourite place in the neighbourhood,

> Because the reason we have everything there (sic). We have market, we have clinic, and we have stationary shops. And we have provision shops. We have everything we need. Everything is there, so we don't have to go anywhere (else). And the MRT [mass rapid transit] station is nearby for all the people (living in flats)… I'm staying in Block 452, next door is the shopping centre. It's convenient. Everything is there. And the clinic is there. For the elderly people, the PG[2] clinic is also there so no need to worry; dental is also there. Everything is there. So I don't have to go anywhere (else). (Madam R, Indian lady)

A number of the respondents mentioned that they welcomed the opening of the new Bukit Panjang Market and Hawker Centre at Bangkit Road (this centre opened in late 2015). Some particularly liked the modern design of the building with its green walls (planting on the facade). Although it was to function as a *pasar* (market in Malay language and conventionally a "wet market"), new-built *pasars* like this will operate in a "dry environment" to maintain stricter hygiene levels and to appeal to present day shoppers. The Fajar Market takes on a similar concept of a dry "wet market".

The three local neighbourhood shopping centres—Fajar Shopping Centre, Greenridge Shopping Centre and Bukit Panjang Plaza—are patronised by almost half of the respondents. Some of them remarked that these decentralised, smaller scale neighbourhood malls are preferred by housewives, retirees and seniors as they are close by, air-conditioned and house a good mix of shops. Some even have amenities such as a rooftop community garden as in the case of Greenridge Shopping Centre. Other community amenities that respondents like are the Bukit Panjang Community Library located within Bukit Panjang Plaza as well as the places of worship—churches, mosque and temples—in the neighbourhood.

One respondent conveyed her appreciation of long-time shop owners of various trades in her neighbourhood; these businesses not only offer convenience for residents in the neighbourhood but are a familiar presence within the neighbourhood. There are interesting instances of specific local knowledge that some respondents have, which

---

[2]PG refers to "Pioneer Generation", that is, to the early generations of people (those aged 16 and above in 1965 or those aged 65 and above in 2014) who have contributed to Singapore's nascent years of nation building. Singapore has initiated a PG package for this group of older citizens to help with their medical costs, e.g. a PG card allows the cardholder to receive subsidised medical treatment at designated clinics.

would not be immediately apparent to outsiders. For instance, one respondent talked about a particular vegetable stall in the Bangkit market that sells vegetables grown in his own farm and her periodic haircut sessions at a neighbourhood hairstylist who worked out of her flat.

Moving away from marketplace/shopping activity hubs are the estate amenities that are located in proximity to residents at the void decks on the ground floor of their residential blocks. One may find childcare centres, seniors activity corners, clinics and even traditional Chinese medicine (TCM) clinics. One respondent remarked on the growing popularity of TCM clinics with the older residents and the importance to include this amenity in the local neighbourhood,

> Oh, this is the best. Nowadays, old people like to visit the TCM, and now it's so near our homes. We don't need to wait for the children to fetch them there or travel for long on public transport. We can just walk for a little bit. It is just behind the CC [community centre], Zheng Hua TCM. It is very convenient for the elderly. (Madam SB, Chinese lady).

Another added,

> Before this, the nearest TCM was in Yishun. The doctor's skills are also really quite good. He is proficient at acupuncture and even massage. (Madam AP, Chinese lady)

A couple of the respondents expressed hopes for more senior activity centres to be located at void decks (see next section).

6.3.2 Structured Social Spaces

Structured social spaces are places that are designed especially for community activities and gatherings. They include exercise hard courts, playgrounds and community centres, which provide a variety of venues for sports, leisure and classes for the community to learn new skills such as music, cooking, languages, etc. (Fig. 6.2). A total of 60 (9%) of the photographs collected pertain to this category.

Structured social spaces are common provision in public housing towns in Singapore. These spaces including both outdoors and indoors venues are used mainly for community exercises, activities and events. Regarding the outdoor spaces, respondents talked about the presence of hard courts as a group exercise venue often used for daily exercises organised by the community (e.g. community centre). The sessions are generally well attended, especially by retirees and housewives. Participants would gather in the mornings, which could begin as early as around 6 am, and during dusk for various exercises such as qigong and tai chi (Chinese breathing and stretching exercises), line dancing and brisk walking. One third of the respondents said they attended these sessions regularly as part of their daily or weekly routines.

Some would bring their grandchildren to these exercises, especially where there are other children with whom they could play together. Brisk walkers could look forward to inter-zone mass walks held on the first Monday of each month, which could attract 400–500 people while the evening walks generally attract an average

Fig. 6.2 Neighbourhood structured social spaces (all photographs from respondents)

of 150–180 people. These walks always start off with warm-ups at the hard courts. Respondents remarked that some exercise goers are very enthusiastic and committed to the activity; they would continue to attend despite minor ailments. Others hinted that they are probably drawn to attend these exercises because of the grocery shopping vouchers that they could receive for participation. As this respondent explains,

> This is at Zone 4 brisk-walking, sometimes up to 400–500 people join the walk. It is (organised) under CC [community centre]. Each zone has their own (sic), but every first Monday of the month, the zones will come together, that's where they get the 400–500 people. (For) night walking we can get 150–180 (people). But, can you see only one or 2 Malays only (sic). Many ah ma (elderly Chinese ladies), limping still want to go, 'cos [because] can get the vouchers (sic). (Madam K, Malay lady)

Another ubiquitous community outdoor space are the children playgrounds, which are at times integrated with an exercise corner with exercise equipment for adults and seniors. A couple of the respondents observed that these places are well used by children, suggesting that more of these should be integrated with exercise facilities for older adults. One of the respondents noted that while there is variation in the design of children playgrounds, the elderly equipment provided are mostly the same, repeated in various locations in the neighbourhood. She asked if there could be more variation in older adults' exercise area design. These comments may point to a need to reimagine this space provision, e.g. to explore the co-location of play areas (children and older adults) given the common arrangement for older people to be caregivers of their grandchildren when their adult children are away at work.

On indoor social spaces in the neighbourhood, respondents noted that these are mostly housed within the community centres. These centres offer classrooms for social, learning and community activities such as cooking, basic computer skills, language classes, etc. Some larger centres even offer sports facilities such as indoor and outdoor multi-purpose sports courts, swimming pools, etc. In Bukit Panjang, there are three such community centres: Bukit Panjang, Senja-Cashew and Zhenghua. Senja-Cashew is comparatively the largest community centre in Bukit Panjang and respondents like it for its comprehensive facilities, which include a swimming pool that opens early on Thursdays at 7 am at the request of residents. During weekends, a number of the respondents would go to the community centres with their family or friends to play sports such as badminton, table tennis, etc.

Of the four respondents who spoke of community centres, three had generally good evaluation of these facilities and amenities. For instance, regarding the Bukit Panjang Community Club, Madam SB shares that, "the CC is pretty and has many facilities, and is the go-to place for many residents", and for Cashew Community Club, she feels that "the amenities are great, there is a pool, gym, computer lab and dance studio. It is a good place that people like to go to, and promotes a healthy body and mind."

There are other respondents who feel that the community centres are generally under-utilised and there is a lack of interesting activities or enrichment classes for adults and seniors.

> Recently there has been 2000 free new courses for the public – but none in Bukit Panjang. CCs seem to be making a profit rather than serving the public. Muffin baking course at $80 is overpriced. As a result CCs underutilised… Hope that CC would open classes on healthy living for elderly (sic). (Madam T, Chinese lady)

This respondent continued to explain that other than community centres, there is a lack of senior-specific venues in Bukit Panjang such as air-conditioned TV rooms, gyms, etc., for older people to spend their time and do recreational pursuits. She desired to have more programmes such as courses for older people to keep occupied and learn new skills in their free time, perhaps on topics such as healthy living for the older adults. These programmes should be free or provided at a low cost so that many more can afford them,

> Suggestion for a centre for old folks (sic), a Senior Activity Centre with air-con, TV. In Bukit Panjang there's no senior specific centre or gym for senior specific activities. There are only CCs. All the old folks are at home after their morning exercise activities. Need more events to keep old folks occupied. Keep costs of programmes low. Cannot keep paying high cost for activities. (Madam T, Chinese lady)

In Singapore, Senior Activity Centres (SACs) are recreational centres for older adults, frequently located at the sub-neighbourhood level and shared by clusters of residential blocks. Some SACs include day-care services and healthcare facilities so as to bring such amenities and health services within the communities, closer to homes. Another respondent added that only Zone 8 has a Senior Citizens Corner, but not in Zones 9 and 10. Senior Citizens Corners are simpler versions of the SACs. They are usually recreational corners with TVs, chessboards, karaoke, chair and tables located at the void decks of residential blocks for the older residents to spend their free time.

Places of worship—the neighbourhood mosques, churches and temples—are another category of neighbourhood social spaces that a few of the respondents mentioned frequenting. One of the Malay respondents, for instance, showed a photograph of the lush garden within the Al Iman Mosque premises, which are well used by Malay-Muslim ladies for gathering sessions.

6.3.3 Incidental Social Spaces

Other than the structured social spaces, a significant amount of interactions take place in incidental spaces in the neighbourhood such as the void decks and common corridors at the HDB residential blocks and on routes to key destinations such as the markets, schools and parks within the neighbourhood (Fig. 6.3). A total of 83 (13%) photographs were collected for this category. Around the Bangkit-Bukit Panjang neighbourhood centre marketplace area, it is a familiar sight to see older men who like to sit around the cluster of seating provided or at informally placed and make-shift tables and chairs to play chess, watch others play and chat the day away. As the photographs show, many of these spaces have spill over activities (e.g. goods on display) being located adjacent to the primary community and business activity nodes and busy pedestrian thoroughfares.

While they appreciate the degree of flexibility to use common spaces, a couple of the respondents remarked that they have neighbours who clutter up and obstruct public passageways when they place their overgrown plants, shoe cabinets, clothes-drying racks, etc. in the common areas. Others pointed out that in a high-density living environment and a multi-racial, multi-religious city like Singapore, there are some cultural or religious practices that when carried out in public spaces could create inconvenience and cause unhappiness to others. One respondent gave the example that people sometimes do not burn "paper money"[3] in the designated containers provided but on the open areas like public footpaths. The resulting fire and debris cause obstruction and inconvenience to pedestrians while flying ash/debris can make the estate grounds untidy.

A handful of respondents expressed appreciation of the general landscaping and greenery efforts around the neighbourhoods—along the main roads, canals, traffic islands and junctions (flower shrubs at traffic lights are beautiful). A couple of them mentioned that they would, in fact, prefer to do their brisk walking along the footpaths of the main roads rather than to go to a park as the former is pleasant enough for exercise. Two respondents are delighted in discovering interesting and unknown aspects in their neighbourhood, in particular, a water source of ground/spring water that is continually running, clean and sweet tasting. Such incidental areas can become an informal social area for the "old-timers" who are in the know,

> Madam D: (Referring to a photograph of groundwater) There is a steady flow of water.
>
> Madam A: People go there to collect water.
>
> Madam D: I was not aware that there was something like this till I walked past it that day. The water is very sweet!

---

[3]Burning paper money for the dead is a religious-cultural practice, part of Chinese Taoist ancestral worship.

Informal social spaces near market places where older residents gather

Morning market spills over to informal spaces *An informal garden*

Residents petition to keep these chairs *Burnt "paper money" residue in public space* *General landscaping in Bukit Panjang appreciated* *Natural spring water discovered by residents*

Fig. 6.3 Neighbourhood unstructured social spaces frequented (all photographs from respondents)

Madam A: The water keeps flowing, it never stops. It is like spring or ground water, even when it does not rain there is still water…There was once they sealed it up, and water simply came out from another hole.

Madam D: It is at the foot of the hill.

Madam A: By the large canal.

6.3.4 Outdoor Green Spaces

There are a variety of outdoor green spaces within Bukit Panjang Town, ranging from neighbourhood parks, community gardens to lush landscaping seen all around the neighbourhood (Fig. 6.4). A quarter (167) of the photographs are in this category. The community is found to be involved in the greening and maintenance of a number of such green areas. Among the variety of green spaces, the community gardens rank as one of the favourite and most frequented places. A majority of the respondents reported visiting these gardens regularly, registering a deep level of involvement.

It was mentioned that the Mayor of the area has been championing the provision of community gardens. Some respondents highlighted that a brochure has been produced to show the community gardens trail and specialities in each of these gardens (National Parks Board n.d.). Under the National Parks Board Community in Bloom Programme, many residents are involved in community gardening. Bukit Panjang has more than eleven community gardens, spread across the neighbourhoods in convenient locations close to residential blocks. While edible fruits and vegetables are favoured by many gardeners and are found throughout these gardens, these gardens also have special foci—some are on organic farming, some have medicinal herbs, others have special varieties of fruits trees and flowering plants.

The community gardens are a well-loved place among half (9) of the respondents. Respondents remarked that they are able to have a sense of ownership as they take care of their own plot of garden and choose their preferred plants, vegetables, flowers and medicinal herbs. They can enjoy the "fruits of their labour" as gardeners get to

Zhenghua Park near residential blocks, platform for exercises *Resident at community garden*

Landscaping brings garden and greenery near to residents *Retirees engaged to do landscaping upkeep*

Fig. 6.4 Neighbourhood green spaces (all photographs from respondents)

keep the produce. Many think it is healthy to eat vegetables that they grow themselves (organic food). It also brings back memories of days when they or their families were farmers. One respondent shared that she found it useful for residents to learn about gardening tips. The experienced gardeners would sometimes share gardening tips on the types and names of vegetables, how to plant and harvest various kinds of vegetables.

For some, gardening has become an integral part of their daily routines, something to do before or after their morning or evening exercises, morning trips to the market or before some of them go to work. For them, the community gardens support social interaction. It is a place to meet other residents in the neighbourhood who share similar gardening interests and through that help them make new friends and strengthen the sense of community. They also take part in competitions such as for the best community garden in the neighbourhood. As one respondent proudly explains, "the community garden beside Block 423 is very good, was done beautifully. It is able to win the first place in competitions!".

Other frequently mentioned outdoor green spaces are Bukit Panjang Neighbourhood 5 Park, Zhenghua Park and Bukit Panjang N2 Park. Bukit Panjang Neighbourhood 5 Park, being centrally located within Bukit Panjang new town and adjacent to Senja-Cashew Community Club, is quite well used, especially by the older population who do morning group exercises. A couple of activities draw the residents to the place. There is a programme: 'Sports to Stay Fit' (takes place every morning from 6.30 am to 8.30 am) that attracts the participation of mostly housewives and retirees for group exercises such as qigong. There are basic amenities such as pavilions and seating provided in this park.

Even so, some respondents suggested that the park lacks a variety of floral planting, attractions and amenities for the older users to spend their spare time. The lack of toilets in the park also poses an inconvenience, especially for the older person. One respondent highlighted the difficulty experienced by her elderly mother, "my wheelchair bound mother has to rush back home because there's no toilet in the park. It puts a lot of pressure on the old folks." Another respondent added that many of the people would just go home or go to Greenridge Shopping Centre after their exercise to use the toilet as it is close by the park.

Zhenghua Park, together with Bukit Panjang N2 Park, forms an elongated park that runs alongside the left flank of the Bukit Timah Expressway (on the right is the large Central Catchment Nature Reserve). The Zhenghua and Bukit Panjang N2 Parks are more undulating than Bukit Panjang Neighbourhood 5 Park and benefits from being close to the nature reserve, having a more forest-like environment with mature trees and interesting trails. On its northern edge is the Zhenghua Community Club. These two parks are viewed with mixed comments from the respondents. A few of the respondents found them to be popular, especially in the evenings. The amenities such as benches and children's playground are considered very good while the provision of pavilions and toilets is a convenience welcomed by visitors.

A respondent shared that residents would go for brisk walking in Zhenghua Park on Sundays, which starts with a warm-up session at Zhenghua Community Club. Another respondent opined that portions of the park near the highway are less attractive and less used. Yet, another commented that she disliked the long flight of stairs from the residential blocks to the park (Bukit Panjang N2 Park) as that can be quite tiring for the older adults to walk up to the park even though younger people may enjoy that.

Two of the respondents shared photographs showing some Bukit Panjang residents, mainly retirees who are hired by the town council in the maintenance of landscaping around the neighbourhood (Fig. 6.4). It is understood that these senior citizens could choose the days and frequency in which they want to work. The respondents felt that this is a good option for senior citizens who may want to engage in some work within the neighbourhood and who would otherwise be bored at home. As this respondent explains,

> They are working as gardeners under Town Council… My husband is in charge of the workers there… Some of them, maybe the ladies, (work) only half a day. Only senior citizens (are employed). After retiring have nothing to do, might as well better they go out and work right? (Madam S, Malay lady)

6.3.5 *Quality of Neighbourhood Infrastructure and Outdoor Street Furniture*

This and the later segment on *Transportation and travel journeys* seek to unpack respondents' concerns and what they see in the built environment to be important for the ageing process—the neighbourhood features that they appreciated, those that they thought were inadequate or needed improvements, and those that they desired to have (Fig. 6.5). A total of 49 (7%) photographs were collected for this category.

On the whole, respondents generally expressed appreciation for the conveniences afforded to them within Bukit Panjang town, that good amenities are available within walking distance. One respondent, a long-time resident who has lived in Bukit Panjang for 27 years, attests to the vast improvements and additions in facilities, amenities and shops in Bukit Panjang over the years, making it an outstanding neighbourhood to live in. When asked what she did not like about Bukit Panjang, she replied, "What I don't like?… Bukit Panjang is my home, It's the best place I think, 'cos [because] it's been 27 years we've been staying in Bukit Panjang." She explained that the amenities keep improving, that there are more shops, new facilities—"I think need to thank our MP Dr. Teo." [4]

Infrastructure such as covered walkways to key destinations in the neighbourhood such as market places, LRT [light rail transit] and bus stops, schools, etc. are very well received and so is the increasing provision of barrier-free access ramps. They were highlighted by six respondents. Two other respondents reported that the

---

[4]MP refers to Member of Parliament. Dr. Teo Ho Pin is the MP representing the Bukit Panjang Single Member Constituency and also Mayor of the Northwest Community Development Council. Dr. Teo has been credited with the many improvements in amenities and environment made to the Bukit Panjang neighbourhoods.

AED defibrillators and CCTV at lift lobbies are useful *Some areas not well-lit* *Abandoned bicycles in void deck*

Damaged benches in Bukit Panjang N2 Park *Rubbish thrown into recycling bins, clutter not cleared*

Fig. 6.5 Neighbourhood infrastructure and maintenance issues (all photographs from respondents)

CCTVs installed at lift lobbies make it safer for the community and the AED defibrillators (installed at selected blocks) are a good first-aid feature. Even so, respondents repeated various observations where they found maintenance and good civic behaviours lacking, for example,

- Damaged benches in Bukit Panjang Park—do not have backrest for the older user;
- Uneven pavements should be maintained to prevent falls;
- Repairs and maintenance needed for exposed wires at Block 213 Petir Road;
- Void deck of Block 211 Petir Road has old bicycles that have been left there for a long time; no one seems to claim them;
- Occasionally unsightly litter can be spotted around the neighbourhood;
- Advertisements pasted around the neighbourhood can be unsightly;
- Over-full recycling bins, which may not have been cleared on time;
- Children's playground is not well-maintained; the rubber flooring material of children's playground damaged by fire from candles during celebrations such as lantern festival has not been repaired;
- Corridor flooring is badly water stained, perhaps due to leakage – looks very unsightly.

Respondents discussed the lack of civic consideration and behaviour by some residents that resulted in clutter or inconveniences for others, for example,

- Recycling bins are misused as rubbish bins;
- Littering at areas around letter boxes as some residents would throw unwanted letters/advertisement leaflets on the ground, even when rubbish bins have been provided close by;

- Large pieces of furniture discarded at lift landings cause obstruction and may be a fire hazard;
- Drivers or motorcyclists who park illegally, especially at places where they obstruct pedestrian flows, etc.

Respondents went on to suggest areas for further upkeep or redesign improvements like,

- Relook into those open, all weather sports hard courts that accumulate water puddles after rain; they are fall hazards, especially for children and older people;
- Provide tables and chairs at void deck for the older adults to sit and linger while their grandchildren play;
- Provide some seating near vehicular pick-up areas so that older residents can sit while waiting for taxi or car to pick them up, they often cannot stand for long;
- Design good drop-off/pick-up points that are sheltered so that the older person does not have to go to the roadside to flag down taxis;
- Provide toilet facilities in public parks as these are very well used by older people during their morning and evening exercises; this would reduce significant hassle and stress for older people if they need to use the toilet at any time;
- Ensure that public toilets come equipped with handrails for older people and handicapped, and bidet/spray wash for the convenience of all races. Toilet floors should be anti-slip;
- Ensure that public areas should be well lit at night as many older people cannot see as well in the dark;
- Revisit upgrading programme of neighbourhoods. While it is good that there is upgrading of neighbourhood amenities, the sense of having constant upgrading can be disorienting and does not support a peaceful, quiet environment. A case in point is the area opposite Block 422 Fajar Road where there is repeated noisy construction work; it seems they are constructing a badminton court there and something else in the past;
- Singapore can consider having priority queues for older people in places like supermarkets and polyclinic; in polyclinics, unlike in hospitals, there are no appointment-making systems.

One respondent emphasised that safety and simplicity in design is key in the design of elder-friendly public spaces:

> (I would) like to have a safe environment for us (the elderly people), and not only for us but for everyone. When you want to design anything, you want to make it good for all the people, (with) elderly people in mind – safe and not complicated. Don't make it so complicated that the elderly cannot use. I mean if you have all these features and you don't know how to (use them, it is it useless), just like a gadget…You think all the elderly know how to use all the sophisticated gadgets? They don't know it! Just to make it safe and easy to use for them. (Madam AR, Malay lady)

6.3.6 Favourite or Frequented Activity Nodes Outside of One's Neighbourhood

Of all the photographs submitted, there are 63 photographs (10%) from twelve respondents, which show places outside of their neighbourhoods and town. Respondents would travel outside their neighbourhoods for a range of purposes: to visit relatives and friends, to workplaces, to institutions for learning, for medical care, to cultural and religious activity nodes and for leisure and recreation (Fig. 6.6). There is no particular weekday/weekend distribution for such trips except that travel to workplaces, learning and healthcare institutions typically take place on weekdays while travel to leisure activities with family members generally take place during weekends.

Three of the respondents mentioned that they would travel beyond their neighbourhoods to receive medical care. As there is currently no polyclinic in Bukit Panjang Town, they typically travel to the nearest one in Choa Chu Kang town (an adjoining public housing town). Comments surfaced that while Choa Chu Kang Polyclinic has comprehensive facilities and provides good medical care, it is often very crowded. As such, one respondent is looking forward to having the new Bukit Panjang Polyclinic, which has been announced to be developed in the coming years (Koh 2015). Another respondent remarked that she would visit the Ng Teng Fong General Hospital in Jurong East regional centre (a regional hospital).

A handful of respondents told us that they would travel outside Bukit Panjang to go to cultural or religious activity nodes. One Indian respondent likes going to Little India as she is able to get a vast range of products such as saris and other Indian costumes, bangles, earrings, gold jewellery, ornaments in places such as Jothi flower shop at Campbell Lane. There is also Tekka Market (in Little India) where she can

Sungei Road Flea Market *Goddess of Mercy Temple at Rochor* *Above: Church of Saint Joseph*
Below: Ng Teng Fong Hospital

Fig. 6.6 Favourite or frequented activity nodes outside one's neighbourhood (all photographs from respondents)

find a wide variety of fresh ingredients as well as cooked Indian food. Sometimes, she goes to Little India to worship in the temples. She would go to Little India whenever there are functions in Tekka Market or during Indian festivals,

> My favorite place is Little India… Jothi Flower Shop - there you go and you can get everything. And you go to market, Tekka Market upstairs. And downstairs all the variety, all the types of fishes, vegetables everything (sic)… Then I will also help out in the community in Little India during Deepavali. I will help them out by taking the old folks to the Little India functions and celebrate with them. I love that. I love to cook for the elderly people. (Madam R, Indian lady)

A Malay respondent shared that she likes to go to Geylang Serai market[5] and she would go weekly. She can get groceries on the ground level and do her shopping (for clothes and other sundries) on the upper levels. She could spend a large part of her day there whenever she visits Geylang Serai,

> I go there every Tuesday. I love to go Geylang (Serai). I will go in the morning and I will just eat breakfast first, then go shopping then after that have lunch, go marketing then I go home. Because the market will close very late about 4 o'clock or 5 o'clock so it is like a whole day process. You can just spend your whole day there you know…. so I like to go there. (Madam A, Malay lady)

A Chinese female respondent shared pictures (see Fig. 6.6) of a popular Chinese temple in Waterloo Street, Rochor area—the Kwan Im Thong Hood Cho Temple (or simply known as the Goddess of Mercy Temple)—taken on the first day of the month of the lunar calendar,

> Today is the first day of the lunar month, this (referring to the photograph, see Photo 6.6) is the Goddess of Mercy Temple at 四马路 (Waterloo Street). The place is bustling with energy, and there is a huge crowd. (Madam C, Chinese lady)

This temple attracts worshippers, especially on the first and fifteenth day of each month of the lunar calendar. Located in downtown Singapore, this place is a confluence of cultural and religious attractions with the Chinese temple, an Indian temple, a Malay mosque, a Jewish synagogue all within walking distance. Waterloo Street is perpetually vibrant with commercial activities with many street hawkers lining the streets and with nearby shopping malls like Bugis Junction, Bugis+, Fortune Centre, Sim Lim Square, etc. It is well connected by public transport with over 20 bus services and Bugis and Rochor MRT stations. It is a place often teeming with locals, tourists and visitors. Yet, another respondent talked about going to her home church—Church of Saint Joseph (in the city) with a history of 165 years where she goes to attend mass every Sunday.

For leisure, respondents listed a wide variety of places they had visited. These include: various malls and supermarkets such as Causeway Point (in Woodlands), the 24-h NTUC Fairprice at Jurong Point, Vivocity Shopping Centre, etc.; Sungei Road Flea Market where one respondent noted that the older generation and lower income people like to go to shop; regional gardens and parks such as the Chinese Garden

---

[5]Geylang Serai market is one of the busiest and largest wet markets in Singapore, offering mainly Indian-Muslim and Malay food and products.

and Gardens by the Bay; other attractions such as the Marina Barrage, Marina Bay Sands, Lee Kong Chian Museum at National University of Singapore campus, etc. In general, it can be noted that almost all of these destinations are located either in the western, northern or central parts of Singapore. They are fairly accessible from Bukit Panjang (located in north-western part of Singapore).

6.3.7 Transportation and Travel Journeys

Respondents submitted 122 (18%) photographs on transportation and travel journeys. Based on these photographs (Fig. 6.7), they shared their views on travel journeys and transportation issues—those taken on foot around the neighbourhood and on public transportation.

Journeys on Foot Journeys on foot are taken largely within the scale of respondents' residential blocks or neighbourhood street blocks. A couple of the respondents opined that the sheltered vehicular drop-offs located at the ground floor of residential blocks are very useful for older users, especially when it is raining. Even so, one respondent felt that some of the design features around the residential blocks do not

Sheltered walkways to key destinations *Longer crossing time for the elderly* *Tactile installations could be trip hazards*

Multi-storey car park without lifts *Bus stop lack suitable seating* *Difficult for older person to climb long flight of stairs* *Reserved seats taken up by young people*

Fig. 6.7 Transportation and travel journeys (all photographs from respondents)

contribute to a barrier-free environment. He cited the example of barriers erected to prevent cyclists or motorcyclists to ride on the pedestrian paths. These same barriers could also cause obstruction for pedestrians and wheelchair users. The respondent explained with a Mandarin caption to his photograph that could be translated as, "The barriers are installed to increase accessibility but instead, added an additional obstacle. I would suggest for them to be removed" (Mr. P, Chinese man).

Another issue raised relates to multi-storey car park buildings—many in the Bukit Panjang neighbourhood still do not have lifts and barrier-free access ramps. One respondent remarked that there are "no barrier-free access ramps through doorways in this public car park building…have to use vehicular ramp exit which is dangerous for wheelchair users" (Madam T, Chinese lady).

Against the increasing popularity of non-motorised personal mobility devices, several respondents anticipated that there would be more conflict of use when more cyclists, kick scooters, hoverboard users, etc. and pedestrians share the same footpaths. Four respondents spoke about pedestrian-cyclist conflicts on shared footpaths. One mentioned that some cyclists do not dismount to push their bicycles at traffic crossings even when there are signs telling them to do so. Another female respondent remarked that, "I don't like it that we (pedestrians) need to share a path with bicycles. Now, they (cyclists assume to) take precedence and expect us to give way to them… some are very inconsiderate, they just (cycle past) from behind you, and you cannot see, so very dangerous." These respondents suggested that there should be dedicated paths provided for cyclists. One respondent explained that Singapore "should have dedicated (bicycle) paths, then I feel more safe. 'Cos [because] I think western countries have those. Canada also have, LA (Los Angeles) also have (*sic*). These are the last two countries I have been."

Three respondents discussed another set of conflict regarding tactile installation on the ground for the visually impaired. They felt that these could become tripping hazards for older people and those who push heavy trolley of groceries. As one respondent remarked, "I nearly fell down quite a number of times, especially when you push the trolley, it is really very dangerous. Then sometimes when the trolley is very heavy, it is so difficult. You have to use a lot of strength…to go forward." Another added, "blind people might not be using the steps as often as we think."

At the neighbourhood scale, more than half of the respondents remarked that there is a very good and comprehensive system of sheltered walkways in Bukit Panjang, connecting them to key destinations such as the markets, LRT and MRT stations, schools and community facilities. A couple of the respondents also observed that many barrier-free pathways (ramps) are not sheltered, which is inconvenient for users when it rains. They suggested that shelters be installed for these ramps. One respondent was particularly appreciative of the longer green man time (for older users to activate with their seniors cards) provided at some traffic junctions. Another thought that this should be implemented more widely, covering more traffic junctions. As this respondent said,

> This place (traffic junction) is at Fajar Road at Bukit Panjang Ring Road. This pedestrian crossing, the old people will definitely not be able to cross the road. So if they tap their cards, the timing extends and they are able to cross faster. I have been to many places and I

> have seen few of this around Singapore. So in places where there is a lot of traffic or elderly people, they should definitely install this! (Mr PT, Chinese man)

A third of the respondents conveyed that older people would not prefer to use overhead pedestrian crossings because it can be very difficult for them to climb the many steps. One respondent mentioned that she and her husband much prefer to cross at traffic light junctions. In the absence, they may resort to jaywalking. Often, when overhead pedestrian bridges are provided, they noted that barriers are also installed at the road median to prevent jaywalking. Yet, to avoid using the overhead pedestrian bridges, some older people would try to jaywalk by climbing over those barriers, which makes it doubly dangerous for them. There were a couple of suggestions that these overhead bridges should have ramps, escalators or lifts installed to facilitate ease of use for the older person.

Public Transportation One of the respondents, Madam R (Indian lady), remarked that on the whole, many of the older population when compared to younger people tend to be quite frugal and if they were still mobile, they would choose to take public transport—bus, LRT or MRT, but not taxis as they found them too expensive. She lamented that young people today would simply take taxis without a second thought.

With regard to transit node infrastructure—bus stops, LRT and MRT stations, one respondent commented that the new bus stop design has seats that are slanted, too narrow and too high, which make them difficult and uncomfortable to sit on. Flat, low seats with larger widths are preferred. A few of them felt that LRT and MRT stations should provide escalators for every upriding circulation. The recently renovated Choa Chu Kang MRT station was brought up as an example of weak LRT-to-MRT station linkage where a long flight of stairs is provided instead of upriding escalators. The lifts are not well-located and commuters have to walk a distance and tap out (exit) and re-enter to make their transit connection. As this respondent explains,

> Bad connectivity from LRT to CCK MRT and vice versa. Inconvenient journey, especially for elderly and wheelchair bound or those pushing trolleys as there is no lift. No back-to-back connection. The lift can only be accessed from outside the gantry after tapping out. (Madam LT, Chinese lady)

The quality of bus service provision was felt to vary on different service routes. One respondent observed that for the buses serving the Bangkit area, bus services 963, 966, 190 and 187 are frequent, coming every five minutes, but bus services 171 and 960 are too infrequent and often come late. Madam AR showed her photographs taken on her buses and train trips to convey that it is quite common to see young people taking up the reserve seats for the older adults or needy. She suggested that reserve seats could come with a tap card system that allows the older riders to use their senior citizens card to tap on those seats to be released for their use. She also felt that all public buses should be equipped with automatic ramps, which does not require the driver to get down to assist passengers on wheelchairs on/off the bus.

6.4 Conclusion

The photo-taking and sharing activity offers a snapshot of respondents' experiential perspective of their daily outdoor environments. It provides a way of crowd-sourcing and capturing personal observations from the community itself. Even though subjective, the information collected has a potential role to inform and improve neighbourhood outcomes, in terms of analysing community wants and needs as well as encouraging residents to share their individual experiences of the neighbourhood and the city.

Respondents in the study shared their favourite places within the Bukit Panjang Town. These include in descending order,

Places of amenities (such as market places, shops, *kopitiams*, etc.). Respondents liked the convenience and proximity of these places to their homes. They also liked how these places provide opportunities to meet and interact with others and to get to know local businesses and shop owners. There was, however, a single negative comment by one of respondents who disliked the new Fajar Market for the lack of variety of vegetables and the cooked food sold in the food centre that is not tasty. She noted that there are still a number of shops that are vacant. For these reasons, she still prefers to frequent the marketplace at Bangkit-Bukit Panjang neighbourhood centre.

Green spaces (community gardens, neighbourhood parks, general landscaping around neighbourhood, etc.). They liked the community gardens as these gardens provided them a space to spend time gardening as a hobby, to grow and enjoy healthy vegetables planted by them as well as to meet and interact with their community. On the other hand, the neighbourhood parks were appreciated as venues for their morning/evening exercises, to have nature close to their homes and to enjoy the greenery and variety of planting in the parks. Notwithstanding, a handful of respondents talked about their dislikes of various green spaces. For the community gardens, while those who are involved have very good comments about the space, others who are not involved felt that the community gardens are not inclusive enough. One respondent felt that even if she does not have a plot in the garden, she should be allowed to purchase the vegetables grown in it. A couple of the other respondents did not like the selection of plants used for landscaping in the parks, citing reasons that they were too dull or the smells were not pleasant. One female respondent commented that the Bukit Panjang N2 Park and Zhenghua Park were too undulating with long staircases that are too tiring for seniors to access. But, she also acknowledged that younger population might like the natural terrain undulation and climbing the long flights of steps are not a problem for them.

Incidental social spaces (e.g. void decks, common corridors, spill over areas of marketplaces, etc.). They liked these spaces for their opportunities to meet and interact with neighbours and community on their way to their regular destinations or activities in the neighbourhood; they liked being able to use common corridors for planting (having nature close to home). Incidental social spaces received a few negative comments, mainly regarding the lack of tidiness of the estate, which are at times caused by residents who lack civic-mindedness.

Structured social spaces (community or exercise gathering spaces, e.g. hard courts, community centres, places of worship, etc.). They liked the hard courts where they could do their communal exercises; the playgrounds that incorporate elderly exercise equipment as venues where they can meet and spend time with other older adults who bring their grandchildren to play; and the community centres where there are recreational and sports facilities and classes that they could take. Even so, a handful of respondents found that venues for social/recreational activities are lacking.

Elsewhere, on neighbourhood infrastructure and street furniture, respondents liked the CCTVs and AED defibrillators at lift lobbies, which provide a sense of security. Some others complained, mostly on issues of maintenance and upkeep of common spaces and outdoor furniture in the estate. In transportation and travel journeys, respondents liked having the comprehensive network of covered walkways to key destinations in the neighbourhood and the barrier-free access ramps. Outside of their immediate neighbourhood and town, respondents' frequently travelled to places include ethnic/culture-related spaces, e.g. religious places, ethnic marketplaces/quarters (like Chinatown, Little India, Geylang Serai), and medical/healthcare-related services (like hospital, polyclinic).

As revealed by respondents' comments, many spaces are viewed with mixed responses, both positive and negative. These may reflect both good and weak aspects of a particular space. To illustrate, public parks and their landscaping are often viewed as attractive places but the lack of toilet provision is a negative issue, causing inconvenience to older people and detracting from their park experience. Another example is the extensive provision of covered walkways, which is considered a good feature but shared walkways for pedestrians and cyclists are not safe for pedestrians, especially for older people.

Mixed responses may also arise because of different types (and scale) of spaces within the same category. For instance, the category of green spaces includes not just parks but also community gardens—the community gardens are largely viewed positively (well-liked because they are accessible and allow the community to be involved in growing their own plants) whereas some parks may raise negative comments (lack of sufficient amenities e.g. toilet, less accessible to older people because of distance and long flights of steps to get to the park).

In aggregate, the amenities and facilities provided in Bukit Panjang are generally well commented on—11 respondents (65%) have only positive comments (no negative comments) on this category of spaces, 1 with mixed response (positive comments as well as suggestions for improvements) and the rest gave no response on this category of space. Respondents' positive comments are generally about the ease of accessibility, diversity of market places and shops provided, range and affordability of groceries and other necessities available, and opportunities to socialise and mingle with the community.

Informal spaces seem to score well with eight positive responses and three mixed responses. These spaces such as common corridors, void decks and landscaped spaces around the residential block are spaces that are immediately outside of the homes and thus, have a direct impact of the quality of the living environment of the residents.

These spaces are mostly appreciated as spontaneous places where respondents meet their neighbours or pass time with friends within the community.

Outdoor/green spaces received six positive responses and seven mixed responses. A closer look into the responses reveals that a great majority of the positive responses (even within the mixed responses) pertain to the community gardens. In fact, all nine respondents who are actively involved in community gardens have only positive comments to offer. The main parks—Zhenghua Park, Bukit Panjang N2 Park and Bukit Panjang Neighbourhood 5 Park—received mixed responses. They are liked for their beautiful landscaping and as a place for community exercising such as qigong and brisk-walking. The parks are already well used by retirees for their morning and evening exercises but many felt that more can still be done to further improve their park experience. Some respondents were concerned about a lack of amenities (in particular, toilet provision) and accessibility features (e.g. long staircases) that make the place less elder-friendly, especially for seniors and those with disabilities.

A higher proportion of negative responses are garnered for the categories of Outdoor infrastructure/street furniture and Transportation and travel journeys, with six such responses for each of these categories. The main negative responses for the former category pertain to lapses in maintenance and lack of elder- and disabled-friendly design of public space infrastructure. For the latter category, the lack of elder- and disabled-friendly design, conflict of uses between pedestrians and cyclists compose most of the negative responses.

What Respondents See in the Built Environment to be Important for their Ageing Process

In the area of provision of amenities, though there is already a high level of satisfaction by the respondents, a handful of them noted a desire to have some of the amenities (such as Traditional Chinese Medicine and 24-h clinics, Senior Activity Centres) that cater specially for older people to be located even closer to them. For green spaces like the Zhenghua Park and Bukit Panjang N2 Park, a couple of the respondents commented that there could be more amenities to attract older users to visit the parks; a basic amenity often mentioned is the convenience of public toilets in the park.

In structured social spaces, a couple of the respondents found that such spaces or activities for the elderly are inadequate. One opined that venues for social/recreational activities tailored for older people (such as Senior Activity Centres) are lacking while another suggested that activities in the community centres do not attract the older person. A couple of the respondents liked the elderly exercise equipment that is integrated into the playground space for children. Could this point to a growing preference for age-inclusive and activity-inclusive public spaces with amenities that encourage people of different ages to share the space and interact?

In neighbourhood infrastructure, a handful of respondents are concerned about safety issues such as ponding in places after rains, causing fall hazards, especially for older people and children. One respondent commented that it is important to use non-slip flooring materials in public spaces and fitting out public toilets with handrails for older people. Another concern is about ensuring public spaces at night are adequately lit for older people with poor eyesight. In the area of transportation and travel journeys, respondents' photographs and conversations often pointed to aspects of urban environment that facilitate or are obstacles for the elderly to navigate in the neighbourhood.

Safety for all is an important consideration as several respondents felt the current situation of cyclists sharing the same paths as pedestrians to be problematic. Several respondents felt that a good network of covered walkways that connects key destinations in the neighbourhood is a good supportive infrastructure provision for older residents and for residents in the estate in general. A handful of them felt that the increasing number of barrier-free ramps is important to support an ageing population. Some others commented that more barrier-free access such as covered ramp-ways, obstacle-free pathways, etc., need to be provided for a greater proportion of public buildings and spaces to better support the older person's mobility needs.

The respondents made many suggestions in what they think to be important for the ageing process. Some of them relate to specific problem areas in the neighbourhood, e.g. fall hazards, maintenance issues while others identified gaps in the provision of more intergenerational communal spaces such as co-locating senior activity centres with childcare centres, etc. The focus on ease of accessibility and simplicity in urban design with the older users in mind are repeated themes that emerged.

Acknowledgements We gratefully acknowledge the following for their generous support and assistance with the See and Snap Urban Audit Survey: our colleagues Cheong Sue Wei and Parvathi Nair for facilitating the group discussions with respondents; Bukit Panjang Town residents who participated in the survey and generously shared with us their responses and experiences of their urban environment. And our colleague Adithi Moogoor for assisting with the photo presentation in this Chapter.

References

Collier, J. (1967). *Visual anthropology: Photography as a research method*. New York: Holt, Reinhart and Winston.

Harper, D. (2002). Talking about pictures: A case for photo elicitation. *Visual Studies, 17*(1), 13–26.

Hurworth, R. (2003). Photo-interviewing for research. *Social Research Update, 40*(1), 1–4.

Koh, V. (2015, March 13). New facilities to make healthcare more accessible. *Today Online*, http://www.todayonline.com/singapore/new-facilities-make-healthcare-more-accessible. Accessed August 1, 2016.

Kong, L. (2007). *Singapore Hawker centres: People, places, food*. Singapore: Straits Times Press.

Lai, A. E., Collins, F. L., & Yeoh, B. S. A. (2012). The kopitiam in Singapore: An evolving story about migration and cultural diversity. In A. E. Lai, F. L. Collins, & B. S. A. Yeoh (Eds.), *Migration and diversity in Asian contexts*. Project MUSE.

Ministry of Health. (2014). *Creating senior-friendly communities: Tips and tools for the City of all ages project*. Ministry of Health, Singapore. https://www.moh.gov.sg/content/dam/moh_web/Publications/Educational%20Resources/2014/MOH-CFAA%20guidebook%20(final).pdf. Accessed July 12, 2017.

Mitra, R., Siva, H., & Kehler, M. (2015). Walk-friendly suburbs for older adults? Exploring the enablers and barriers to walking in a large suburban municipality in Canada. *Journal of Aging Studies, 3,* 10–19.
Moore, G., Croxford, B., Adams, M. D., Refaee, M., Cox, T. J., & Sharples, S. (2008). The photo-survey research method: Capturing life in the city. *Visual Studies, 23*(1), 50–62.
National Parks Board. (n.d.). *Bukit Panjang community garden trail*. National Parks Board. https://www.nparks.gov.sg/gardening/gardening-resources/caring-for-plants/other-useful-links/bukit-panjang-community-garden-trail. Assessed July 1, 2016.
Novek, S., Morris-Oswald, T., & Menec, V. (2012). Using photovoice with older adults: Some methodological strengths and issues. *Ageing & Society, 32,* 451–470.
Project for Public Spaces. (n.d.). http://www.pps.org/, Accessed March 10, 2016.
Prosser, J. (1998). *Image-based Research: A Sourcebook for qualitative researchers*. Routledge.
Ronzi, R., Pope, D., Orton, L., & Bruce N. (2015). Towards healthy ageing in urban environments: Engaging older people and stakeholders in strengthening respect and social inclusion in cities using photo-voice participatory methods. *Journal of Epidemiology and Community Health, 69*(Supplement 1), A 64.

Chapter 7
Investigating Space, Activities and Social Dynamics

Belinda Yuen and Parvathi Nair

Abstract The On-site Observation, one of three Urban Audit tools, observed and recorded the various ways in which a range of public outdoor spaces are used by older and other users from the community through a weekday and weekend day. Interest is with investigating the relationship between spatial characteristics, activity patterns and the social dynamics of urban public spaces in residential neighbourhoods. The analysis showed that design features and time of day could impact the activity patterns of these public spaces.

7.1 Introduction

The On-site Observation, one of three Urban Audit tools, aims to study and record the various ways in which public outdoor spaces are used by older and other users from the community through a weekday and weekend day. In order to include a range of spaces, four public spaces were surveyed based on different characteristics:

A. A successful (well used) [1] senior activity centre site: a precinct square;
B. An unsuccessful (not well used) senior activity centre/senior care centre: senior residents' corner at a void deck[2];
C. A site which presently does not have any elderly service facility: a void deck; and
D. A neighbourhood park, which is the site for community design workshops.[3]

[1]Carmona et al. (2010) have argued that successful spaces are spaces that support and facilitate people's activities.

[2]Void decks can be found on the ground floor of apartment blocks in public housing. They have been purposefully left vacant (open space) and typically used for communal activities.

[3]See Chap. 10.

B. Yuen (✉) · P. Nair
Singapore University of Technology and Design, Singapore, Singapore
e-mail: belinda_yuen@sutd.edu.sg

P. Nair
e-mail: parvathi.snjv@gmail.com

B. Yuen (ed.), *Ageing and the Built Environment in Singapore*,
https://doi.org/10.1007/978-3-319-92444-1_7

Our interest was with investigating the relationship between spatial characteristics, activity patterns and the social dynamics of urban public spaces. The methodology used was one that combines Whyte (1980) work on field observation and Goličnik (2011) activity maps as a means of indicating user behavioural patterns physically on a two-dimensional map.

This Chapter outlines the methodology and elaborates the key findings of this observation survey.

7.2 Methodology

The On-site Observation, similar to Whyte (1980) methodology, seeks to look at movement patterns 'in the field' and extract empirical knowledge based on activities that are taking place within the space, considering the number of people involved, the number of older users, gender, access to the provisions on the site within the given day, time and weather conditions.

Trained observers were stationed on the ground, on shifts to record the frequency, intensity and nature of activities that occurred in the study public outdoor spaces. Three types of data were collected:

- Actual numerical account of users and their activities;
- Activity map that traced the path taken by the users; and
- Photographs of the site taken at various times of the day.

The observers were given standard data collection forms with guidelines to note the data along with easy to read maps of the study sites so that the type of recording was uniform. Observers were stationed at vantage points to the public spaces that allow for non-obstructed observation of space and its activities without appearing as intruders. Data was recorded every hour over 15 h per day over a two-day cycle: one weekday and one weekend day in March 2015. Each full-day observation was divided into three shifts, five hours each.

Despite the somewhat structured nature of the observations, it is acknowledged that much of the data collected is subjective and dependent on individual observer's perceptions and views of events on the site as well as their personal discipline and diligence in recording these observations. As there is no direct contact with the users, there could be human error in locating and recording users on site, e.g. the older person is defined through cumulative visual cues such as appearance, gait and speed of mobility. Notwithstanding the inherently subjective nature of this method, the on-site observation is one of the most effective tools for discovering the functioning of spaces in the literature (Whyte 1980; DeWalt and DeWalt 2002). It is integral to understanding the breadth and complexities of the human experience, providing additional information to participants' subjective reporting about their public space usage.

In order to set a common language to study the four selected and distinctly varied sites, the physical characteristics of the four sites are evaluated based on Kevin

Lynch's wayfinding elements: edges, nodes,[4] landmarks and paths, for a successful public space. This classification reminds the importance of bridging the objective urban elements and subjective existential experiences of lived space.

The four study sites are[5]:

Site A. 256 Bangkit Road—*site with successful senior activity centre*

This site was chosen as it presents a successful seniors' public space that has bustling activities around it throughout the day. The site, a precinct civic plaza *cum* children's playground, is nestled between public housing blocks on all four sides. Analysing this site as per Lynch's elements:

- *Edges*—Ground floor shops in surrounding public housing (HDB[6]) blocks, a bustling pedestrian off-street and neighbourhood centre (Bukit Panjang Neighbourhood Centre) form the immediate edges of this site. There are over 50 shops and fresh food market stalls, a supermarket and other commercial amenities like coffee shops, bakery, bank, day-care, tuition and enrichment centre for children, town council office, etc.
- *Nodes*—This site is provided with primary seating in the form of steel benches on the eastern and western sides that offer nodal points for users who pass through this site. The benches are arranged in clusters (each around a table) and also in a straight row. The benches are all fixed to the ground. The children's play area in the centre of the site acts as a major nodal focus for children and the adults who accompany them.
- *Landmark*—The senior fitness corner at the north-western corner is a landmark for older users. It is surrounded by elevated landscaped area that further defines the boundaries of the fitness corner. Importantly, the provision of elderly exercise equipment in such a location—away from the main road and at the centre of a bustling neighbourhood marketplace and close to (but appropriately separated from) a children's play area ensures safety, security and socialisation, encouraging the older residents to use the space more often. The accompanying seating provision near the fitness corner and around the periphery of the civic space enables the older adults to rest after their workout or to meet friends and chat.
- *Paths*—This site has two barrier-free access ramps at prominent and easily accessible locations that enable universal access. This provision will enable those with walking aid and mobility issues to access this outdoor recreation space.

Site B. 434 Fajar Road—*site with unsuccessful senior activity centre/senior care centre*

This site was chosen as it contains an under-used senior resident centre. Even though Block 434 has a very large (long) void deck that runs the full length of the block, only

[4]For the purpose of analysis in this study, nodes will also refer to the physical provision on the site that encourages the usage of the space.

[5]For site maps, see Chap. 5.

[6]HDB is the abbreviation for Housing and Development Board, the public housing agency in Singapore. Public housing is popularly referred to as HDB housing/block/flats.

a section—the senior residents' corner and its immediate adjoining hard court—is observed. The observation area has two distinct parts with different characteristics: a senior residents' corner at the void deck and an open exercise area at the adjoining hard court. To study the site in detail, the site was observed separately as: the Void Deck and the Hard Court respectively.

The Void Deck

- *Edges*—The void deck space at 434 Fajar Road, with the senior residents' corner is part of a larger void deck with the community centre office at one end and a bird singing corner (that functions only on weekends) at the other. On its external edges stand a car park situated on the northern edge while an open hard court forms its southern edge. The Fajar Road light rail station (LRT) is across the road from this building block. The edges of this study area are not seen to generate much activity that would aid the use of its senior residents' corner.
- *Nodes*—The fixed seating located within the residents' corner forms the nodal points in this space. Being a fairly large space, the zone is provided with two types of seating - rows of stone seats in front of a TV console and a couple of stone tables located close to each other, each with six stone stools (no back). The residents' corner is equipped with a toilet, wash area and a vending machine. The area is under CCTV surveillance.
- *Landmark*—This location does not seem to possess a prominent landmark element that attracts the attention of the older users. Its on-site provision—the senior residents' corner with its mentioned facilities—seems to suggest a design for generating activities within the void deck of this building block. But, the present poor condition (locked up TV console, broken cabinet doors/missing doors, etc.) seems to indicate a general abandonment, a lack of animation in terms of physical design and programming to attract older users (understand the site has been earmarked for renewal).
- *Paths*—This space does not have any clearly defined pathway. It is open and traversed by many (in many different directions) due to its location abutting the car park, main road (Fajar Road), proximity to LRT station (across Fajar Road) and bird singing corner within the void deck, which is used during weekends.

The Hard Court

- *Edges*—HDB building blocks form three edges of this adjoining hard court. One edge is abutting Fajar Road, across which the Fajar LRT station is located. The edge abutting Block 434 is lined with trees, with seating along its shadow area and an exercise/play corner at the far end.
- *Nodes*—The seating along the northern edge (abutting Block 434) is a nodal point of this hard court, offering a gathering space for the older and other residents.
- *Landmark*—As a hard court, the space is designed as an open (concrete) space that can be adapted for a multitude of community uses including ball games, mass exercise, etc. The exercise corner and its adjoining (small) children's play equipment would constitute a landmark element. Even though many people cross

the hard court daily, this space is observed less used; these provisions do not appear to be an active landmark for this location.

- *Paths*—The paths taken across this hard court are dependent on the desired destination of the user. Most users are seen crossing this space to walk to the main road (Fajar Road) and from there to the LRT station or to bus stop further along Fajar Road.

Site C. 415 Fajar Road—*site that presently does not have any elderly service facility*

This site is a regular void deck space that is commonly found on the ground floor of most HDB blocks across Singapore. Analysing this site as per Lynch's elements:

- *Edges*—The space is bookended by other HDB blocks on one side and a Chinese temple on the other; its northern edge faces Fajar Road (the main access road) and the eastern side opens onto an open surface car park.
- *Nodes*—Two sets of permanent seating, fixed to the ground, offer the nodal points—a bench and a set of stone table and seats at one end of the void deck and a set of stone table and stools near the lift lobby. Residents have placed additional movable chairs (including armchairs with cushion) to augment the seating provision at the far end of the void deck, away from the lift lobby.
- *Landmark*—Like most void decks of public housing in Singapore, aside from some seating and the usual paraphernalia of lift lobby (like letterboxes and notice board), no significant landmark element is found in this void deck.
- *Paths*—No specifically defined pathways. People use their most preferred paths to walk through the block and/or to access the lift lobby and letterboxes.

Site D. 201 Petir Road, Bukit Panjang N2 Park—*site for community design workshops*

This site, a neighbourhood park, is used as a reference site for the project's community design workshops.[7] The park is designed as an outdoor recreational/green space for residents of the neighbourhood. It is built on the pre-public housing undulating landscape and hillock. The observation area covers an area of the active recreation zone at the top of the park where two pairs of pavilion and some fitness equipment are located. Analysing this site as per Lynch's elements:

- *Edges*—The northern edge is bounded by a service driveway and a community garden for the residents. The other sides are bordered by jogging/walking tracks that run parallel/adjoining the surrounding HDB blocks.
- *Nodes*—There are several nodal points at this site, comprising amenities such as exercise corner, multi-purpose hard court and two community gardens at the eastern and western extremes of the site.
- *Landmarks*—An elderly exercise corner with fitness equipment and the community gardens act as major landmarks of this site.

[7] See Chap. 10.

- *Pathways*—Being a recreational green space, this park consists of various pathways meandering through its undulating, hilly terrain. These footpaths are designed as jogging tracks as well as recreational walks. But, these walkways are not always shaded and are at times littered by outgrowth of vegetation from the secondary forest in the centre of the park. Many older users we spoke to expressed concerns over sections of these footpaths.[8]

Even though the observers were in unobtrusive places, their long presence attracted the attention of residents in three of the four sites (Sites A, B Void Deck and C), especially among the regulars (mostly older people) who engaged the observers in conversation to find out what they were doing there. Perhaps, because of the proximity of these spaces to the residences, people were alert to the presence of strangers lingering in the place for an extended time.

7.3 Findings

The observations from each site are analysed as a function of day, time and activity under *Activity Level* and *Activity Typology*.

Activity Level Analysis studies the results from the numerical data in a quantitative manner. It is acknowledged that activity levels could vary from day to day and the pen and paper field observation may have missed recording all of the activities, especially during busy times. Notwithstanding, the results offer a window into how the study spaces are being used, creating empirical knowledge on how people are using different public spaces in local neighbourhoods.

The activity patterns within the study sites are studied in terms of two major types of user engagement—active and passive—as categorised by Carmona (2010). Primary form of passive engagement includes sitting and people watching. While there is no need to become actively involved with the environment, this form of engagement does require some appropriate design features, e.g. providing sitting choices. Active engagement represents an active experience within a place and the design of a space can create such opportunities, e.g. provision of play equipment, running track. This form of engagement can often result in social interaction among the involved people. Carmona argues that successful people places provide opportunities for different levels of engagement. This section of the analysis will attempt to answer the following:

- What is the proportion of older adult users?
- Which time of the day is the space most used and what time of the day are these older adults found using the space?
- What proportions of the users are engaged in passive and active activities?
- What proportions of the older adults are seen using walking aid to move around these spaces?

[8]See Chap. 10.

- How do these trends vary over weekdays and weekends?

Activity Typology Analysis explores the mapping of the various activities in a qualitative study. In order to further analyse the activity-physical design features relationships within the study spaces (the relationship between daily uses of the space and its physical design features), the study will identify not only how many people (especially older users) were within the spaces but also what they were doing, focusing particularly on the influence of physical design features on activities.

Two types of activity-place relationship analysis were conducted. The first is Gehl (1987) urban space design that investigates people's activities in terms of 'necessary' (what you have to do), 'optional' (what you like to do and get tempted to do when given the right circumstances) and 'social' activities (meeting fellow residents). Gehl argues that any good, popular and loved public spaces will always have some necessary activities plus many optional and social activities. He further concludes that necessary activities are influenced only slightly by the physical space quality because they have to happen anyway, optional activities are a direct barometer of space quality as they only occur when conditions are optimal while social activities happen whatever the physical context though their quality and intensity could be influenced by both the number of people and quality of the space. The second is Pinto and Remesar (2012) analysis of public space network cohesion through possible polarising effects of nodes to 'anchor' activities and movement patterns. The analysis of activity mapping and site photographs will attempt to answer the following:

- Where does each activity take place?
- What activities are the older adults involved in?
- Do the presence of nodes/physical provisions generate physical activity?
- If so, what are the physical design features that facilitated the activities?
- What effect does the weather have on the number of users and activity on site?

In order to explore the abovementioned themes, each site is analysed in detail followed by site comparison. The comparison of different spaces and the provisions in each site will address the following:

- How are outdoor activities influenced by the quality of space?
- How do people's activities relate to the physical characteristics of a public space?
- What outdoor physical setting is most attractive to older adult users?

As it is not possible to explore all the nuanced groups and activities within and between the different spaces, descriptions (based primarily on aggregations of data) of the more prominent uses and users, especially older users, are presented.

7.3.1 Site A—256 Bangkit Road

Over 1500 users were observed on each of the two days during the observation hours (6 am to 9 pm). Only 20–30% of these users were older people. Given its setting amidst surrounding shops and marketplaces, there were numerous people going through the site, making it difficult to count at times.

More people (5.4% more) were found using this public space on a weekday when compared to a weekend day. This trend was more marked among the older users. There were more older users on the weekday (543) than weekend day (290). This could be because they would stay at home to spend time with working family members and children during the weekend. On both the observed weekday and weekend day, the number of older people visiting this public space with walking aid was found to be low (less than 30 people) despite the provision of two barrier-free ramps to the site. Several of the older users were accompanied by one or two people, particularly young grandchildren, domestic helpers or other family members.

Figure 7.1 illustrates how the flows of people vary during the observation times. There is a lot of movement to and through the space, e.g. users walking across the site to other destination or just loitering around. A variety of activities, both passive (e.g. sitting) and active (e.g. walking, exercising, playing) was observed, with walking being the most frequent activity on both weekday and weekend day.

Walking is not just a necessary mode of transport, to get from point A to B. It is also a social process—to hear, see and meet other people. It is an activity that self-reinforces life in public spaces—it is easy to change to other types of activity while walking, e.g. to stop to look at something interesting, to talk to someone or to sit down for a while. Even though some activities are common on both days, there appears to be more walking on weekday while talking, sitting, eating and shopping seem to happen more during weekend day, suggesting that perhaps people have more time for such leisurely activities during the weekend.

In order to understand these trends in detail, the user type and activity frequency of weekday and weekend days are further analysed.

Weekday

User frequency—People's space use differs throughout the day. On the weekday, more people (48.5% of total users) were recorded using this space in the morning (from 6 am to 11 am) followed by 29.9% towards the evening (from 4.01 pm to 9 pm) and 21.6% during the afternoon (from 11.01 am to 4 pm). Majority of the older users were recorded during the morning hours—324 older users during 6 am to 11 am as compared to 120 recorded during 11.01 am to 4 pm and 99 between 4.01 pm and 9 pm. The highest numbers of users to this space including older users were found between 9 am and 10 am. With regards to timing, the most populated time was in the morning while the afternoon was the least populated time—dipping to a trough just after lunch, around 1 pm, before rising in the mid-afternoon, around 3 pm and 5 pm and into the night (after 7 pm). The number of older users, however, remained almost consistently low (<30) from 1 pm onwards, after peaking in the morning (around 9 am).

In terms of group size and composition, more individuals and small groups of two to three were recorded in this space during the morning and afternoon. They were found mostly engaged in social activities (e.g. talking to each other), anchored at the seating areas and exercise corner. Larger groups of three or more people were

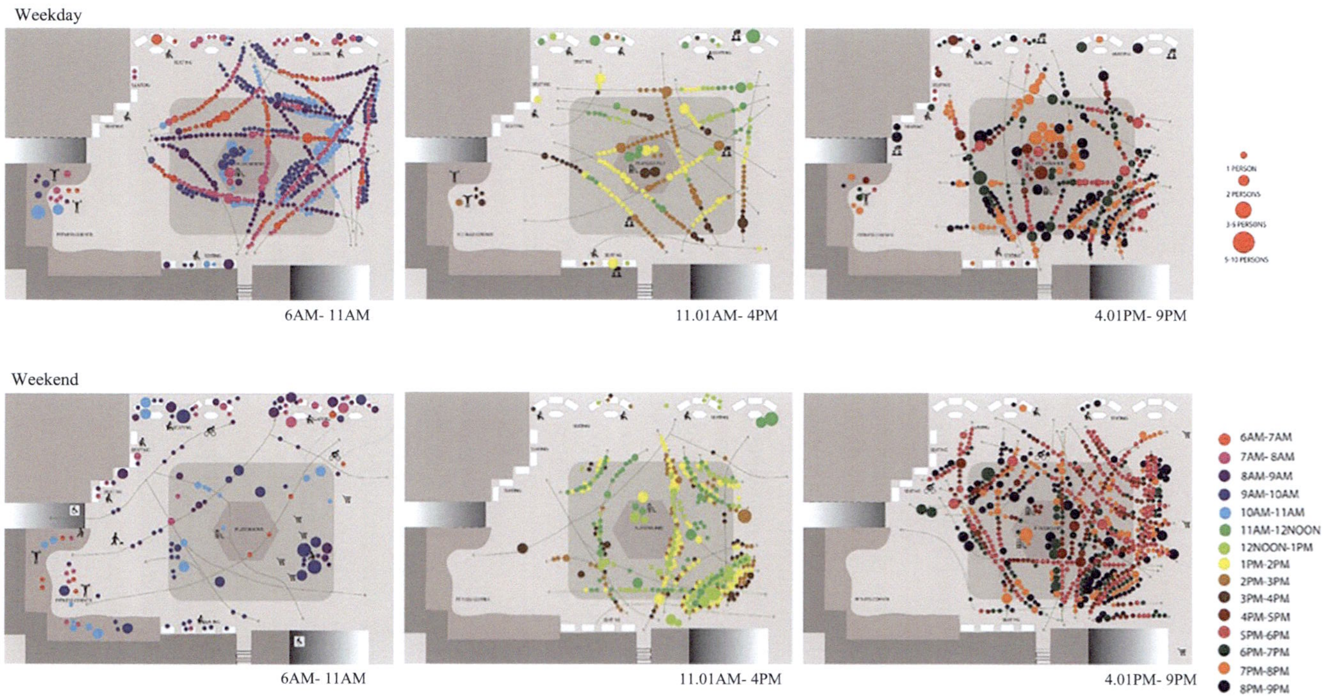

Fig. 7.1 Activity mapping—Site A

observed to gather in the evening than other times. During the morning, the groups tended to be mostly women. By afternoon, groups of men were also observed though often sitting apart from the women groups and occupying different sections of the same space. The groups seemed to co-exist in the same space without paying much attention to each other. During the evening, even though there were many more groups, more came in pairs and more adult and older users were observed using this space alone. Groups of children were observed using the central play area along with their caregivers and friends.

Types of activities—Over 85% of the users of this public space were engaged in active activities like walking, playing and exercising. The volume of walking was highest during the morning, from 8 am and peaking around 10 am (when the surrounding marketplaces were in full swing and at their busiest). This activity would tend to reduce during the afternoon before increasing again in the evening hours (when people returned from work or went out for meals). A smaller number of users were found in passive engagement such as resting at the seating area, sitting on the benches to read newspapers or talk, show and tell about shopping purchases, etc. Sitting and talking were found through most of the day, peaking in the morning (around 8 am), noon and again towards evening hours (from 8 to 9 pm). Talking seemed to happen through the day (among friends and sometimes with strangers), peaking in the afternoon even when other activities (apart from walking) were at an ebb. Observers were sometimes approached and asked what they were doing in the space. The activities were not always singular happenings; sometimes, two or more activities would occur at the same time, e.g. walking and talking, sitting, talking and eating.

A second categorisation of activities indicated that most users were involved in necessary/functional activities like walking to destinations within (such as its elderly fitness corner and children's playground) and beyond the observed space throughout the day. About 80% of the users were walking (taking shortcuts) across the study site (Fig. 7.1). The majority of these anchored movements happened during the morning and evening observation hours. A large number of people were observed using this site as a transit pathway from and to work/destination in the morning and evening. Such movements were carried out by all age groups and genders and found to be taking place all over the site, depending on the user's origin and destination points.

Different areas were accessed at different times. During the morning and afternoon hours, most walks were to the amenities/shops around the site. During the evening, the children's playground would become animated, attracting the most number of users, especially children and their caregivers. As a public space set amidst surrounding mixed uses (housing, car park, shops, supermarket, town council office, children learning enrichment centre, etc.), random, dispersed activities such as those that occur as isolated events without a specific pattern of occurrence were observed through the day.

Although in smaller numbers, optional activities (like sitting on the benches, reading, watching the crowd, etc.) and social activities (like meeting and chatting with friends and passers-by, playing games, etc.) were happening too, mostly during the morning and evening, with fewer users in the afternoon (Fig. 7.1). Such optional and social activities were carried out by older users in the morning (mostly women) and afternoon (a mix of men and women). By evening, all age groups including children were engaged in these activities. The frequency and location of these activities were often influenced by the sun's path and shadow from surrounding buildings. Users were found to sit in areas that were under shade, especially during the afternoon. The sunny weather and lack of shade could have affected recreational use of this space as indicated by the smaller number of users during the afternoon.

The location and provision of on-site facilities like exercise/play areas, benches and shade seem to play an important part to initiate/support these activities. The elderly fitness corner and seating provisions are observed to be major action spaces, becoming nodes of attraction among the older users while the presence of the children's playground paves the way for children and their caregivers to gather, play and socialise. These elements and their co-location in a single space (appropriately distanced and buffered) are key to the social appropriation and dynamics generated by the observed space.

Spaces were adapted to use where appropriate; sitting was not limited to the seats provided. People would sit on play equipment, edge of flowerbeds or wherever a group can have a conversation. In the absence of dedicated seating, wheelchair users and their caregivers were observed to sit and group on (e.g. sit on the equipment) and around the fitness equipment. Foreign workers would occupy the edge of flowerbeds, sitting in a space that no one claimed rather than on the provided seats. Some seemed to adopt parts of these areas as personal meeting places, e.g. some women meeting their friends around a particular table and benches close to the shops, domestic helpers gathering around the fitness equipment with their elderly charge.

The presence of magnets like supermarket, coffee shops, market stalls and children learning enrichment centre on the edge of the observed space is yet another important anchor to attract users of the neighbourhood and of other areas of the town/Singapore. The various provisions and the activities they offer draw people out and to the site. The presence of people further becomes a welcoming factor to attract a secondary crowd of users to the site—to watch people and the world go by (Whyte 1980).

Weekend

User frequency—Weekend data reveal a different pattern to that of weekday. One in two (51.4%) users were recorded during the evening hours. There were fewer older users at the site during the weekend than weekday; about one in five users were older adults. The highest numbers of users to this site were found around noon and between 5 pm and 6 pm while the highest numbers of older users were recorded between 9 am and 10 am and around 5 pm.

There were more large groups of people (five and more) on the weekend day than a weekday, especially during the morning and afternoon observation periods as compared to the evening when group size was smaller, mostly two people, taking leisurely walk or crossing the site to another destination. During the morning, the groups were mostly women. By afternoon, men as well as women were found seated in groups, though separately whereas in the evening, there were more groups of male users.

Types of activities—The majority (about 85%) of users to the site were engaged in active activities with walking being the most prominent active activity. Most of these activities occurred during the afternoon (around lunch time) and towards the evening (from 5 pm onwards). Among the other activities, there were more sitting and talking activities during the morning but these seemed to somewhat diminish by the afternoon (perhaps because of the afternoon heat).

During the weekend, the site was most busy during the evening. The provision of various nodal functions and socialisation spaces such as seating and fitness corner were key reasons for this trend. The active edge (supermarket and retail stores) of the public space was another draw factor, drawing in users to the observed space throughout the day. The connectivity and permeability of the site to its surrounding uses are important conditions in the movement of users and the experience of the space. Majority of the older users would use the exercise corner and seating areas in the morning. Many of the users during the rest of the day were observed to move along their various chosen paths across the site (shortcuts and alternative routes to a destination beyond, e.g. the supermarket, etc.) and around its facilities (to avoid space furniture, playground, etc.), which condition them to move along certain directions.

There were more optional and social activities during the weekend day than the weekday. During the morning, optional and social activities (e.g. exercising, sitting on benches in small groups) seemed to be higher than the necessary activities. As with the necessary activities, the social and optional activities were carried out by all age groups and genders though they appeared more prevalent among adults and older users in the morning (both men and women) and mostly men in the afternoon. There were fewer optional and social activities during the afternoon—some individuals, mostly men, were observed seated on the benches on the eastern side of the site that was then under shadow of the adjoining HDB block.

In the evening, more users were found walking through the site to other destinations, especially to the adjoining marketplace and eating-places. The only optional and social activities seemed to involve groups of children and their caregivers at the play area. The cloudy weather with drizzle in the early evening hours did not seem to have much apparent impact on the activity pattern or number of users; people just went on with their activities.

On the weekend day, the space design features like exercise corner, benches and children's playground as well as the proximate location of the supermarket on one edge and the presence of users in the observed site continued to act as anchor spaces to draw users to the site—to sit on the benches (to rest, relax or just watch passers-by), exercise or play at the children's play area. Majority of these anchored movements and activities happened during the morning and evening hours when the weather was cooler, making the space more pleasant to use.

Active Zones in Site A

There seems to be a direct relation between the physical architecture and atmosphere of the public space and its activity and social dynamics. At this site, due to its design as a site of community recreation, catering to both the old and the young in the neighbourhood as well as its location amidst mixed uses, the precinct civic plaza has become a site of public gathering in and passing through. The predominant activity was walking—the large number of people walking through the space to either use its facilities or to reach other destinations on its edge and beyond the space. The path taken through the site is the result of its open setting coupled with the number of easy access points that connect this site to its surroundings.

The designated central play area was a major activity zone for children and the adults who accompany them (Fig. 7.2). Senior users were observed to congregate largely within the elderly fitness corner, especially during the morning hours. A majority of them were found seated on the benches provided nearby—some to rest between their walk/workout, some for relaxation and socialising with friends/others and some just to watch the world go by. The shops that line the periphery of the site, add to the attraction, with goods spilling over at times to the pedestrian spaces (especially on weekends) to generate centres of consumption and interest around them.

Fig. 7.2 Site A—active zones

The benches provided around the site are important activity hubs. They are seen to attract users throughout the day, especially during the morning hours though there are a couple of seats that were less used, e.g. those which did not have a good view of the playground or are located too close to the shop and its activity to afford sitting comfort. People tend to prefer to sit where they can view action, but out of the way of the flow of pedestrian traffic or where they can sit comfortably for conversations or be alone. The seats, apart from being functional nodes, are used by people in diverse ways, e.g. to create 'talkscapes', as a meeting or waiting place, for people watching or to while away a few minutes and to just be as Gehl (1987) observed a long time ago on public space use in Europe.

International evidence suggests that public benches are an important element of inclusive design; without them, many older people may remain in their homes, unable to travel to the local shops or visit friends on foot (World Health Organisation 2007). By giving people the opportunity to sit down creates possibilities to spend longer periods of time outdoors and the conditions for social interaction. The ability to attract people to stay is the hallmark of successful public spaces as the presence of people creates a certain rhythm of life in public spaces (Alexander et al. 1977).

Rhythm and use are not static but change with time and on a daily basis, e.g. weekday is different from weekend day. These variegated space-times of aggregation can in turn generate more activities and may later influence the space design. For example, if there were a large number of people viewing the pedestrian traffic, there will be a need for more seating. As is, aside from existing seating areas, some people were observed to sit on the play equipment and the edge of flowerbeds to converse or watch the world go by (Fig. 7.2). This could indicate an unmet need that requires further examination.

7.3.2 Site B—434 Fajar Road, Void Deck

The activity level was observed at Site B void deck on a weekday and weekend day. Over 200 users were observed on each of the observation days; about half of them were older people.

More people (25.3% more) were found using the senior residents' corner on a weekday when compared to a weekend day. An opposite trend was observed among the older users. There was a marked increase in older users on the weekend day (65.5% of users) than weekday (45.3%). The number of older adults visiting this space with walking aid was generally low; 6 users on the observed weekday and 10 on the observed weekend day, perhaps because more working family members had the time during the weekend to accompany these older adults out of the house. Majority of the older users were with one to two other people, mostly peers of similar age groups. Larger groups of three to five older persons were observed congregating in the morning (especially on weekend day) and during evening hours (especially on weekday).

In terms of activities, walking was the most popular activity on both weekday and weekend, followed by talking, sitting and play. While these activities were common on both days, there appeared to be more walking on a weekday. Talking, sitting, play, eating and shopping seemed to occur more on the weekend. The activity maps in Fig. 7.3 illustrate the visitation during the duration of the observation sessions.

The user type and activity frequency of weekday and weekend days are further analysed as follows.

Weekday

The day of observation was mostly cloudy with intermittent sunshine from 7 am to 10 am. Sunshine from 7 am; dark by 8 pm.

User frequency—On a weekday, most users were recorded at this site during the hours of 11.01 am to 4 pm (38.7%) and 4.01 pm to 9 pm (37.3%). The highest number of users to this space was found around meal times, around 12 pm and again around 6 pm. Most older users were recorded during the morning, around 9am and in the evening around 6 pm; 95% of male and 79.3% of female users in the morning appeared to be of older age. The afternoon was the least popular time.

Larger groups of five people were gathered in the evening while smaller groups of two to three were most often found in the morning and afternoon. During the morning and evening, the groups were mostly female whereas in the afternoon, there were small groups of female and male, some sitting together at the tables, talking and playing cards and games.

Types of activities—The most popular active activities (about 65% of users) in Site B void deck were walking and play/exercise, e.g. playing with cards and chess. The latter activity seemed a common pastime during the afternoon and evening. As a void deck and given its position between the car park and the hard court (exercise area), LRT station and adjoining Fajar Shopping Centre, there were often people walking through/by the space, especially in the morning and evening, e.g. children and young people going to and from school, adults going to and from work or about their daily businesses like markets, shopping, etc. The space was used by different age groups as a shortcut (through the void deck) to get to the main road and public transport-shopping nodes beyond. Many of these random flows were observed on the periphery of the site.

A smaller volume of users was engaged in passive and optional or social activities like sitting and talking. Sitting was common through the day. For the better part of the day, some individuals as well as small groups of users (less than five), usually older people, would be sitting at the void deck, at the provided seats or along the low parapet walls on the fringe of the space. Around meal times (11 am–12 pm and 5 pm–6 pm), their number would increase, peaking around 6 pm, perhaps knowing that there would be others at the site. They were engaged in a range of activities—sit and gather at the provided seating, reading (e.g. newspapers), chatting or using the vending machine, wash area and toilet from time to time.

By their clothes, it would seem that the users were local residents (living in the block) and regulars using the space as an alternative or extended living room area. Talking was most observed in the morning (around 9 am) and early evening

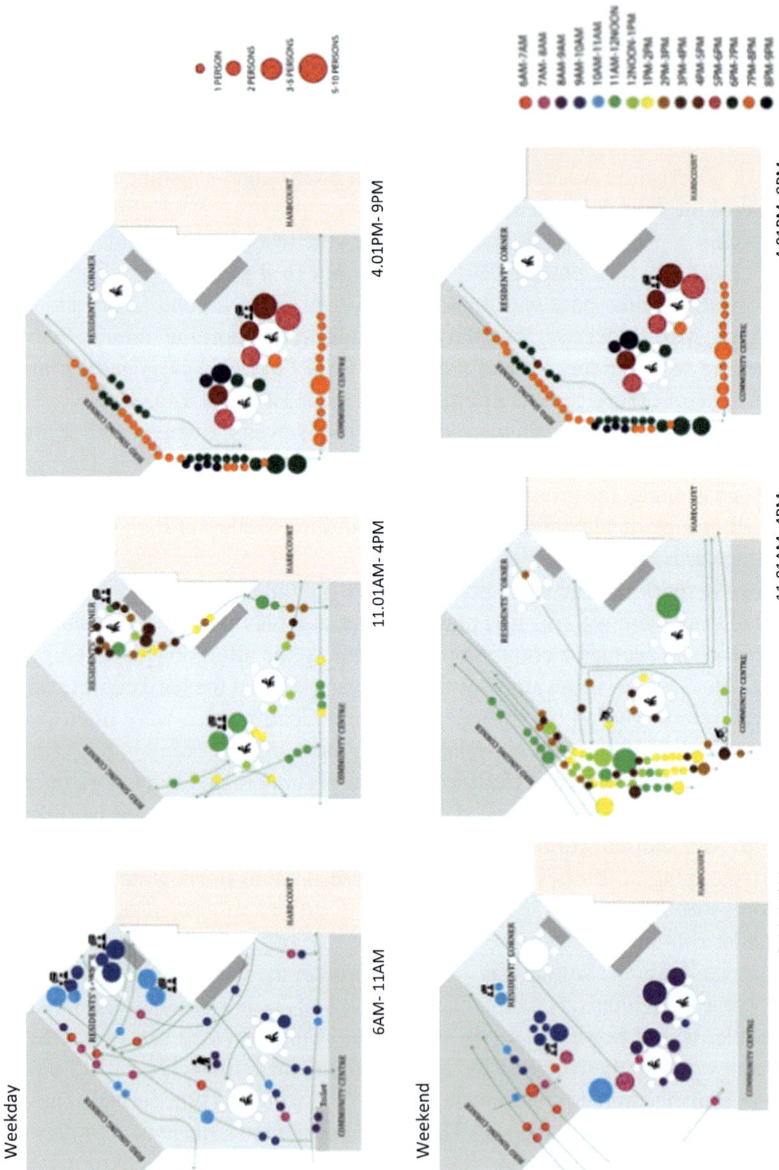

Fig. 7.3 Site B void deck—activity mapping

(around 6 pm). As regulars to the space, they were interested in finding out what the observers—strangers—were doing in their space for an extended time.

Several of the older users (along with their caretakers in some cases) seemed to know each other (perhaps through frequent meetings in the space) and would chat and play games together. Many of these activities were anchored by the provision on site (such as seating, vending machine and toilet), acting as a destination for the older users and from which more activity can propagate. Even so, the space did not appear to be busy. A small number of people were observed to be carrying out maintenance works at the void deck like repair works in the morning (around 10 am) and clearing garbage in the evening (around 8 pm). Perhaps, may be because of these maintenance works, there were fewer visitors.

Weekend

User frequency—On the weekend, most users (47.2%) were found during the evening observation hours (especially between 6 pm and8 pm). Older users were fairly constant through the day (an aggregate of about 40–50 people were recorded during each observation period), except between 10 am and 12 pm when there were more of them, perhaps going out to lunch with family. Over 50% of users were older people. During the afternoon and evening, one large group of seven people was observed to stay through most of the day, from morning to afternoon. An increase of older users with walking aid and wheelchair (4.4% of users) was seen when compared to weekday.

Types of activities—A large majority (about 75%) of the users were engaged in active activities like walking and play on the observed weekend day (Fig. 7.3). Walking occurred throughout the day and peaked in the evening, especially after dinner, 7 pm–8 pm. In the afternoon, part of the walking along the edge of the site could be traced to the adjacent activated Bird Singing Corner. Even though it occurred through the day with spurts of activity at certain times, play (e.g. with cards and chess) as a social (group) activity was most concentrated around lunchtime (12 pm) and the evening hours, 5 pm–6 pm.

A number of users were in passive forms of engagement like sitting and talking through the day, with most intensity between 10 am–12 pm and 5 pm–8 pm. Many of these activities were enabled by the space provision like tables and seats. Some people, usually older people, were observed to sit on the low parapet walls on the edge of the space and to strike up conversation with passers-by who may then stop, sit and talk for a little while before walking on. Some were interested to see and talk to the observers to find out what they were doing in the space.

Compared to the weekday observation, there were less users walking across/through the site on a weekend morning. This could be because the users who traversed the site on the weekday used it as a shortcut to get to work or transport nodes while people were sleeping in during the weekend day. Morning activity was relatively sedentary, with groups of two or three people (usually older people) gathered around the seating provision. As with the weekday, most activities within this space were optional and social activities, largely anchored to the location and provisions within the senior activity corner itself.

The dedicated seniors' activity area seemed to offer a meeting place for some groups of older people (including a large group of seven or eight users) to meet regularly over longer periods. They seemed to be the only people around the place most of the time, staying to talk to each other or passers-by or to play table games and using the toilet and pantry from time to time. While there seemed to be a regular group, this space did not appear to be busy. Even though the area is under the protection of CCTV surveillance and located adjacent to the residents committee office, it seemed somewhat underused. Perhaps because of the decrepit condition (broken fixtures)/lack of facilities/programming, it did not seem to present an inviting environment. Similar to the weekday day, workers were found to be carrying out certain odd jobs around this void deck. Maintenance and repair works were undertaken in the morning (from 7 am to 11 am) and the garbage clearance was done in the evening (about 7 pm).

Active Zones in Site B Void Deck

This void deck space is designed and designated as a Senior Citizens' Corner. Besides being a space for people to walk through to other amenities (shops and transport services), its major activity zone was clustered around its seating provision. Some older residents were found occupying the seats through the better part of the day (Fig. 7.4). They were involved in leisure and social activities like having a chat, reading newspapers and playing games. They seemed to use the space as their extended living room, a place where they know they can meet other frequent visitors. Being a sheltered space, change in external weather conditions would not affect the activity within this space. Perhaps because of this, its space features and residential location (on the ground floor of a residential block), about one in two users were older adults (45.3% during weekday and 65.5% during weekend day).

The prevalence of older users over other users suggests two things: (a) that the void deck is a convenient and therefore, a popular social gathering space among the older adults as it is relatively easy to access from their residences; (b) the designation of a 'seniors' space for the older adults, despite the poor maintenance of facilities, works to encourage their participation. Of importance, the provision of facilities such as a toilet, wash area and vending machine supports and sustains a longer stay at the site. It lends support to the findings from the study's other research instruments[9] that toilets are vital for older people to maintain a level of independence outside of their homes. A place to buy or make drinks like coffee for cheap is also largely welcome.

7.3.3 Site B—Fajar Road, Hard Court

The activity level was observed at Site B Hard Court (adjoining the previously discussed void deck) on a weekday and weekend day. On each observation day, 1200–1500 users were observed at this site; about 20% were older users.

[9]See Focus Group Discussion in Chap. 2 and Community Design Workshop in Chap. 10.

Fig. 7.4 Site B void deck—active zones

There were more people (17.3% more) on a weekday than the weekend day. There were more older users on the weekday (324) than the weekend day (271). On both the observation days, the number of older people using walking aid was generally low (less than 10). Most of the older users were either alone or in small groups of 2–3 people. Their favourite time was mostly during the morning hours and their common space seemed to be around the fitness equipment. Many were accompanied by domestic helpers, friends or a partner.

On both weekday and weekend day, the most frequent activity at this hard court appeared to be walking, followed by talking, play/exercise and sitting. Some activities like sitting, talking, play/exercise appeared more during the weekday than weekend day. Eating and walking were observed more on the weekend day (97% of users were walking as compared to 93% during the weekday). The maps in Fig. 7.5 illustrate how activities vary by number of people during the observation sessions.

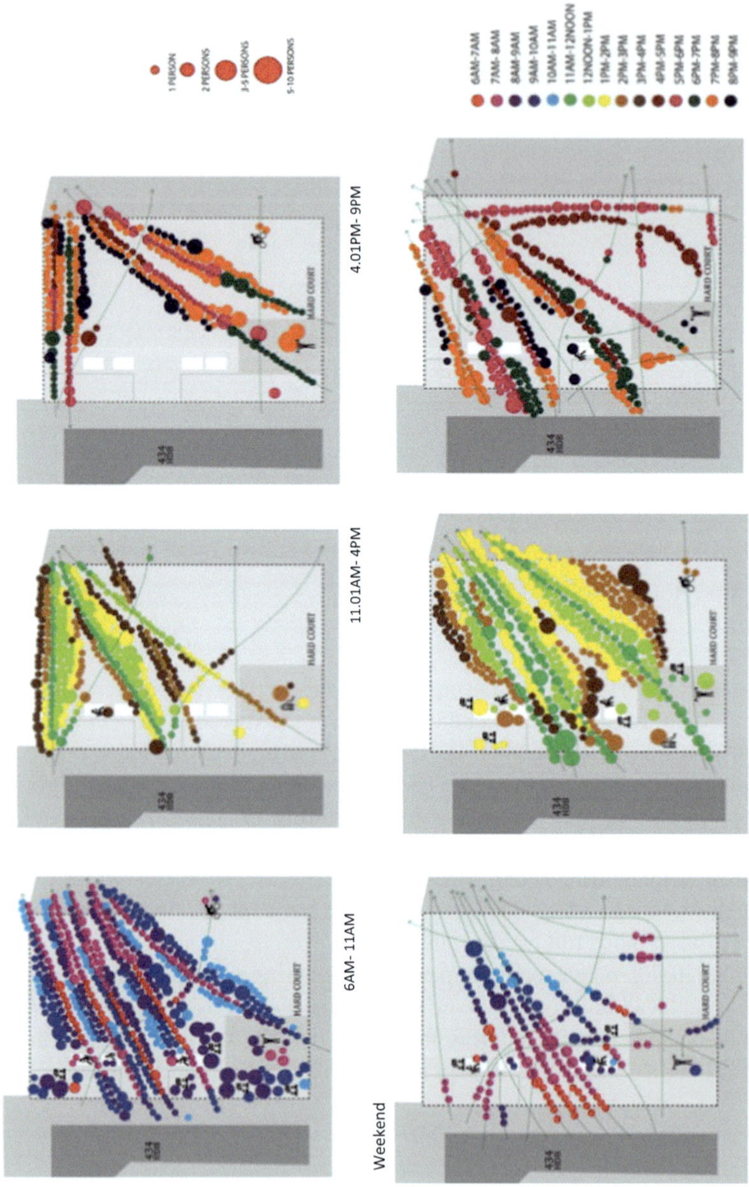

Fig. 7.5 Site B hard court—activity mapping

Weekday

User frequency—On the observed weekday, the number of users was highest (48.8%) in the morning, peaking between 8 am and 10 am. About one-third of the users visited the site in the afternoon while 18.2% came in the evening. Most older users were recorded during the morning observation hours (212 between 6 am and 11 am as compared to 50 during 11.01 am–4 pm and 62 during 4.01 pm–9 pm). Their highest numbers were recorded around 10 am, dipping thereafter and remaining consistently low through the rest of the day.

Types of activities—About 90% of users were engaged in active activities like walking and play/exercise. Such active activities were generally at their highest in the morning, 8 am–11 am, peaking between 9 am and 10 am. Aside from some who walked to the site for its facilities (e.g. fitness corner, play equipment/space), most people were walking across the site to a further destination, i.e. between the housing blocks and the bus stops, LRT stations, shopping centre, etc. For play/exercise, there were two peak periods of activity—around 9 am and again at about 8 pm in the evening. Some workers were seen doing maintenance works during the afternoon. Unlike the adjoining void deck, comparatively fewer users were in this space for passive activities like sitting though it was not uncommon to see some among those exercising to also sit before/after their exercise/play and chat with others present. These activities seemed more prevalent in the morning than the rest of the day. All activities were found to considerably reduce during the course of the day with the reduction in user numbers.

As with Site A, due to its location, the necessary activity of walking across the site to other destinations (such as to the main road, nearby bus stop and Fajar LRT station and adjoining shops) was a common occurrence on this site. Such movements were carried out by users of all gender and age groups, some carrying umbrella to provide shade from the hot sun. From the activity mapping of this hard court, it can be observed that movement patterns were spread across the entire hard court during the morning. But, as the day progressed, these lines of movement became more concentrated around certain areas, largely influenced by the shadow and shade across the hard court.

A small volume of optional and social activities was recorded. Social activities included meeting other users, usually in small groups and chatting. These activities were generally observed to occur in and around the play/exercise facilities and at the edge of the hard court, largely where seating is provided. These activities were more common during the morning hours and mostly among older women. Observed optional activities included sitting, reading and chatting, mostly in groups of two or three by older users. These activities were largely anchored at the row of benches located at the edge of the hard court (closest to the void deck) and at the fitness corner. Even though the fitness corner was not used, its nearby benches provided a meeting point (mostly female older users) to sit and gather in the morning. Their numbers were not many.

Several reasons may be offered. Although the benches gave people the opportunity to sit down and watch the world go by, there was little activity to see. Also, the location of the benches, all lined in a straight row, did not seem conducive for a group to have conversation. The global literature on seating design in public spaces suggests that the location and placement of benches are important considerations in how they will be used (WallJasper 2007; Whyte 1980). The lack of shade could be another reason why activities were not recorded, especially during the afternoon. The presence of dark cloud cover (hence, possibility of rain) in the early evening hours could have further discouraged people to use the space during that time.

Weekend

User frequency—During the weekend, the highest number of users was observed in the afternoon (50.2%), peaking around 12 pm–2 pm, followed by evening (34.7%) and then morning (15.2%). About one in five users was an older person (an aggregated 50 to −110 older users during each 5-hobservation periods) except around 9 am, 2 pm and 6 pm when slightly more older people were observed.

Types of activities—Similar to the weekday, a large majority of users (about 97%) were observed walking (active activity) at/across the hard court. The activity was most popular during the afternoon and evening hours, with the highest intensity between 12 pm and 3 pm. The next most common activity, though far smaller in scale, was talking. This seemed to continue through much of the day but, especially during the evening (5 pm–8 pm) and in the morning (9 am–10 am). Compared to the void deck, there were fewer users sitting at this site. Most seemed to do this in the morning (around 9 am) and towards the evening (4 pm–6 pm) when the weather was cooler. Some were, however, observed sitting in the afternoon (1 pm–2 pm), perhaps taking a lunchtime rest. Some of them were observed eating (12 pm–2 pm). While walking, talking, sitting, playing/exercise remained proportionately the same, eating was nearly 5 times more frequent during the weekend day than weekday.

Despite it being sunny, the hard court was most active in the afternoon and evening. Most people were walking (usually individuals) across the hard court to destinations beyond the site; the peak hours were 12 pm–2 pm, perhaps out for lunch, errands, etc. Exercise was not limited to the exercise corner. Due to the size of the hard court, in the evening, several users were observed to walk/jog round and round its periphery for a workout. Some older users were engaged in social activities like gathering and sitting at the benches and chatting. These activities, largely anchored to the seating and fitness corner, were observed mostly in the late afternoon and also in the morning. Despite it being the weekend, only one group of children (three children and two caregivers) was seen playing at the children's play equipment—in the evening (5 pm–6 pm).

Active Zones in Site B Hard Court

A couple of active zones may be discerned. The most active zones were the diagonal walking paths (undefined) across the hard court, connecting people from the HDB blocks to destinations beyond the hard court like shops and transport nodes—car park, bus stop, LRT station, main road, etc. Walking constituted over 90% of user activities. Even though it was often used as a throughway between the housing block

Fig. 7.6 Site B hard court—active zones

and the nearby activity nodes (like LRT station and shops), this space did not seem to generate as much animation and secondary activity of people-watching as Site A, which had active frontages, high volumes of activities and users.

Another active zone was the seating and exercise corner (Fig. 7.6). These spaces, though used, did not seem crowded. They were largely used by older people (about 20% of users). All day long, only two groups of children and caregivers (of 4 and 7 users) were recorded for the weekday and one group (of 5 users) during the weekend day. This could be due to a combination of reasons: (a) the current provisions are not attractive to local residents, (b) the locations of these provisions are somewhat isolated (on the edge of the hard court) and not coupled with other nodal points, and (c) a lack of shade and shelter that did little to encourage people to stay on in this space. Perhaps, because of its design and function as a hard court, much of the site is open and exposed to the weather.

7.3.4 Site C—415 Fajar Road, Void Deck

The activity level was observed at Site C void deck on a weekday and weekend day. Some 300–700 users were observed on each of the days; less than one-third were older people.

Nearly twice as many users (87.7% more) were found at Site C void deck on a weekend than a weekday. Similarly, there were more older users on the weekend day (202) as compared to weekday (108). The number of older users with walking aid was low (2.3% on weekday and 1.4% on weekend day). Majority of the older users were accompanied by one or two people, particularly by domestic helpers or young grandchildren and other family members.

The most common activity was walking, followed by talking, sitting and playing/exercising. Although these activities were common on both days, a higher volume of most activities was observed on the weekend day, probably because of the larger user numbers on that day. There was more walking (99.7% more) but less playing/exercising (19.2% less) on a weekend day than weekday, suggesting that people were going to other destinations on the weekend. The maps in Fig. 7.7 illustrate how usage varied during the observation periods.

Weekday

User frequency—On the weekday, the highest user counts were recorded in the morning (57.2%) as compared to 33.9% in the afternoon (11.01 am–4 pm) and 8.9% in the evening (4.01 pm–9 pm). The general usage of the site peaked in the morning around 8am when most people were leaving home for work, market or errands. Another *albeit* smaller wave of users was recorded around noon (lunch time), after which user numbers decreased quickly and remained low until around 7 pm onwards when user traffic climbed as the evening working crowd returned home. As the ground level of a residential block, the void deck is a must-use space for residents and their visitors when getting in and out of their homes. It is the ground entrance/exit to the flats located above. It is where the lift lobby and staircases, noticeboards and letterboxes are located.

About 5–20% of users were older users. Their largest number was in the morning (13.3% of all users during 6 am–11 am), peaking around 9am (around 8 am–9 am for older female and 9 am–10 am for older men) before declining to 9.9% in the afternoon (11.01 am–4 pm) and 5.0% by the evening (4.01 pm–9 pm).

Types of activities—About 88% of users were engaged in active activities like walking and playing/exercising. These activities peaked in the morning (highest around 7 am–8 am) and then gradually reduced during the course of the day. By the evening, around 6 pm–8 pm, there was almost no more playing/exercising observed at this void deck as most people returned home for dinner and the night. Through the better part of the day, a small proportion of users were engaged in passive activities like relaxing and resting at the seating area, sitting on the fixed benches to read, chatting with friends or family and occasionally strangers (observers were asked

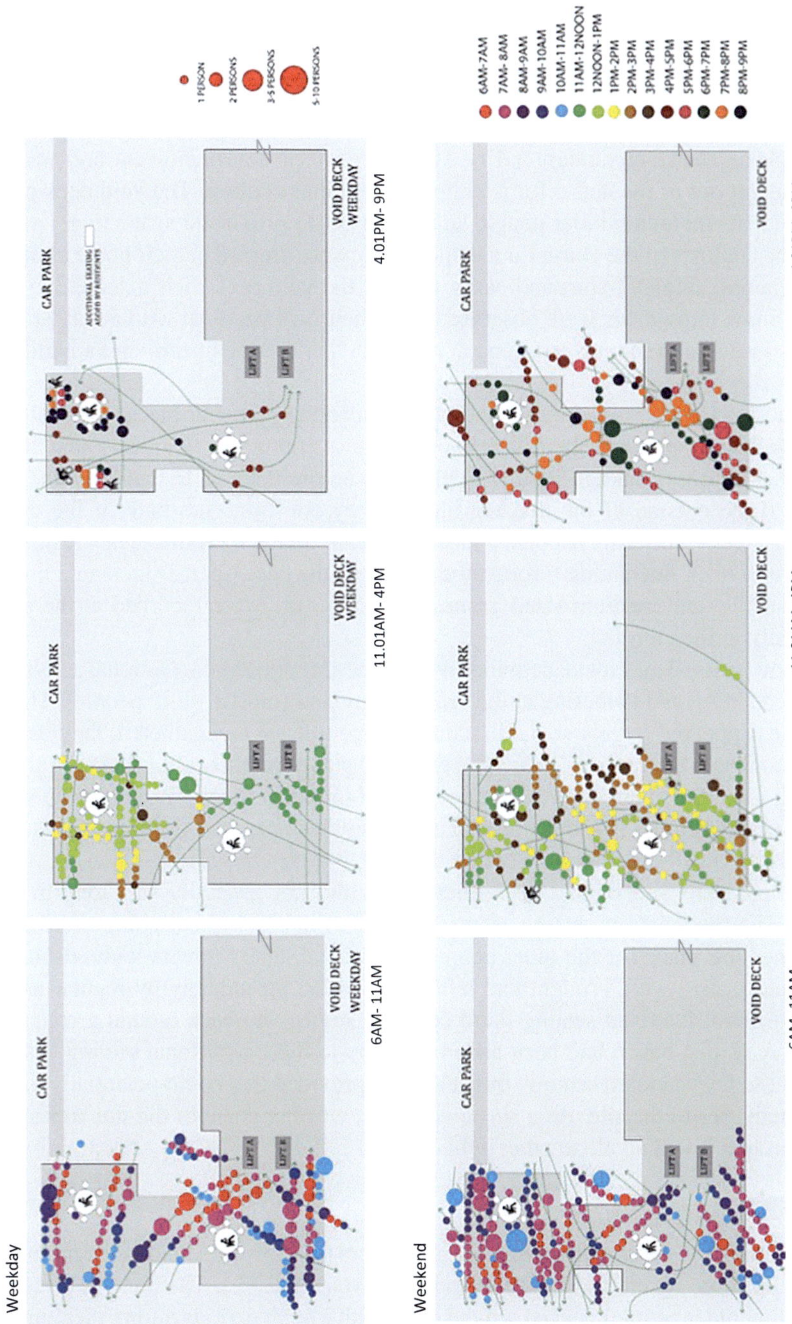

Fig. 7.7 Site C—activity mapping

what they were doing in the space), collecting and reading mail retrieved from their letterboxes.

Sometimes, people would say "hello" or acknowledge each other with a nod or smile as they move through the void deck. Others would pass by avoiding eye contact. Older people who occupy the space were usually in passive engagement, lingering at the benches. These activities seemed to pick up towards the night, around 8 pm–9 pm as people finished their dinners and used the space to enjoy the night scene and breeze, rest and get out of the house for a while before going to sleep. The void deck gave the residents including older people an easy to get to proximate space to get away from the confines of the house for a while. Many were dressed in their home clothes, e.g. pyjamas, singlet/T-shirt and pants, making the void deck their extended living room. Some individuals were observed doing their own personal work/study at this space—one adult (about 8 am), one domestic helper (about 11 am) and a child (at noon).

Importantly, the void deck served as a connecting space to places beyond the housing block, particularly in the morning and afternoon. Residents and their visitors—people of all age groups would walk across the site, to and from the lift lobby to the outside of the housing block. They constitute the bulk of the daily necessary activities. This happened most frequently in the morning (when residents leave for work), fluctuating through the course of the day. Apart from single users, some small—and medium-sized groups of 2–5 people were recorded at the site, especially in the morning.

A few optional and social activities were sometimes observed. Optional activities included sitting and gathering at the seating provided (mostly older people). These activities appeared largely social in nature. As people sat and gathered, they began to talk amongst themselves. Some played board games (e.g. chess), others did show and tell (showing their shopping). They tended to gather at the far end of the void deck, away from the lift lobby area. These activities appeared popular among the older users, mostly in groups of two or three.

Compared to Site B void deck, these users did not generally stay long in the space. The provision of seating seemed to be an important focal point, offering a local meeting place for the older people. Additional chairs (chairs with cushions, arms and backs) were brought and left in the space, presumably by regular users to supplement the fixed seating (hard concrete stools—no back around a concrete table). A garden bench had been added close by to these additional seating. Aside from these few standard seating, little else was provided that could potentially draw more activities to the site. As a sheltered space, weather changes did not appear to affect usage; it was an all-weather public space.

Weekend

User frequency—During the weekend, most users (47%) were found in the morning hours, followed by afternoon (28%) and then evening (25%). The number of users (including older people) peaked around 8 am with trough periods during mealtimes, 12 pm and 7 pm, suggesting that perhaps more people were staying home for their meals. The afternoon was less busy than the morning though slightly more people were observed in the space around 3 pm and 5 pm. A similar trend was noted among the older users—the highest number was found in the morning (16%), decreasing to 7% in the afternoon and further to 5% by the evening observation period.

Type of activities—Similar to the weekday, about 85% of the users were found walking across the void deck. The next most popular active activity was talking (chatting with others) followed by playing/exercising. The most common passive activity was sitting (at the void deck seating provision). All four activities were more prevalent during the morning hours. Walking and talking peaked around 8 am while sitting and playing/exercise peaked around 11 am.

The void deck was most active during the morning hours. Compared to the weekday, there were more users—a constant stream of activity (especially walking) during the afternoon and evening hours as people came and went out of the block on their various businesses. The space served as a path to and from the lift lobby, car park, nearby transport nodes or other HDB blocks and beyond. Necessary activities were carried out by all age groups.

Compared to the weekday, there were slightly more optional and social activities—sitting and gathering, chatting or playing board/table games, mostly by the older people. Most of these occurred during the morning, around 3 pm and after 7 pm. Sitting and chatting were not limited to the fixed seating areas. They could take place by the low parapet wall near the lift lobby, a primary area of walking through the space. A group of young children were found playing here around noon (11 am–12 pm). During the evening, a few residents were seen using the seating, having a conversation though they did not seem to stay long. A small group of older people was also found doing self-motivated mild exercises in the evening (4 pm–5 pm). Aside from this, there seemed little else to support social activities on site.

The sit, talk and chat activities were largely carried out by older users, mostly in groups of 2–3 people sitting at the provided seating area and often using their own chairs. Most of the older users were single and at other times a few small groups of two to three people, usually in the morning and early evening hours. It was as if they knew there would routinely be someone at the void deck to talk to or play a game of chess/card with at a certain time of day—a local meeting place.

Active Zones in Site C

The most activated zones of this void deck were the randomised walking routes taken by residents and passers-by to and from the lift lobby, letterboxes and other HDB blocks to transport nodes and other destinations beyond the block. There was a constant stream of such movement across this space through the day. The space was otherwise concrete and empty-looking as the term ‘void deck’ would suggest. Its only amenity—the fixed seating (Fig. 7.8), though used did not seem altogether popular—the users of the seats did not usually stay long and the regulars had added their own chairs to the space. The fact that some people had occupied the space with their own chairs suggests a sense of ownership and perhaps a certain rhythm of life in public housing—provided a quick and easy to get to near-home respite place. Aside from the seating, there was little else to attract users and activate the zone effectively.

Fig. 7.8 Site C—active zones

Compared to Site B void deck (of similar typology though Site B was a much larger space with more facilities), the proportion of older users to this space was much lower (28.2% on weekday and 28.1% on weekend versus 45.3 and 65.5% respectively for Site B void deck). Aside from resident population, the difference in usage levels between the two sites may be due to the lack of provision and programming—Site B had a dedicated though somewhat decrepit Senior Citizens Corner, more seating and auxiliary facilities such as a toilet, a pantry/wash area and a vending machine. These elements could be the missing link, indicating that some programmatic and infrastructural features are favourable to public outdoor space usage amongst older people. The exact provision would have to take account of people's needs and preferences since there is no Mr. Average older person or resident.

7.3.5 *Site D—201 Petir Road, Bukit Panjang N2 Park*

Like the other sites, the activity level at Site D was observed on a weekday and weekend day. About 350 users were recorded on each day; one-quarter were older people.

Slightly more users (5.6% more) were recorded on the weekend when compared to the weekday. There were slightly more older people on the weekend day (85) than weekday (81). Only a very small number of the older users (<2%) were with walking aid. There were even fewer of them on the weekend day (2 as compared to 6 on weekday).

The most popular activity on the site was walking. This was followed by playing/exercising (e.g. children playing in the open space, jogging or exercising at the fitness corner), talking, sitting and others—the most prominent of which was working in the community garden. Although these activities occurred on both observation days, only walking appeared more on the weekday (25.0% more). Playing/exercising at the fitness corner, talking and sitting occurred more during the weekend day. Figure 7.9 illustrates how usage varies by the number of people during the observation sessions. The afternoons appeared less used, perhaps because of the heat.

Weekday

User frequency—On the weekday, the number of users was highest in the morning (47.2%) and the evening (38.9%) while there were less people in the afternoon (13.9%). This could be attributed to the lack of shaded paths in the afternoon. The trend was similarly pronounced amongst the older users—this group constituted 11.2% of all users in the morning, 12.1% in the evening and a mere 0.6% in the afternoon. After the morning peak at about 9 am, usage dipped to a trough, the lowest was around 1 pm before picking up gradually but remaining low till after 5 pm. There was nearly no activity between 12 pm and 1 pm. Visitation only picked up again after 5 pm, hitting a second high around 7 pm before declining once again as night and darkness fell. Even though the park was lit at night, user number declined rapidly with nightfall. The most popular time for older visitors was around 8 am and

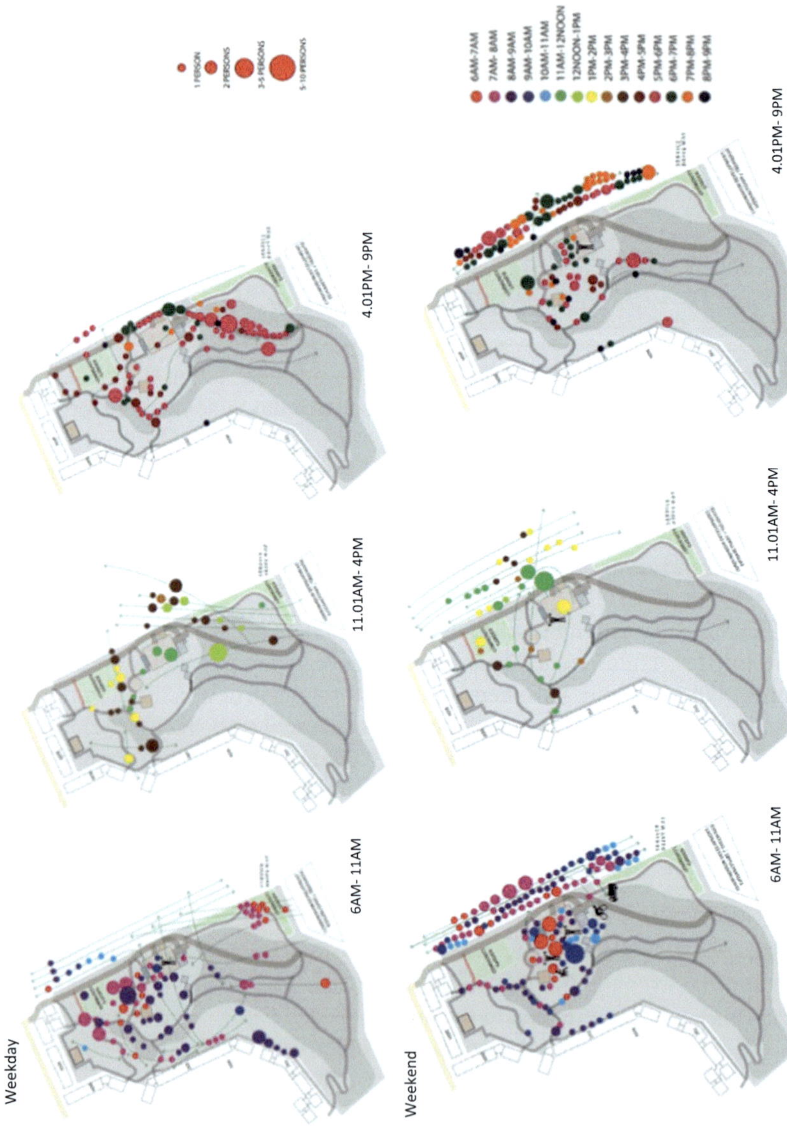

Fig. 7.9 Site D—activity mapping

between 5 pm and 7 pm while the least popular time was between 12 pm and 5 pm (the hottest period of day). Many were local residents, going by their home clothes.

Types of activities—About 90% of the users of this park were engaged in active activities like walking, jogging, exercising or working in the community garden. Site D was most active in the morning and evening. Morning activity was spread across the park but concentrated around the community gardens, exercise area and jogging path on the park periphery adjacent to Block 201. Most walking and exercising took place during the morning (especially 8 am–10 am, some started exercising before dawn around 6 am) followed by evening (6 pm–8 pm) when the outdoor weather was more cool and pleasant. Morning walking had many benefits and could be easily integrated into one's routine.

The activity of community gardening was carried out from late morning onwards, sometimes into early afternoon (10 am–3 pm), indicating the dedication of the gardeners. The availability of a portable toilet nearby (at the hard court) made the extended stay possible. During the afternoon hours, there was minimal or no physical activity in the park. The reduced activity during the afternoon highlighted the shortage of shaded zones within the park.

A small volume of passive activities like sitting around the exercises corner/hard court/pavilion was observed through the day. Some foreign workers (usually an individual) were sitting/sleeping on the bench in the pavilion, generally in the afternoon. From the pavilion (especially the upper set), people could see who was approaching and though in a very public place, they could be relatively private. One could command quite a good view of the activity areas of the park; it also provided shelter from the rain and sun.

As with the void deck, some people tended to adopt parts of the park as personal meeting places, e.g. women meeting by the community garden, older users to exercise at the hard court. In contrast to the other sites, only a small volume of necessary activities was observed at this park. These included users traversing paths along the perimeter of the park to get between the HDB blocks on the western side of the park, residents walking along Bukit Panjang park connector[10] on the eastern periphery of this park that provided a green and traffic-free path from Petir Road to the shops and marketplaces at 256 Bangkit Road-Bukit Panjang neighbourhood centre.

Majority of the activities were optional and social in nature. These included exercises at the elderly fitness equipment and IPPT[11] fitness areas, group fitness classes at the hard court, walking and jogging along the various paths, running and playing with pets in the grassed area (like dwarf rabbit) and gardening at the two community gardens. Some of the social activities included group exercise (mixed gender at the hard court in early morning), gathering among friends at the pavilions after the group exercise, chatting and working in groups at the community garden. Majority of these

[10]This 1.4 km park connector links Bukit Panjang Park to Zhenghua Park and Central Catchment Nature Reserve via the heartlands of Bukit Panjang, passing by Bukit Panjang Ring Road, Segar Road and Saujana Road.

[11]Refers to the Individual Physical Proficiency Test (IPPT) set by the Singapore Armed Forces, Singapore Police Force and Singapore Civil Defence Force to test the basic components of physical fitness and motor skills of their service people.

optional and social activities were observed during the morning hours and then again in the evening, largely anchored around the facilities provided in the park. All age groups participated in these activities, e.g. younger users, usually men in IPPT exercise or running, older people, mostly women, in community gardening. Generally, the groups—older and younger people were in the same place at the same time but not interacting or talking to each other unless they were of the same family or group in the first instance.

The older users engaged in these activities (especially gardening) mostly during the morning. A majority of them were found at the two community gardens located at the north and south sides of the park. The second largest group of users were at the fitness area exercising (elderly exercise equipment). Medium-sized groups of three to five people were found congregating for most part of the day, especially in the morning and evening, in the community gardens and fitness corner. These groups often comprised of older residents.

In the afternoon, the space was sparsely used. Only a few users were found accessing the community garden and the hard court (where the only toilet in the park was located). This could be exacerbated by the lack of sheltered spaces within the park to escape the afternoon sun. In the evening, the site came alive again, with most users coming to the park to exercise, perhaps after work. They occupied the fitness area (e.g. IPPT area) and many used the various pathways up to the top (highest point) of the park.

Weekend

User frequency—On the weekend day, the highest number of users was recorded in the morning (54.2%) followed by evening (28.2%) and then afternoon (17.6%). The usage pattern was similar to that of the weekday but during the weekday, the peaks were sustained over a longer time. Older users were most active during the morning (19.3%) with minimal to no activity in the afternoon (1.7%) and evening (2.8%). Peak usage was between 7 am and 9 am. Beyond this, user number dropped drastically reaching its lowest point around 3 pm. Usage was observed to increase again from 3 pm onwards, rising to a second high (half of the morning peak) around 6 pm before retreating as night and darkness advanced.

Types of activities—During the weekend day, the park was most active during the morning hours. Morning activity was centred at the exercise area and users were mostly older people. Afternoon and evening users were fewer when compared to the weekday, with many people exercising in the fitness area during the evening hours.

About 90% of the users were engaged in active activities such as walking and exercising. Although walking remained the dominant activity at the site, exercising/playing significantly increased during the weekend day as compared to the weekday. Walking and exercising were at their highest during the morning, peaking around 9 am and again around 6 pm. There was perceptibly less use of the community gardens, almost none in the morning with a handful in the afternoon and evening. There was a noticeable increase in passive activities like sitting together to play games and chatting at the seating near the exercise corner and hard court when compared to the weekday. Talking peaked around 9 am.

Differing from the weekday, there were more necessary activities at the site during the weekend day. High foot traffic was observed in the morning and the evening along the park connector, more than during the weekday. Another necessary activity was walking through the park between HDB blocks (e.g. Block 210 to 203) and to and from different nodal points within the park such as the exercise equipment area and community garden (mostly in the afternoon).

Even so, the majority of activities were optional and social in nature. Similar to the weekday, such activities were largely anchored to the physical provisions such as the elderly exercise equipment, IPPT fitness area, various pathways and the two community gardens. Optional activities included walking and jogging along the various footpaths in the park and exercising in and around the different fitness equipment. Some of the social activities included group exercise (tai chi at the hard court), gathering after exercise and chatting. Majority of these optional and social activities were observed during the morning hours followed by evening, mostly older users. Groups and older users were mostly observed in the morning. There were far fewer older users on the weekend day than weekday. As a public park, there was hardly any litter around the place except for leaves litter.

Active Zones in Site D

Located in a residential neighbourhood, this park offers free and open access all day and every day. Over 300 people used this park on a daily basis, almost all in home or sports clothes. The most active zones in this site were clearly its facilities—the community gardens, exercise area and hard court (Fig. 7.10).

These were people places where most users could be found, generating further social and optional activities like talking, meeting other residents, exchanging and learning about gardening. Different groups would claim and occupy different parts of the park at different times, e.g. the gardeners the community gardens, sports enthusiasts the fitness equipment area, the tai chi exercise goers the hard court. Oftentimes, the individuals and groups would co-exist with little interaction between them. The park is a local everyday space these users visit regularly, that contains important social and functional use.[12]

In contrast, the forested part of the park seemed to be hardly used, probably because of the unkempt vegetation and fallen trees. The dense and overgrown nature of the forested area seemed to evoke a sense of fear at times and potentially a threat to usability; users might feel unsafe to walk through such areas, especially in early morning or evening when illumination and visibility were low.[13] It is important to always provide places where people will desire to visit, avoiding public places that are 'doing nothing' because there is no one present. It bears reminding that people make places, more than places make people (Mean and Tims 2005).

---

[12]Everyday space is a space in which all life occurred—the casual, daily, functional uses. For further discussion, see Lefebvre (1947); Mean and Tims (2005).

[13]See Chap. 10.

Fig. 7.10 Site D—active zones

There seems little difference in the number of users (including older people) over the weekday and weekend day with the possible exception of the use of the park connector, which saw more users during the weekend day. The implication is that most users came to this site as a destination in itself, due probably to its intrinsic qualities (nature, open space, view, etc.) and provisions (the facilities, activities and programmes it offers), and not as an *en route* pathway to other destinations. An important factor could be its location and topography.

The site is enclosed by adjacent HDB blocks and new condominium developments and hidden to other users, apart from residents and locals who know the place (the park was developed at the same time as the first HDB blocks in the area decades ago). The park facilities (e.g. exercise equipment, community gardens) could only be accessed by climbing a long flight of staircases or walking along a long and upward going barrier-free access pathway (developed a few years earlier)—one must first negotiate these before reaching the facilities. In consequence, we saw few older people with walking aid, especially when they could find other public spaces and seating areas at the foothill of the park, closer to the residential blocks.

The location of this park could prove to be advantageous (offering quietness, privacy and safety from noise, traffic, etc.) but also unfavourable at times (tucked away from common sight, lack of visual accessibility causing the space to be less known to non-regulars). Its accessibility could become a consideration on usage, depending to a large extent on who are its users—their age, gender and time of day/week, among others. International research has shown that successful parks require innovation and imagination, recognition of the diversity of users and access to resources to provide the needed infrastructure (and relevant activities to encourage participation) and their continuing maintenance (Tate 2015).

7.3.6 Site Comparison

A comparative study of the activity patterns of the various sites is carried out to understand better how older people use and adapt to these outdoor recreational spaces. The different sites are compared by *activity level across time* and by *activity typology across design elements*. This part of the analysis is aimed to pinpoint the provisions and deficits in the built environment that encourage/discourage the use of a public space by the older and other users.

7.3.6.1 Activity Level

The attempt is to compare user frequency of public space usage at different times of the day (how many, when and who), outline the different activities (what) they were seen to be engaged in at the respective locations (where) at the sites of observation.

User frequency

Most active sites—The largest number of users was recorded at Site A followed by Site B hard court on both the weekday and weekend day. Higher usage of these two sites could be the result of a combination of factors: their larger size, provision of physical infrastructure (e.g. benches, exercise equipment), accessibility, openness, linkages and connectivity within and beyond these locations via transport nodes (e.g. car park, bus stop, LRT station) and surrounding amenities (e.g. shops, hawker centre, market).

Of importance is access and linkages, especially in Site B hard court as this site has comparatively fewer on-site infrastructure provisions than Site A. Like Site A, much of its rhythms of use are connected to things outside the site—business, retailing, transportation in their immediate surroundings. Aside from being a destination in itself, these sites serve as important *en-route* places (shortcuts) to other destinations. Site D (Bukit Panjang N2 Park) is also comparably large and open, certainly has more physical provisions of seating, fitness area, etc. than Sites A or B. Notwithstanding, this site does not seem to attract as many users—it appears more as a destination in itself than as a connector between places.

Peak hours—The peak usage time of the day was morning (6 am–11 am) for Sites A and B hard court on weekday. This could be related to their activity patterns as

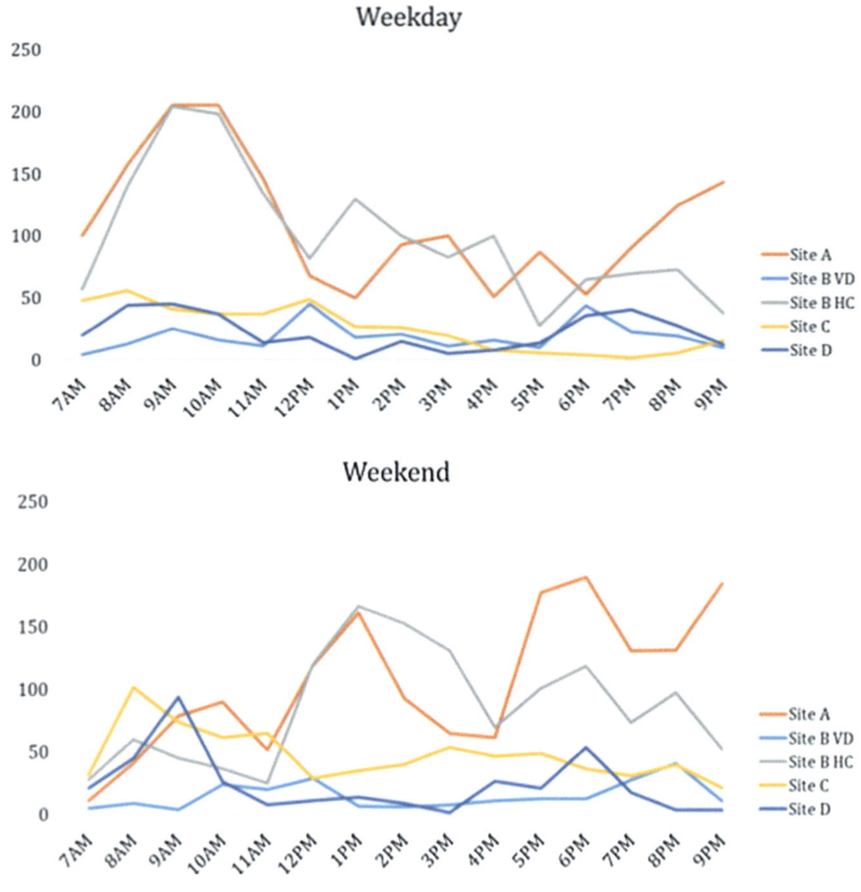

Fig. 7.11 Users across 4 sites

users go about their business, retailing and working day. The usage at site B void deck (a Senior Citizen Corner) peaked around 12 pm and then again around 6 pm, suggesting these to be common times of the older residents when they know they can meet another person or two and have a conversation, especially if they are all alone in their homes and the rest of the family are away on their respective chores. On weekend day, however, there appeared to be varied peak time usage for all sites (Fig. 7.11). Sites A and B void deck were most used during the evening while Sites C and D in the morning and Site B hard court in the afternoon. This variation could be due to the fact that this is a non-work day and change in routines of working people on the weekend; they have leisure time and free time to do things or go out with family and older members.

Variation of usage on weekday and weekend—When weekday and weekend day usage is compared, there were more users on the weekday than weekend at Sites A (86 more), B void deck (58 more) and B hard court (221 more), indicating their banal but important function as a go-through place in people's daily routines during the working week. The trend seems reversed for Sites C (336 more, almost twice as many than weekday) and D (19 more), which saw more users on weekend than weekday. The notable increase for Site C could probably be the result of people traversing this void deck to access the adjacent Chinese temple on the weekend. Of the 4 sites, Site B void deck had the least number of users at all times of the day during the weekday and weekend day. This trend is possibly because this void deck is a designated Senior Citizens Corner, hence less people other than the older residents would use this space, especially when people could traverse this void deck at other points (this space is part of a very long void deck).

Older Users

Most active sites—Similar to the general trend, more older users were found at Sites A and B hard court on weekday. On weekend day, most numbers of older users were recorded at Sites A, B hard court and C.

Peak hours—On the weekday, all sites of study, except Site D, were most used by older people during the morning hours. The usage at Site D (Bukit Panjang N2 Park) peaked from 6 pm to 7 pm. Older users at this site were almost nil from 11 am to 5 pm, suggesting a lack of shaded zones and activities to attract the older people to the park at those times. On the weekend day, Sites A, C and D were still most used in the morning by older people whereas Site B void deck was most used in the afternoon and Site B hard court in the evening.

Variation of usage on weekday and weekend—When compared to weekday, more older users were found on weekend day at Sites B void deck (20 more), C (94 more) and D (4 more). More older adults used Site A (253 more) and B hard court (53 more) on weekday than weekend. This could be because the older users prefer to access the amenities (e.g. market, bank, LRT) that are located on the edge of these sites on weekdays. Unlike the weekday, there was no significant spike in users at Site A and B hard court during the morning hours of the weekend day (Fig. 7.12), suggesting that perhaps older people prefer to spend their weekend indoors with working family members or they visited other places during the weekend. For Site B hard court, there was more activity during the weekend afternoon and evening than on a weekday. For the rest of the sites, the number of older users remained relatively low at around 20 people or less per hour. On weekend day, older people's usage of Site B void deck remained more or less the same as weekday while Site D showed more usage in the evening on the weekday and more usage in the morning on the weekend. At Site C, however, almost double the number of older users was recorded on the weekend day than the weekday.

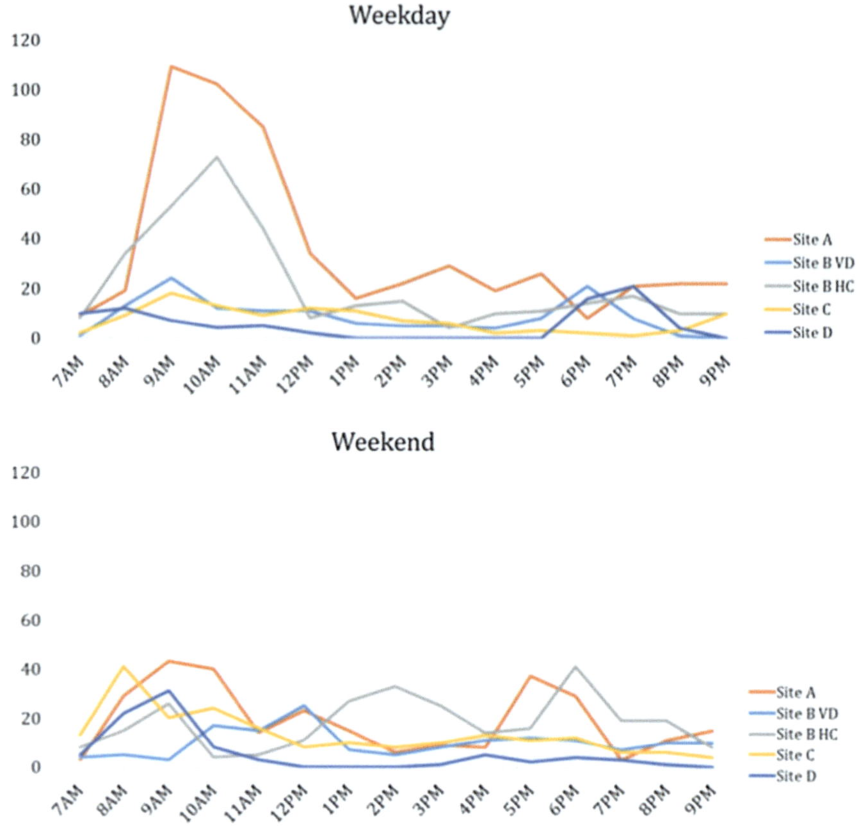

Fig. 7.12 Older users across 4 sites

7.3.6.2 Activity Profile

Table 7.1 summarises the types of active and passive activities at the study sites and the locations (where) within the site that support such activities.

From Table 7.1, it can be noted that most of the activities, be it active or passive, happen at locations that were purposefully designed. The nodal points (with design elements) at each site, e.g. seating, fitness area, children's playground were the main activity generators.

7.3.6.3 Activity Typology

In order to understand the activity pattern across the study sites, the activities in each site are compared based on their nature and typology, drawing on public space literature, particularly from the works of Carmona (2010), Gehl (1987) and Whyte

Table 7.1 Active and passive activities across 4 sites

	Types of activities	Site	Location at site
Active activity	Walking	A, B void deck, B hard court, C and D	• Site A—all across the site • Site B void deck—all across the site • Site B hard court—paths that are the shortest routes to adjoining main road/bus stop/LRT station • Site C—across the site from and to lift lobby • Site D—to and from the community gardens, along jogging tracks
	Exercising/playing	A, B void deck, B hard court, C and D	• Site A—at fitness corner and children's play area • Site B void deck—at seating area with tables in Senior Citizens Corner • Site B hard court—at elderly fitness corner and children's play area • Site D—at elderly exercise corner and adult fitness corner, along the jogging tracks and hard court
	Talking	A, B void deck, B hard court and D	• Site A—at seating provided along the periphery, walking in/through the space • Site B void deck—at seating areas, standing around the Seniors Citizen Corner • B hard court—at seating provided along periphery adjoining HDB block, walking in/through the space • Site C—at stone seating provided at void deck and on additional chairs brought in by regulars, walking in/through space • Site D—at seating provided and pavilions, walking or doing other activities in the park
	Gardening	D	• Site D—at two community gardens within the park
Passive activity	Sitting	A, B void deck, B hard court, C and D	• Site A—at seating provided along the periphery • Site B void deck—at seating provided at Senior Citizens Corner • Site B hard court—at seating provided along periphery adjoining HDB block • Site C—at stone seating provided at void deck and on additional chairs brought in by regulars • Site D—at seating provided and pavilions
	Reading/eating/or any activity carried out alone	A, B void deck, B hard court, C and D	• Site A—at seating provided along periphery • Site B void deck—at seating provided at Senior Citizens Corner • Site B hard court—at seating provided along periphery adjoining HDB block • Site C—at stone seating provided at void deck and on additional chairs brought in by regulars • Site D—at seating provided or pavilions
	Watching the view	A, B void deck, B hard court and C	• Site A—from seating provided along periphery • Site B void deck—from seating provided • Site B hard court—from seating provided along periphery adjoining HDB block • Site C—from seating provided • Site D—from seating provided, especially the pavilions

(1980). Here, the activities are categorised simply and broadly as necessary, optional and social activities based on the way the users interact with the public space.

Since necessary activities, by their nature will take place regardless of the physical infrastructure (participants have no choice), Gehl and many others have argued that high quality outdoor public spaces are those where necessary activities and optional activities occur together (the latter are especially dependent on exterior conditions, they will only occur when place and situation invite people to stop, sit, eat, play, etc.). In other words, in a good outdoor environment, a broad range of human activities is possible—instead of hurrying home, people linger on and the resultant social activities usually increase substantially. In other words, the presence of only necessary activities in a public space though important is not sufficient; it must also have optional and social activities. The proportion of optional and social activities is an indication of what the space has to offer and how the users are influenced by the design features (Gehl 1987).

As Fig. 7.13 indicates, the largest proportions of activities in all four study sites are necessary activities. The void deck space at Site B has the largest proportion of optional and social activities (72.2% on weekday and 85.5% on weekend) followed by Site D (27.1% on weekday and 66.3% on weekend) and Site A (19.7% on weekday and 40.5% on weekend). This proportion is largely dependent also on the total number of users of the sites. The void deck at Site B has a much lower number of people going through the space compared to the other sites, as it is a designated Senior Citizens Corner.

As for the activity typology of Sites D and A, the higher proportion of optional, recreational and social activities when compared to the other sites could be a direct impact of the design elements like provision of seating, pavilions, community gardens, etc. and other programmed activities like fitness corners, jogging tracks coupled with the location and nature of these sites. One is a neighbourhood park with mature trees and greenery while the other is a neighbourhood centre with a bustling marketplace (lots of activities and people to watch through the day). These typologies in themselves could have attributed to the large volume of said activities. Perhaps, because of these locational contexts, the physical provisions in the other study spaces like Site B hard court are not sufficient to attract users to do more optional and social activities despite high necessary activities. Notwithstanding, the proportion of necessary activities is still higher for these two sites (A and D) when compared to optional and social activities. This opens up opportunities to improve the quality of these public spaces, to expand the conditions to encourage people to spend more time in these spaces and more optional and social activities to take place.

The contrast in usage between the void deck spaces at Site B and Site C is another clear indication of the role of physical elements and programming in activating public spaces. Although similar in typology and spatial design, Site B has higher optional and social activities due to interventions like a designated Senior Citizens' Corner and additional physical provisions like washing area, toilet and vending machine. The programming and auxiliary facilities have visibly led to the largest proportion of older users (46–66% of total users) at this site among the four study sites.

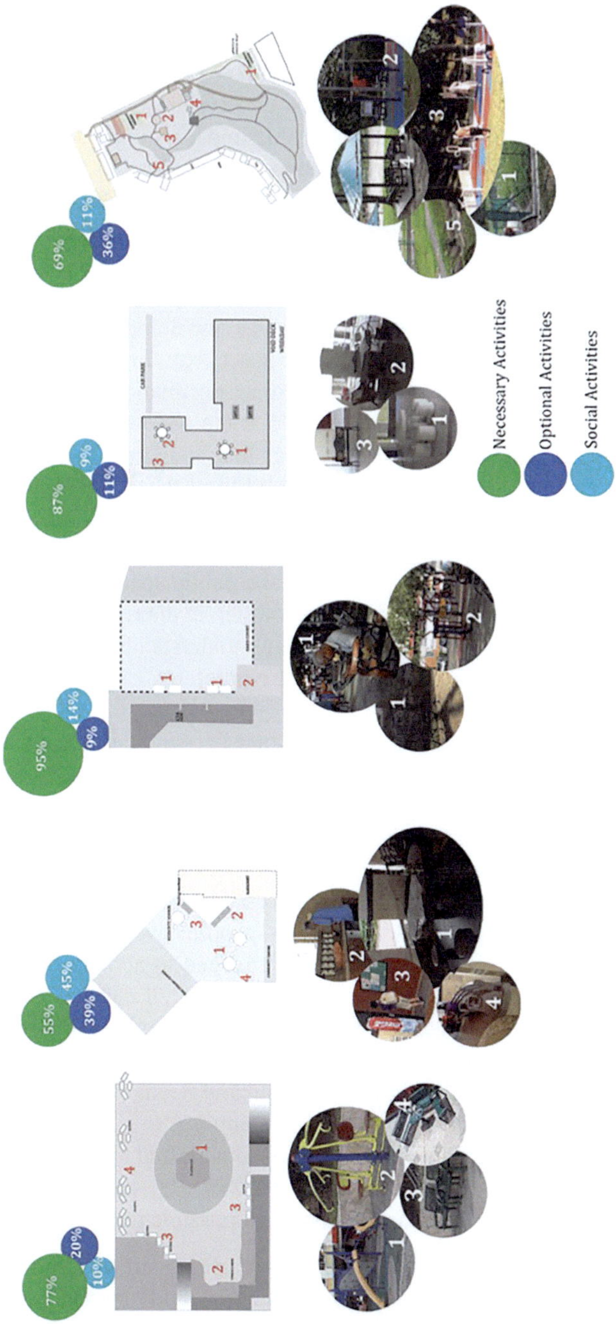

Fig. 7.13 Activity typology across 4 sites

7.4 Conclusion

The onsite observation aims to offer a glimpse into the daily usage levels (across the span of a day during *albeit* one weekday and weekend day) of the four public spaces examined. Through the study of their activity levels and activity typology, we learned that design features have discernible impacts on the activity patterns of public spaces. As the analysis has shown, the public spaces examined, though all located in the same town, have diverse characteristics and were used in different ways by the local population, by different segments of the community.

In all of them, the older users were always present though their numbers might fluctuate from under 20% to over 50% of total users, depending on space characteristics and features, time of day/week. Their usage and activities are clearly related to their personal needs and perceptions of the space, e.g. the large numbers that go through Sites A and B hard court *en route* to other destinations (using these spaces as shortcuts), the children and their caregivers at the children playground at these spaces or the older residents at Site B void deck and Site C.

This points to the importance of the availability of these close to home spaces for the community, especially an ageing population. Compared to other segments of the community, the old and young are less able to travel further for their leisure and recreational needs. The policy direction to make public spaces in neighbourhoods more age-friendly and attractive for its users is a step in the right direction. This call for close study of outdoor activities and a greater understanding of the needs of the users: their activities and play habits, their seating preferences, among others. There are many case studies to draw from.

Gehl (1987) has posited that as a prime concept, everyday life must form the centre of attention and effort. The consequential requirement is to invest public spaces with desirable conditions for the three basic activities—necessary outdoor activities; optional, recreational activities; and social activities. The Project for Public Spaces (2009a) has highlighted that great public spaces share four qualities. These are: access and linkages—the space must be accessible, both visually and physically, easy to get to and through; uses and activities—people have something to do in the space, a reason to come to the space and are engaged in activities there; comfort and image—the space is comfortable, presents itself well and has a good image, and sociability—It is a place where people meet each other, take people when they come to visit and fosters social activities. Ongoing maintenance of the public space is equally important, especially considering changing needs (Madanipour 2010). A practical way is to begin with small-scale interventions, to experiment and further improvise schemes and strategies to arrive at sustainable and adaptive solutions with community inputs (Project for Public Spaces 2009b).

The on-site observation data reinforces several possible interventions:

Creating more 'sittable spaces'—Sitting and talking are the most frequently recorded optional/social activities among users at the study spaces. Sittable spaces as termed by Whyte (1980) are the most essential components in the design of public spaces. Whyte (1980) described these spaces as comfortable physical and social environments where people can choose to sit, presenting the opportunity to spend more time in the public space. Such comfortable and flexible choice of seating is currently insufficient in the sites studied, especially within the void deck spaces (many are fixed seating, often in arrangements that do not facilitate group conversation). As the literature shows, seating could be primary elements like benches or secondary seating like ledges, landscaped lawns or edge of a design feature (like fountain, sculpture, etc.) (Whyte 1980). The location of such sittable spaces is also significant. Studies have proved that users prefer to sit at locations that have pleasant views to offer (can watch the world go by). Additionally, due to Singapore's tropical climate (hot sunny or raining and high humidity), these sittable spaces need to be located under shade or shelter (e.g. naturally tree-lined shelters). This could improve the microclimate and encourage the older adults and other users to use the spaces at their own convenient time, any time of the day.

Introducing 'landmarks'—Landmarks, as pointed out by Lynch (1960) are navigational features of the built environment. These features could be especially relevant to the older users, as they will require a significant, memorable feature to mark their path/location to easily orientate themselves in the outdoor environment. The physical manifestation of a landmark feature for outdoor spaces like Sites A, B hard court and D could be as simple as a strategically located tree, a unique pavilion or even a special colour bench. For semi-open spaces like the void decks in Sites B and C, these could be a differently painted wall, column or bench. In other words, landmark elements need not be expensive, but can be organic and creative (from the people or community), quick and cheap, yet will generate a strong identity for the public and eventually attract users of all ages.

Introducing 'triangulation' of activities—This is a commonly discussed concept of placemaking, as a process by which "external stimulus prompts strangers to talk to each other" (Whyte 1980, p. 154). In the practical sense, it refers to the clustering of activities together so that these primary activities would generate more activity (e.g. social activities) within the place. The primary activities need to be selected and located such that they encourage the users to interact with each other. Triangulation can enable integration and segregation of activities based on the desired programming (Gehl 1987). In order to achieve the triangulation of activities, design features have to be located strategically. For example, a seating area can become a more desirable congregational spot in combination with landscaped flowerbed, shading and a waste bin rather than each of these elements located in isolation. This careful process of planning public spaces can go a long way in improving user interaction and social activity.

Programming of public spaces—This refers to a process of designating specific function/use to a public space. These utilities may be defined based on the context of the space (e.g. location, size, residents' demography, etc.). Public spaces can serve many different functions, e.g. recreational, fitness, relaxation, escape, etc. Studies show that the presence of activity programming in parks and public spaces increases the usage of spaces (Project for Public Spaces, N.A.). Activity programming is being practiced in many nursing homes and care facilities today (such as planned scheduled events games, self-care activities etc.). Such programming could revolve around existing (and future) provision, e.g. recreational offerings like tai chi, qigong and other kinds of exercises, dance, community gardens, etc. Some of these are already happening in the study sites, e.g. tai chi and community gardens in Site D, tai chi/qigong in Site A. With more older residents wishing to age in place, there is opportunity to do more to keep them active physically and socially. Programming options of public spaces should not be limited to elderly fitness zones but also extend to other interest areas like drawing on some older adults' skills of storytelling, developing a network of volunteers of young and old, e.g. to offer bird-watching, garden walk, etc. The aim is to draw on the creative energy of the community to generate more widespread use of public spaces and strengthen the sense of positive use.

Adopting participatory design—In order to plan and design spaces for the ageing individual, it is important to understand their needs and aspirations as well as that of the rest of society as they will together be using the spaces. The best way to understand their perceptions is by involving the older adults themselves in the design process. Immediate and first-hand responses to design can help to uncover previously unknown needs, empower them and promote a greater sense of ownership towards the space. It bears reminding that there are two fundamental dimensions in public space design: a physical dimension, which involves the designed elements that are part of the space (e.g. seating, shelter, amenities, etc.), and a social dimension, which elucidates the appropriation of physical space by the users; these are able to generate new functions and activities within the same space (Mandanipour et al. 1998).

Acknowledgements We gratefully acknowledge the following for their generous support and assistance with the On-Site Observation Survey: all students at Singapore University of Technology and Design (SUTD) including Architecture and Sustainable Design (ASD) 2015 class of 20.016 Urban Analysis who helped us record the observations at the four sites of study; student intern, Lim Aunn Ning, for meticulously compiling the observation data digitally; our colleagues Cheong Sue Wei and Emily Yongxu Soh for verifying the data entry; Laura Jasmine for coordinating the observation data collection and Adithi Moogoor for assisting with the photo presentation in this Chapter.

References

Alexander, C., Ishikawa, S., & Silverstein, M. (1977). *A pattern language: Towns, buildings, construction*. USA: Oxford University Press.

Carmona, M. (2010). *Public places, urban spaces*. Routledge.

Carmona, M., Tiesdell, S., Heath, T., & Oc, T. (2010). *Public places urban spaces*. Oxford: Elsevier.

DeWalt, K. M., & DeWalt, B. R. (2002). *Participant observation: A guide for fieldworkers*. Walnut Creek, CA: AltaMira Press.

Gehl, J. (1987). *Life between buildings*. New York: Van Nostrand Reinhold.

Goličnik, B. (2011). Analysis of patterns of spatial occupancy in urban open space using behavior maps and GIS. *Journal of Urban Design International, 16*(1), 36–50.

Lefebvre, H. (1947). *The critique of everyday life*. Paris: Grasset.

Lynch, K. (1960). *The image of the city*. Cambridge: MIT Press.

Madanipour, A. (Ed.). (2010). *Whose public space? International case studies in urban design and development*. New York: Routledge.

Mandanipour, A., Cars, G., & Allen, J. (1998). *Social exclusion in European cities: Processes, experiences and responses*. London: Jessica Kingsley Publishers.
Mean, M., & Tims, C. (2005). *People make places*. London: Demos.
Pinto, J., & Remesar, A. (2012). Public space networks as a support for urban diversity. *Open House International, 37*(2), 15–23.
Project for Public Spaces. (2009a). What makes a successful place? http://www.pps.org/reference/grplacefeat/. Accessed April 20, 2016.
Project for Public Spaces. (2009b). What is placemaking? https://www.pps.org/reference/what_is_placemaking/. Accessed July 11, 2017.
Tate, A. (2015). *Great City Parks*. New York: Routledge.
WallJasper, J. (2007). *The great neighbourhood book*. Canada: New Society Publishers.
Whyte, W. (1980). *The social life of small urban spaces*. Washington, D.C.: The Conservation Foundation.
World Health Organisation. (2007). *Global age-friendly cities: A guide*. Geneva: WHO.

Chapter 8
Real-Time Data Analysis Using a Smartphone Mobile Application

Sumudu Hasala Marakkalage, Billy Pik Lik Lau, Sanjana Kadaba Viswanath, Chau Yuen and Belinda Yuen

Abstract Information about older adults' everyday journeys and outdoor activities can provide insights into their travel habits, leisure preferences and other daily activities. In this research, a smartphone-based mobile application was developed as a tool to better understand such daily activities of older persons who reside in Bukit Panjang Town, Singapore. The Chapter delivers a detailed methodology of the mobile application data collection, which includes device location information (GPS coordinates), background noise, background light level, and device battery level. Using analysis in MATLAB, heat maps were used to visualise the gathered insights. The findings showed that older participants tend to visit and spend their time mostly in their local neighbourhood whereas the visits to places outside of it are mostly done for shopping, religious and healthcare purposes.

8.1 Introduction

Previous research has found that older adults have the tendency to travel in their leisure time (Capella and Greco 1987). This brings forth a related set of research questions: Where do they go? Which are their preferred places of visit? When and how do they travel to these places? To gather such information, we need to reach the older individuals and communicate with them to collect the required information.

S. H. Marakkalage (✉) · B. P. L. Lau · S. K. Viswanath · C. Yuen · B. Yuen
Singapore University of Technology and Design, Tampines, Singapore
e-mail: marakkalage@mymail.sutd.edu.sg

B. P. L. Lau
e-mail: billy_lau@mymail.sutd.edu.sg

S. K. Viswanath
e-mail: sanjana@sutd.edu.sg

C. Yuen
e-mail: yuenchau@sutd.edu.sg

B. Yuen
e-mail: belinda_yuen@sutd.edu.sg

B. Yuen (ed.), *Ageing and the Built Environment in Singapore*,
https://doi.org/10.1007/978-3-319-92444-1_8

Rather than conducting verbal or written surveys, which are based on recall, the more convenient and perhaps, more precise approach is to utilize the increasingly ubiquitous usage of smartphones among older people as a resource to collect real-time data from the heterogeneous sensors available in smartphones. Previous studies by other researchers show that older people in Singapore have positive attitude towards smartphones and have found these phones to be useful and entertaining (Pang et al. 2014).

Location tracking using smartphone applications is well established in research. There are many previous studies on tracking user locations using smartphone applications. Viswanath et al. (2014) used smartphone location data to track tourist travel locations along with surveys to collect feedback on tourists sites. Others study location tracking in the context of smart homes to support the older adults who stay at home. These include research on proactive environment change for older people (Helal et al. 2003), cloud-based real-time activity tracking in home (Fahim et al. 2012), remote monitoring of older adults who stay at home (Yassin et al. 2016), RFID-based location tracking at home to ensure the safety of older adults in smart home environments (Kim et al. 2013), among others.

Others, e.g. Arab et al. (2013) have researched on how to support older adults to use smartphones conveniently and developed an adaptable interface for smartphones. Mirri et al. (2014) suggested accessible urban paths using crowd-sourced data for people who have limited mobility due to disabilities. Another interesting application is to transfer the knowledge and experience of older adults who stay at homes to younger generation by a passive information gathering system using mobile terminals such as smartphone applications, emails and phone calls (Hiyama et al. 2013). Yet, another development is to support older adults to stay active in life and improve their healthy living conditions using a web-based platform (Ghiani et al. 2013).

Taking such studies as our point of departure, we have included the development of a smart phone based mobile application as a tool to better understand the daily activities of an older person. This Chapter presents the methodology and key findings from the smart phone mobile application.

8.2 Methodology

A user-friendly smartphone application that runs on Android platforms was designed and developed by the research team to collect travel journey data from older people. The app collected device location information (GPS coordinates), background noise (it does not record sound, only the noise level), background light level, and device battery level. The intent was to find out where older people travel to and how much time they spend at each of those places.

More than 100 people, the majority aged 55 and above, who reside in Bukit Panjang Town participated in the study by installing the application on their mobile phones. Bukit Panjang is the study site for a number of other study instruments in

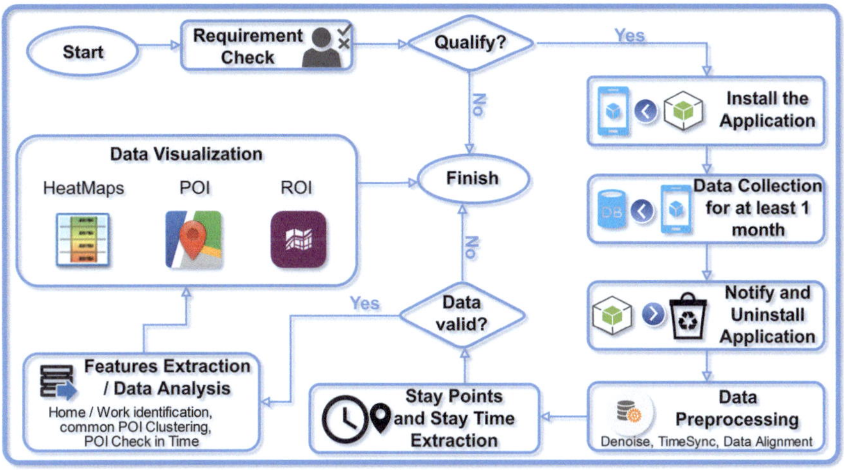

Fig. 8.1 Elderly data collection, analysis and visualisation flow chart

this project.[1] The participants were recruited at community centres, religious sites (e.g. temples) and community events. The data was collected for a minimum period of one month from each user.

In this Chapter, we use data from only 50 participants whose collected data is consistent throughout the data collection period. That is, the participant data was available nearly continuously for a minimum period of one month without any huge data loss in-between.

There are three broad stages to the smartphone application development:

1. Mobile Apps Development;
2. Data Collection from Older People;
3. Big Data Analysis.

Figure 8.1 shows the data collection, analysis and visualisation process.

8.2.1 Mobile App Development

The app is published on Google Play Store (for Android) as “City”. It has three key data components—demographic information, a map showing travel route and a short survey that asked about the user’s travel and place experience. The mobile application runs automatically in the background when the user turns on the phone. If there is no available internet connection, the app will store the data in the device until it connects to internet to upload the data to the server.

[1] See, for instance, Walk and Talk Survey and On-site Observation in Chaps. 5 and 7 respectively.

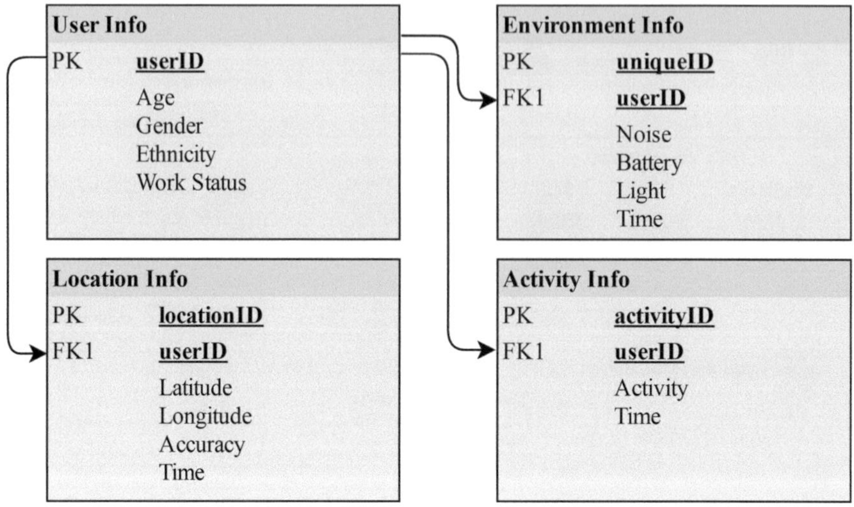

Fig. 8.2 Database model

The data collected from the older person's smartphone are sent to store in a central server at the Singapore University of Technology and Design (SUTD). The database model is shown in Fig. 8.2. The database has 4 tables/types of information pertaining to User, Location, Environment and Activity. The types of data stored in each table are as follows:

- **User Information** This table contains basic information about the older adult such as his/her age (years), gender (male/female), ethnicity (Chinese, Malay, Indian, and Other), and work status (working/non working);
- **Location Information** This table contains the location information collected through Android location API. It gives the location parameters such as latitude, longitude, accuracy (in metres), and timestamp in which the location is acquired;
- **Environment Information** This table contains basic device information and its surrounding environment information. Parameters such as environment noise level (dB), device battery percentage, light level (Lux), and the timestamp in which the data is acquired;
- **Activity Information** This table contains the activity information received by Google API. 4 types of activities are stored in this table such as Still, On Foot, On Bicycle, and In Vehicle.

When the app is installed for the first time, users are directed to the "Registration Page" and prompted to input their demographic details like name, phone number, age, gender, among others. The personal information is mainly for the researcher to contact the participants for the collection of incentive at the end of the study.

When the app is opened, the "Apps Homepage" shows the day/calendar where the user can select "Map View" for a particular day. Map View displays the locations the users had visited that day and also the paths taken. Users can also choose "Map View" for one week of activity. Under this option, all the locations and paths taken for that one whole week are shown.

Users may also participate in the "Survey" to share more information about a place. The survey questions include the purpose of visiting a location, how he/she feels about the place, how he/she gets to the place, how often he/she visits the place, and whom he/she visits the place with. Users can elect to share their activity on Facebook through the "Share" button, e.g. number of places visited, number of hours at work place, time spent on commuting.

The "Settings" tab is used for advanced setting/debugging purposes (users can view their stay points on a Google map for a given stay time value in minutes).

Challenges The main challenge with the mobile app is the trade-off between energy consumption and data quality. To obtain a better accuracy or more detailed information, higher sampling rate is required. However, a higher sampling rate will lead to higher energy consumption. The most common feedback that we received from users is that after the installation of the app, the battery drained faster. This was especially obvious in the cases of those older persons who are new to smart phone technology, and have been using their smart phone as a traditional phone, i.e. only for texting or making phone calls. Going forward, several features of the app can be improved, e.g. energy efficiency, transport mode detection. We can also introduce new features to the app by sharing analysed information with the users.

8.2.2 *Older User Data Collection*

As mentioned above, the recruitment of target participant group for the study (the older person who use Android smartphones and reside in Bukit Panjang Town) was done in public locations around Bukit Panjang. The team performed recruitment of 101 eligible participants over the span of seven months at Butik Panjang. In most cases, the team helped the participants to install and register the app.

Three participants' phones transmitted only limited location information. This is because we received some other sensor data from their phone and it was later discovered that certain brand of smart phone (Xiaomi) had a very different configuration than typical Android phone that resulted in missing location information. After the incident, Xiaomi phones were excluded from the study.

All the data were collected for a minimum period of one month with written consent from every participant in the study. Participants' personal identity was used only during registration. The data collection and analysis was done through a unique auto-generated device ID to ensure privacy of the participants.

While there is a built-in survey in the mobile apps for the users to provide further information on the places that they have visited, hardly any user answered the survey through the app. Only one male participant answered the survey for one to two locations on day one. As an alternative, the team decided to print out hardcopy of their summarised data and conduct face-to-face survey when the participants collected their token of appreciation at the end of the 1-week data collection. This increased the survey response rate.

In the survey, the purpose of visit was classified using the following pre-tested categories:

B Bank
C Community Club
D Dining
E Exercise @ Garden/Playground
F Friend/Relative house
G Grocery/Shopping
H Home
M Medical/Clinic/Hospital
S School/Class
T Temple/Mosque/Church
W Work
O Others (Please Specify)

Challenges Several challenges were encountered during this phase of work. Despite using smartphones, some older adults did not have data plans for Internet access. For example, they only get online when they go to the community centre. As a result, we were not able to receive their data as they are hardly online. Some participants did not turn on location services (GPS) or they turned it off after they went back home. Hence, we were not able to get all their location information. Others ran the smartphone in power saving mode, which resulted in turning off the GPS. Again, we were not able to get their location information. Our app has problems in collecting data from selected smart phone model, e.g. Xiaomi, and Xiaomi appears to be a popular phone among the older participants due to its affordability.

8.2.3 Big Data Analysis

Acquired data from the mobile phones of the older participants need to be analysed to gather meaningful insights into their travel journeys and places visited. But, first the data had to be cleaned and prepared for the analysis. Based on the collected data (but, in some cases, subject to availability, some information may not be complete for certain users), the research team tidied and organised the information for a particular location visited by a particular user in the following way:

- User ID (every participant is identified by an ID to protect privacy)
- Work type (e.g. full time or part time)
- Age
- Gender
- Location (latitude, longitude)
- Google suggestion for location

- Foursquare suggestion for location (top 5 suggestions)
- Purpose of visit (collected from the survey)
- Number of times they visit this place
- Noise (Total noise, average noise and variance of noise)
- Duration (Total duration, average duration and variance of duration)
- Total period of participation (calculated for each participant).

Data pre-processing was done in two major stages using: (1) Denoise, which removes duplicate data records and anomalies occurred due to location accuracies, and (2) Time sync, which aligns data from different tables in relevant time slots. After pre-processing stage, the raw location data were clustered using k-means clustering technique and generated a summary database for each user. A total number of 1781 clusters were generated from raw data of all valid 50 users.

Several data analysis including Voronoi zoning and place analysis were then conducted to identify the popular regions/places visited by the participants. Data analysis was done using back-end Java application and MATLAB software package.

8.2.3.1 Voronoi Zoning

In mathematics, a Voronoi diagram is a partitioning of a plane into regions based on distance to points in a specific subset of the plane. That set of points (called seeds) is specified beforehand, and for each seed there is a corresponding region consisting of all points closer to that seed than to any other. These regions are called Voronoi cells. A MATLAB programme has been developed to divide a region (e.g. Singapore or Bukit Panjang area) into several zones based on the Voronoi diagram, perform statistical analysis on the summary results generated from the processed data and present the results based on the Voronoi cells for given seeds (see Sect. 8.3).

Regions of interest (ROI) were analysed using Voronoi diagrams. In a Voronoi diagram, a plane is partitioned into n points in such a way that each polygon contains one seed and every point in each polygon is closer to the seed inside each polygon than to any other one. These polygons are called Voronoi zones. ROI analysis was done in three categories as follows:

- Identifying the percentage of home stay duration of each user out of their total participation time;
- Identifying the ROI across user neighbourhood, which is Bukit Panjang, Singapore;
- Analysis was done across Singapore to identify island-wide ROI.

8.2.3.2 Place Analysis

MATLAB was used to analyse the usage pattern on places. For a given location, we performed the statistical analysis based on the following:

- Histogram of first check-in across 24 h of a day (i.e. the first arrival time of the participants at the location);
- Total number of check-in across 24 h of a day (i.e. how many participants at the location on a particular hour);
- Duration spent by a person per calendar day;
- Total number of check-in per calendar day (i.e. how many participants visited the location on a particular day);
- Duration spent by a person based on each day of a week;
- Number of check-in based on each day of a week (i.e. how many times the participants visited a certain location on a particular day of a week).

After the data was analysed, tools such as Google Maps, Voronoi Diagrams and Heat Maps were used to visualise the gathered insights to identify the Regions of Interest (ROI) and Points of Interest (POI).

8.3 Findings

We examine the key findings under two major sections:

- Regions of Interest (ROI);
- Points of Interest (POI).

In all cases, the home and work records were excluded (i.e. their stay in home/work is not considered in the count) because the analysis' intent is to identify the number and places visited outside home. The inclusion of such records would affect the number of visits and time spent in the overall analysis. We have separate statistics on the stay home duration (see Sect. 8.3.1).

8.3.1 Home Stay Duration

Figure 8.3 shows the percentage of time each participant spent at home, out of his/her total participation duration for data collection. It also shows the age and gender of each participant. The oldest participant was 75 years old and the average age of participants was 60 years. There were 28% male and 72% female participants. It can be observed that the majority of participants spent a large amount of their time at home,

44% of participants stayed at home for more than 21.6 h a day on average;
22% of participants stayed at home between 19.2 and 21.6 h a day on average;
18% of participants stayed at home between 14.4 and 19.2 h a day on average;
16% of participants stayed at home less than 14.4 h a day on average.

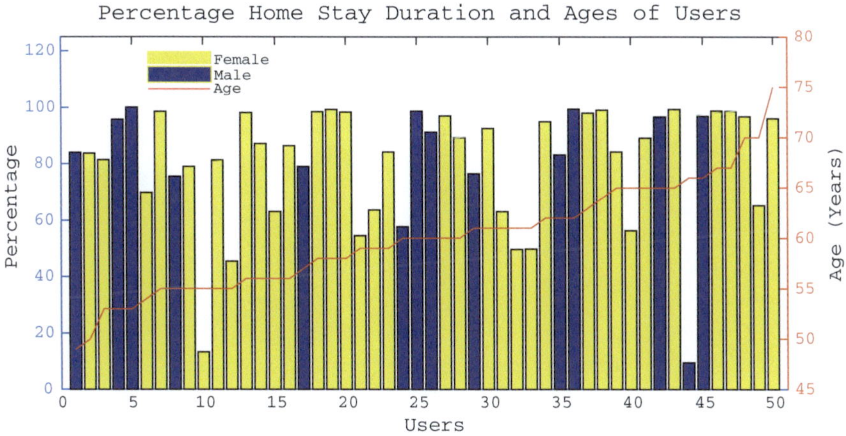

Fig. 8.3 Respondents home stay duration

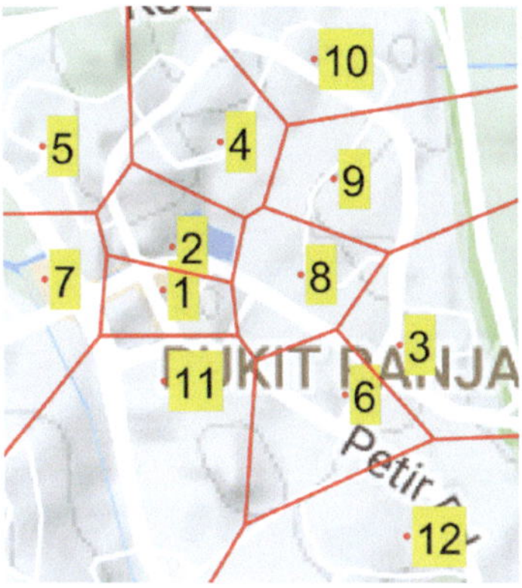

ID	Location	Homes
1	Bukit Panjang Plaza	5
2	Senja Cashew CC	2
3	Masjid Al-Iman	8
4	Greenridge Shopping	2
5	My Family Clinic	5
6	Bukit Panjang CC	4
7	Junction 10	0
8	Zhenghua School	6
9	Fajar Market	5
10	Sheng Siong	2
11	Kimly Seafood	3
12	Michaels' Residence	8

Fig. 8.4 Voronoi zones and participant homes across Bukit Panjang

We were able to notice that participant ID 10 and 44 have substantially lower home stay duration, when compared to others. One possible reason for such a result is that these participants switched off their mobile phones at night.

8.3.2 *Across Neighbourhoods*

We divided Bukit Panjang town into 12 Voronoi Zones to get an understanding about participants' home region since all of them are residents of Bukit Panjang. The Voronoi zone distribution, zone number and number of participant homes in each Voronoi zone is shown in Fig. 8.4.

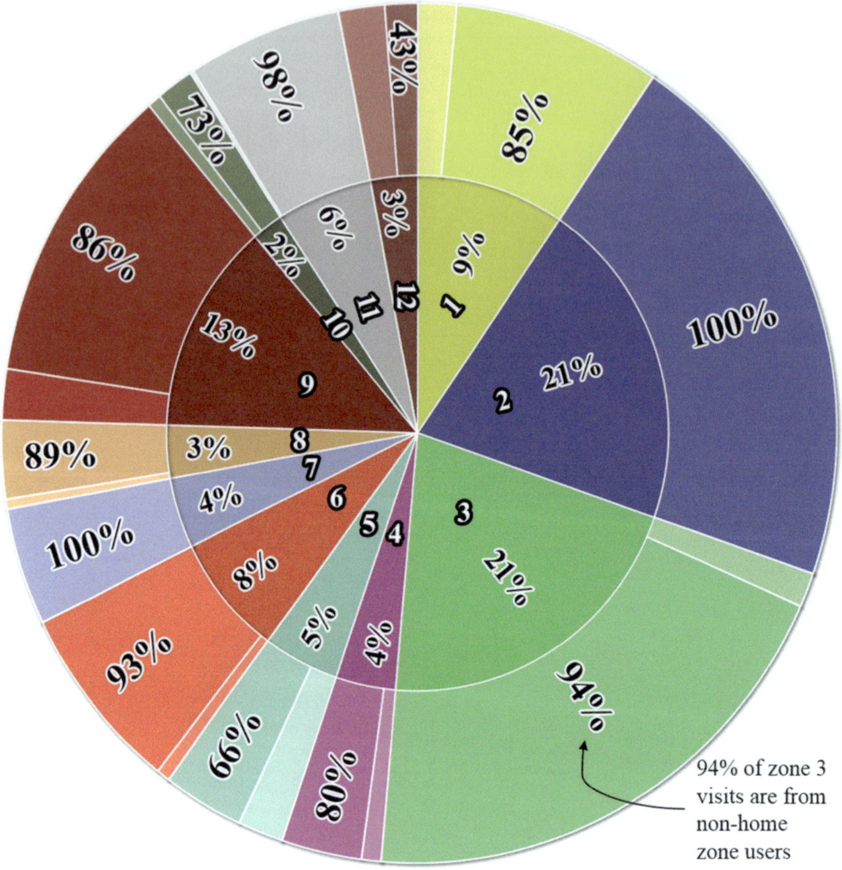

Fig. 8.5 Voronoi zones in Bukit Panjang town and non-home zone visits

Figure 8.5 shows the percentage of visits in each region while Fig. 8.6 shows the percentage of durations in each region. In both Figures, the inner circles display the Voronoi zone number and percentage of visits/duration accordingly. The outer rings show the percentages contributed by non-home zone participants. From the Figures, the following patterns of non-home zone visits and durations are observed,

- Zone 2 (Senja Cashew CC) has 100% of non-home zone participant visits and durations, indicating that all the participants who visit and spend time in that zone are residing outside that zone;
- Zone 2 shows the highest percentage of visit duration;
- Zone 3 (Masjid Al-Iman) shows the highest percentage of visits;
- Zone 3 has higher percentage of visits when compared to percentage of duration. A possible reason for such a result could be the location of the Bangkit LRT station inside zone number 3;

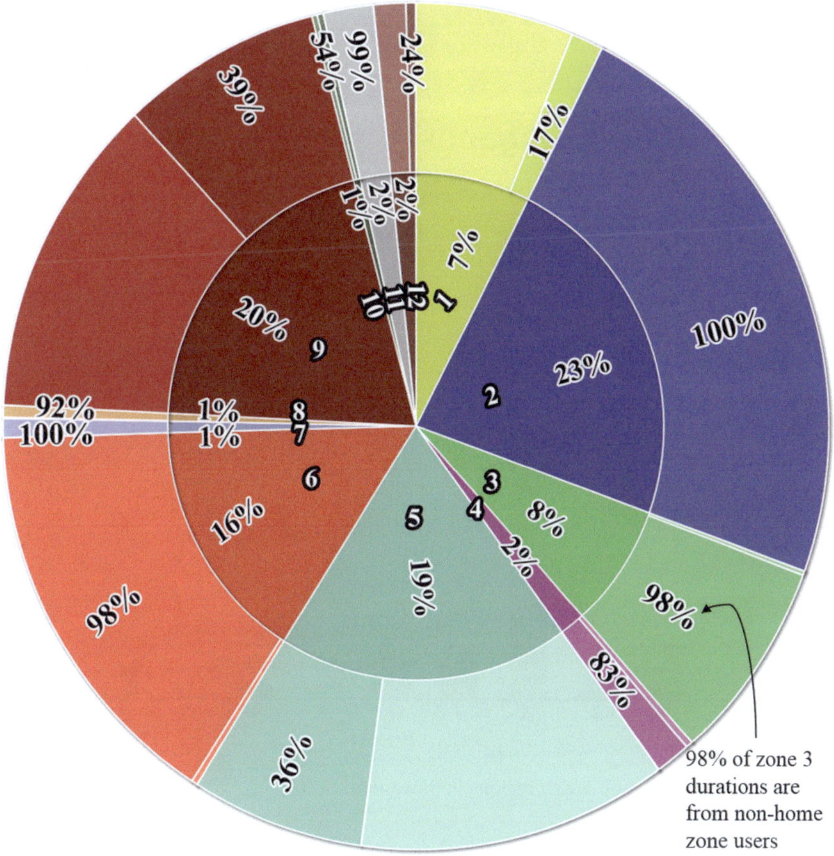

Fig. 8.6 Voronoi zones in Bukit Panjang town and non-home zone durations

- Zones 2 and 6 are popular among non-home zone participants, in terms of percentage duration;
- Zones 5 and 9 are popular among home zone participants. The reason for such popularity could be those zones' POIs such as Bukit Panjang CC (zone 6), Senja Cashew CC (zone 2), and Fajar Market (zone 9).

8.3.3 Across Singapore

The map of Singapore was divided into 9 Voronoi zones in order to understand what regions across Singapore are of interest to the participants in terms of their number of visits and duration of time spent there. The Voronoi seeds were selected to cover

ID	Region
1	Outram Park
2	Punggol
3	Bukit Panjang
4	Jurong East
5	Clementi
6	Tampines
7	Ang Mo Kio
8	Yishun
9	Woodlands

Fig. 8.7 Voronoi zones across Singapore

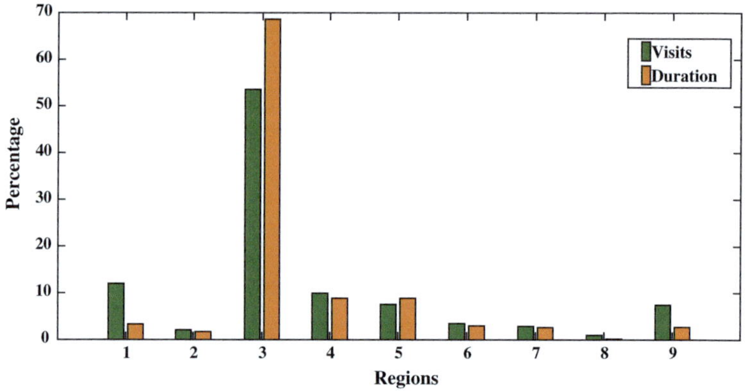

Fig. 8.8 Visits and duration across regions in Singapore

the main regions of the entire Singapore. The Voronoi zone distribution and zone names are shown in Fig. 8.7.

Figure 8.8 shows the percentage of number of visits and durations for each region across Singapore. The following can be observed from Fig. 8.8,

- Region 3 (Bukit Panjang) is the most popular among participants, in terms of both visits and duration. In fact, 53.5% of visits and 68.6% of durations belong to that region. The reason for such popularity is because all the participants were residing in Bukit Panjang Town and they tend to visit places near their homes;
- Regions 1 (Outram Park), 4 (Jurong East), 5 (Clementi) and 9 (Woodlands) are popular regions. However, Regions 1 and 9 tend to have shorter visit duration than number of visits.

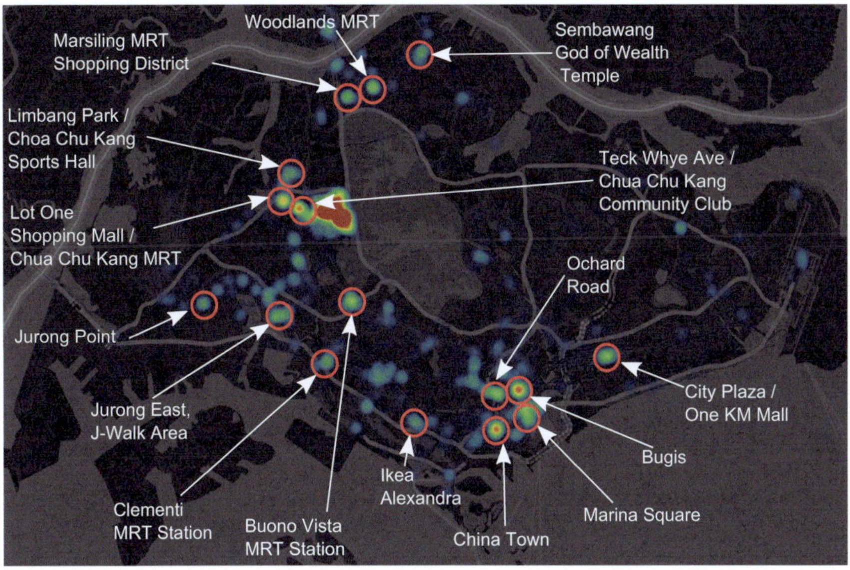

Fig. 8.9 POI heat maps across Singapore

8.3.4 *Points of Interest (POI)*

Next, we analysed in more detail each specific POI. For each POI, we performed POI extraction and ranked it according to the number of participants and time stayed (duration). For privacy reasons, we removed participants' home and work places. We clustered common POI among the participants using density based clustering with radius of 100 m (hence, the POI indicated later include also near-home related POI). The clustering process yielded 980 POI from the collected data.

8.3.4.1 Correlation of POI with Participant Numbers and Stay Time

We plotted heat maps for the participants based on the number of check-in across Singapore as displayed in Fig. 8.9. The heat zones are shown in red colour while yellow colour indicates the places where fewer participants visited. High heat area is circled and pointed out in the maps. We did not include Bukit Panjang Town as it is near to the home of participants. We highlighted a few high heat zone areas using circle and labelled it accordingly. In Table 8.1, we present the top POIs and time spent per visit for the participants (the ranking is evaluated based on the number of participants checked in). The POI highlighted in orange indicates that these are located within the Bukit Panjang Town whereas the blue colour indicates those within five kilometres of Bukit Panjang Town (nearby POI). Finally, the purple colour indicates the POI located outside Bukit Panjang Town.

Table 8.1 Top POI and participants time spent per visit

No.	Point of interest	Percentage of participants visit	Average hours spent per participant per visit
1	Bukit Panjang Hawker Centre	68%	3.7
2	Bukit Panjang Plaza	56%	2.9
3	Bukit Panjang Community Centre	34%	1.9
4	Fajar Shopping Centre	30%	2.5
5	Senja Cashew Community Club	28%	4.4
6	Holland Bukit Panjang Town Council	28%	2.2
7	Block 125 Neighborhood Stores	26%	3.0
8	Lot One Shopping Mall	24%	1.7
9	National Healthcare Group Pharmacy - Choa Chu Kang	22%	1.7
10	Chinese Temple at Bencoolen Link	22%	0.9
11	Greenridge Shopping Centre	20%	2.5
12	Ten Mile Junction Mall	20%	2.2
13	Plaza Singapura	18%	2.6
14	Causeway Point	18%	1.6
15	Block 172 Neighborhood Stores	16%	2.4
16	Zheng hua Community Club	16%	2.0
18	Teck Whye Ave Neighborhood Stores	16%	1.3
19	Temple Street at China Town	16%	1.2
20	Phoenix Road Shop Lots	14%	2.6

Orange - Within Bukit Panjang Area, Blue - Within 5KM range of Bukit Panjang, Purple - 5KM away from Bukit Panjang

Some key findings:

- The top POI consists of locations in the Bukit Panjang Town. Hawker centre and community centre were the most commonly visited places among the participants, where at least 1 h 55 min per visit time spent was recorded;
- Next are the shopping mall and neighbourhood stores. Note that a high-density HDB public housing block can be a multi-purpose place such as indicated by Rank 7, Block 125, which has a grocery store and traditional medical centre at the first storey of the building. This is not an uncommon situation in Singapore's high-density new town development;
- Participants spent an average of 2.5 h in the Bukit Panjang shopping mall area (Bukit Panjang Plaza, Fajar Shopping Centre, Greenridge Shopping Centre, and Ten Mile Junction Mall) per visit.

Next, we discuss the POI located outside Bukit Panjang Town. As shown in Fig. 8.8, there are a handful of frequently visited locations outside of Bukit Panjang area. They are distributed across the central area (Chinatown and Bugis), southwest area (Jurong and Clementi), Bukit Panjang nearby area and north area (Woodlands). A top common area among the older participants was the National Healthcare Group Pharmacy, which is located some two kilometres away from Bukit Panjang Town.

Other POI includes mostly shopping malls and temples, which can be related to participants' cultures and religion.

In a nutshell, participants had the tendency to visit nearby amenities for daily convenience. This can be highly related to the transportation accessibility and travel time. However, when it came to shopping, religious, and medical purposes, participants were more likely to travel further, outside Bukit Panjang Town to access these facilities.

8.3.4.2 Correlation of POI with Check-In Time and Time Spent

After obtaining the POI ranking, we investigated the check-in time and time spent for each POI. The results are displayed in Figs. 8.10 and 8.11. We divided the analysed data into two categories, which are check-in time during different hours of the day and time spent on different days of the week. Several analyses were performed.

First, we analysed the top POI from the previous section—Bukit Panjang Hawker Centre and Market. Based on the data, we can consider these as morning places—-morning (8am–10am) indicated the most check-ins compared to other hours. When combining the time spent on different days of the week, Monday appeared to have less away from home time spent compared to other days. The main reason could be the operating commercial hours of local market where Monday is the rest day (market is closed). Reviewing the information collected, Bukit Panjang Hawker Centre appeared as a place where most participants went to either secure their groceries or for other personal purposes. Note that participants could have gone to the Bukit Panjang Hawker Centre for different purposes (we only considered the majority purpose collected from the mobile app survey).

Next, we discuss the shopping complex that older participants checked in during the data collection, after Bukit Panjang Hawker Centre. There are two shopping complexes studied here: Bukit Panjang Plaza and Fajar Shopping Centre. The pattern for both shopping complexes is diverse. The pattern for Bukit Panjang Plaza is in Gaussian distribution shape and Fajar Shopping Centre is in two peaks shape. This would seem to suggest that checked-in time might highly correlate with the location of the shopping mall. For example, Bukit Panjang Plaza is located at the mass rapid transit interchange, which presumably has more checked-in participants compared to the Fajar Shopping Centre. We noticed one or two participants that frequently spent their time there and contributed a lot of hours.

Another common POI for the participants would be the community centre. It is observed that both Bukit Panjang and Senja Cashew Community Centres had similar check-in patterns where morning and evening display high checked-in time for the participants. However, by comparing the number of hours spent over a week, there is a significant variance between Bukit Panjang Community Centre and Senja Cashew Community Club. Bukit Panjang Community Centre had more check in on Monday and Sunday. We believe this could be linked to the activities that the community centre offers and transportation accessibility to travel there.

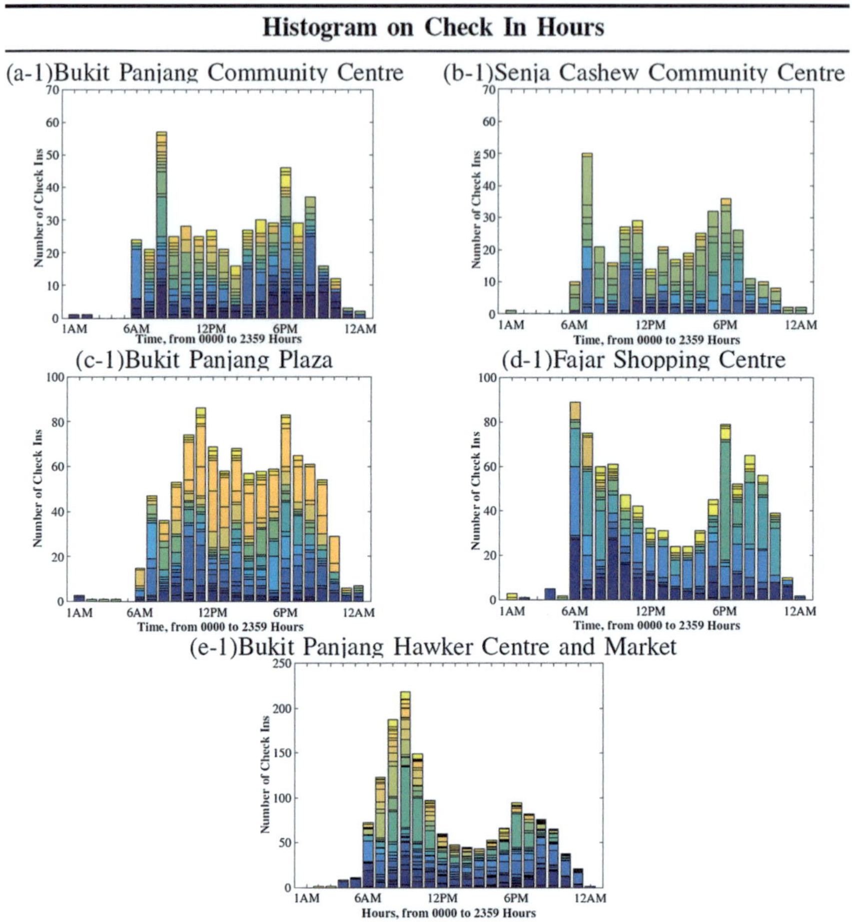

Fig. 8.10 Histogram of check-in during the day

In conclusion, we notice that participants are likely to go to Bukit Panjang Hawker Centre and Community Centre in the morning. Such visit pattern can vary and is subjected to commercial operating hours and activities offered. However, in contrast to community centre and hawker centre, the visit pattern of participants in the shopping mall seems to highly correlate with the location of the mall itself.

8.3.4.3 Correlation of POI with Gender and Employment

The heat maps for different gender and employment status are presented in Fig. 8.12. The highly active gender-employment is highlighted with red bold boxes. Several observations are derived.

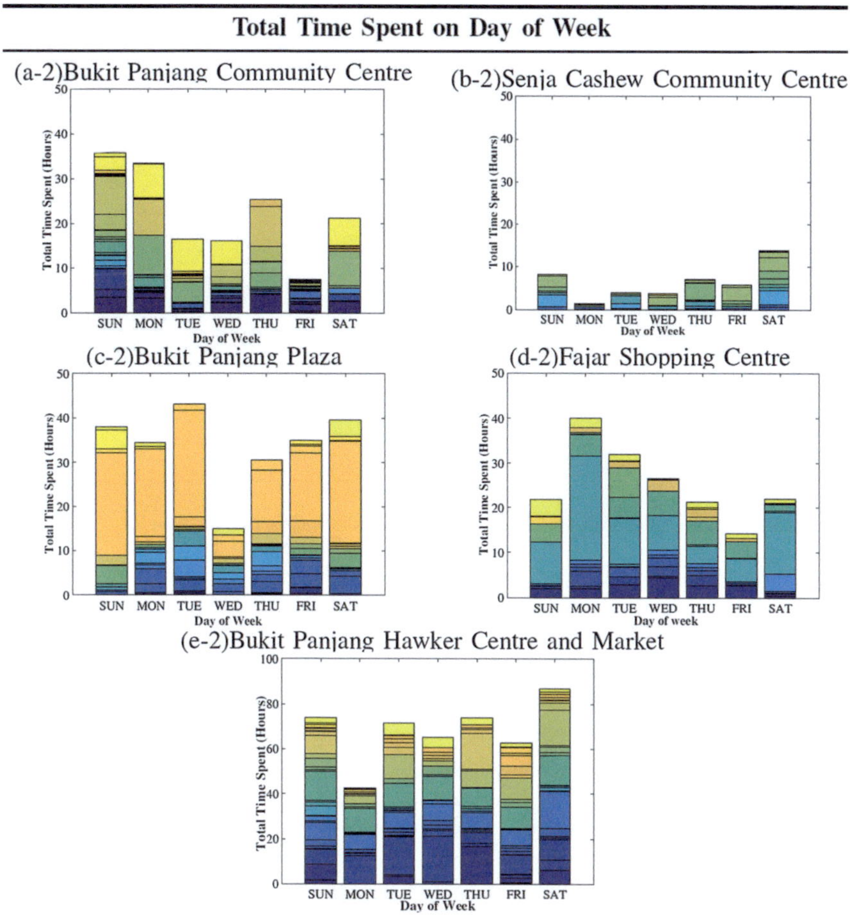

Fig. 8.11 Total time spent on POI during the week

First, we observe the non-working group's travel. The non-working female participants seem more active than the non-working male older participants. The map shows that non-working female older participants have a tendency to go to the city centre area for shopping purposes. These are a few common spots for the female participants: Bugis, Marina Square, Chinatown, Orchard Road, among others. Based on previous data, we can draw a conclusion that the majority of the POI in the central area largely consists of shopping malls, and the like. In contrast to female participants, non-working male participants seem to travel less and mostly spend their time in Bukit Panjang's nearby POI. However, we noticed that participants, both male and female, were willing to travel more than 10 km for religious purposes.

In contrast, working male participants are more active than their female counterparts in terms of travelling around Singapore when they are working. This may

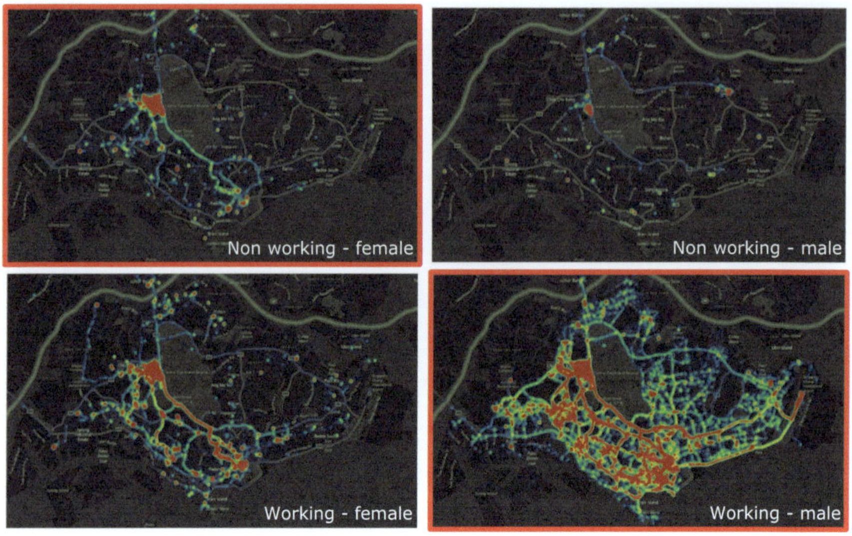

Fig. 8.12 Heat maps based on gender and employment

correlate to the nature and location of work. We have yet to investigate such occurrence and further research is needed to confirm the correlation. As for the female participants, the heat maps indicate a concentration towards the central and southern part of Singapore whereas the male participants' travels are spread more evenly.

A third observation is that the non-working female and working male participants are the most active/travelled groups compared to the working female and non-working male participants. It is hard to determine what is the main factor that separates the behaviour of these male and female participants according to their employment status. An area of future research is to correlate their age and their hobby to determine the actual factor of separation.

8.4 Conclusion

This Chapter introduces an approach using smartphone development application and big data to identify the Regions and Points of Interest where studied older people in Singapore prefer to visit and spend their time. Notwithstanding the challenges in data collection, the study demonstrates that it is possible to collect detailed real-time information with quantitative measurement using smartphone-based mobile application.

The key findings revealed that older participants tend to visit and spend their time mostly in their neighbourhood (in this study, this is Bukit Panjang Town, Singapore). Visits to places outside of the neighbourhoods occurred for shopping, religious, and healthcare purposes. We observed that non-working female participants were generally more active than non-working male participants, and working male participants were more active than working female participants.

Big data research is a relatively new approach and it is acknowledged that there are a number of limitations to our study. One limitation of the abovementioned mobile app method is that it is difficult to figure out the purpose of the visit from the mobile sensor data. This is especially the case in Singapore where most buildings are high-rise and there are multiple businesses on the same building location. A potential solution is to add a query on location and design an incentive scheme to encourage participants to share more information about the purpose/activity conducted at any particular location. We are presently studying this extension development.

Further refinement of the mobile app is needed, especially in terms of transportation mode recognition (walking, taking bus, taking train, for instance). Future research can be extended to profile each participant by creating a mobility profile. The multi-sensor data collected from smartphones make it possible to implement sensor fusion techniques to determine high-level information about participant behaviour.

Acknowledgements We gratefully acknowledge the following for their generous support and assistance with the mobile apps development, recruitment of older participants and analysis of data: Ku Xuefang (Peking University, China), Zhang Wei (Peking University, China), Zhao Jun (Peking University, China), Guo Jindong (Peking University, China), Zuo Jun (Peking University, China), Cao Qiangqiang (Peking University, China), Guo Shaowei (Peking University, China), Wan Yi (South University of Science and Technology of China, China), Wang Ming (BUPT, China), Emma DeSoto (MIT, USA), Herman Li (MIT, USA), Gao Wenxuan (SUTD Undergraduate), Li Wanying (SUTD Undergraduate), Anushka Pakhale (SUTD Undergraduate), Lim Chang Shi (SUTD Undergraduate), Sitoe Wang Yan Samuel (SUTD Undergraduate), Wang Shuo (SUTD Undergraduate), Thirunavukarasu Bala (SUTD), Dr. Mo Ronghong (SUTD), Dr. Wu Jiyan (SUTD).

References

Arab, F., Malik, Y., & Abdulrazak, B. (2013, September). Evaluation of PhonAge: An adapted smartphone interface for elderly people. In *IFIP Conference on Human-Computer Interaction* (pp. 547–554). Berlin, Heidelberg: Springer.

Capella, L. M., & Greco, A. J. (1987). Information sources of elderly for vacation decisions. *Annals of Tourism Research, 14*(1), 148–151.

Fahim, M., Fatima, I., Lee, S., & Lee, Y. K. (2012, February). Daily life activity tracking application for smart homes using android smartphone. In *Advanced Communication Technology (ICACT), 2012 14th International Conference on Advanced Communication Technology* (pp. 241–245). IEEE.

Ghiani, G., Manca, M., Paternò, F., & Santoro, C. (2013). Towards an Architecture Supporting Social, Adaptive and Persuasive Services for Active Elderly. In *CASFE* (pp. 36–41).

Helal, S., Winkler, B., Lee, C., Kaddoura, Y., Ran, L., Giraldo, C., et al. (2003, March). Enabling location-aware pervasive computing applications for the elderly. In *Pervasive Computing and Communications, 2003. (PerCom 2003). Proceedings of the First IEEE International Conference on Pervasive Computing and Communications* (pp. 531–536). IEEE.

Hiyama, A., Nagai, Y., Hirose, M., Kobayashi, M., & Takagi, H. (2013). Question first: Passive interaction model for gathering experience and knowledge from the elderly. In *Pervasive Computing and Communications Workshops (PERCOM Workshops), 2013 IEEE International Conference on Pervasive Computing and Communications* (pp. 151–156). IEEE.

Kim, S. C., Jeong, Y. S., & Park, S. O. (2013). RFID-based indoor location tracking to ensure the safety of the elderly in smart home environments. *Personal and Ubiquitous Computing, 17*(8), 1699–1707.

Mirri, S., Prandi, C., Salomoni, P., Callegati, F., & Campi, A. (2014, September). On combining crowdsourcing, sensing and open data for an accessible smart city. In *Next Generation Mobile Apps, Services and Technologies (NGMAST), 2014 Eighth International Conference on Next Generation Mobile Apps, Services and Technologies* (pp. 294–299). IEEE.

Pang, N., Zhang, X., Vu, S., & Foo, S. (2014). Smartphone use by older adults in Singapore. *Gerontechnology, 13*(2), 270.

Viswanath, S. K., Yuen, C., Ku, X., & Liu, X. (2014, October). Smart tourist-passive mobility tracking through mobile application. In *International Internet of Things Summit* (pp. 183–191). Springer International Publishing.

Yassin, A., Nasser, Y., Awad, M., Al-Dubai, A., Liu, R., Yuen, C., et al. (2016). Recent advances in indoor localization: A survey on theoretical approaches and applications. *IEEE Communications Surveys & Tutorials, 19*(2), 1327–1346.

Chapter 9
Exploring Perceived Neighbourhood Social Capital

Anna Lane

Abstract A survey of older people's perceptions of their social capital was implemented in Bukit Panjang town, Singapore. Complementing the on-site observation study (Chap. 7), this survey was administered at four different public spaces and elicited information on socio-demographics, neighbourhood connections, social and civic participation, helping behaviour, trust, neighbourhood facilities and problems.

9.1 Introduction

This social capital survey complements the On-site Observation study (Chap. 7) that aimed to explore the various ways in which public outdoor spaces are used by older people. The social capital survey was implemented with users at the same four spaces that were subjected to observation.

According to Putnam et al. (1994, p. 167), social capital is a shared property or public good and can be defined as: "features of social organization, such as trust, norms, and networks, that can improve the efficiency of society by facilitating coordinated actions." This definition is often referred to as the social cohesion definition of social capital and it emphasises themes of social networks, trust in others, norms of reciprocity, and social and civic participation (Nyqvist and Forsman 2015, p. 2).

In Singapore, research on social capital as perceived by older people is limited. Also, local studies of social capital have primarily focused on aspects specific to the individual such as living alone, or family and friend networks, rather than neighbourhood (e.g. Lim and Ng 2010).

Social capital within a neighbourhood setting is considered to be an influential determinant of health and wellbeing for older persons. There are a number of ways in which social capital is thought to shape health and wellbeing (Hawe and Shiell 2000). For example, higher social capital is thought to: lead to a better provision of public infrastructure and services that contribute to health, such as health care services or facilities, through the collective action of citizens; encourage healthy

---

A. Lane (✉)
Singapore University of Technology and Design, Singapore, Singapore
e-mail: anna_lane@sutd.edu.sg

B. Yuen (ed.), *Ageing and the Built Environment in Singapore*,
https://doi.org/10.1007/978-3-319-92444-1_9

behaviours through effective social norms and sanctions; and enhance social supports that buffer the negative effects of major life events on mental and psychological wellbeing. Given the import of social capital to health and as Singapore's population ages rapidly, exploration of older people's perceptions of their neighbourhood-based social capital is warranted.

The aims of this survey were twofold. First, to describe perceptions held by older people living in Bukit Panjang Town about social capital in this community. Second, to explore spatial differences in neighbourhood-based social capital as perceived by people aged 55 years and over and measured at four different public outdoor spaces.[1] It was hypothesised that social capital would be higher among people using those spaces that have been successful in connecting older people within the community.

This Chapter outlines the methodology and elaborates the key findings of this survey.

9.2 Methodology

The survey was designed as a structured, personal interview with the help of a questionnaire and administered at four public outdoor spaces at Bukit Panjang Town:

A. A successful (well used[2]) senior activity centre site: a precinct square;
B. An unsuccessful (not well used) senior activity centre/senior care centre: senior residents' corner at a void deck[3];
C. A site which presently does not have any elderly service facility: a void deck; and
D. A neighbourhood park.[4]

A non-probability, quota sampling method was used to achieve a target sample that approximated the national population profile of Singapore in terms of age, gender and ethnicity. Survey assistants approached older people in each of the four spaces and informed them about the study. People who were aged 55 and older and identified as Singaporean nationals or permanent residents were invited to participate in the survey.

The questionnaire comprised items measuring socio-demographics and social capital. Eight socio-demographics questions were included as they are likely to influence the experience of neighbourhood life—age group, gender, ethnic group, educational attainment, marital status, housing type, living arrangement, and length of

[1]For a detailed description of each site, refer to Chap. 7.

[2]Carmona et al. (2010) have argued that successful spaces are those that support and facilitate people's activities. As this space was well used by older people it was deemed to be 'successful' for the purposes of this survey research.

[3]Void decks can be found on the ground floor of apartment blocks in public housing. They have been purposefully left vacant (open space) and typically used for communal activities.

[4]See Chap. 10.

stay at present address. Drawing upon Putnam's definition, respondents' perceived neighbourhood-based social capital was measured as follows:

- Neighbourhood connections—a question about the number of neighbourhood acquaintances as well as questions about the spaces where neighbourhood friends were first met and spaces friends meet up nowadays;
- Social participation—questions about the frequency of participation in five activity types (leisure activities, sports and exercise, adult education activities, religious group activities, and older person activity groups);
- Civic participation—questions about the frequency of participation in voluntary community work and willingness to contribute (time, money and knowledge) to community projects;
- Helping behaviour—a question about how well neighbours help each other out;
- Trust—question about trust for seven different groups of people in the neighbourhood;
- Neighbourhood facilities—question about the quality of five local services and facilities; and,
- Neighbourhood problems—question about how problematic six common neighbourhood issues are in the neighbourhood.

Each interview was approximately 30 min in duration. All participant responses were voluntary and anonymous. The survey was largely administered in English and a small number were conducted in languages other than English (e.g. Chinese or Malay). Survey assistants attended a mandatory training session prior to implementing the survey in the field. During the interview, survey assistants entered participant responses to questionnaire items into a digital device, which were downloaded directly into excel and imported into IBM SPSS Statistics.

Individual responses were aggregated and between-site differences were explored using non-parametric tests; chi square tests (categorical data) or Kruskal-Wallis H tests (ordinal data[5]), where appropriate. To reduce the increased risk of Type I errors (false positive findings due to the number of tests being run), a Bonferroni correction was applied to all tests (Burton et al. 2011; Tabachnick and Fidell 2013). Accordingly, for tests between site and the social capital measures, a p value of less than $0.05/30 = 0.002$ was required to reject the null hypothesis.

9.3 Findings

The findings are presented in two parts. First, the socio-demographic characteristics of the sample are described. Then descriptive statistics are summarised in tables and/or figures for responses to questions about social capital. Although no statistically significant differences in social capital were found between sites, descriptive statistics

[5] Where ordinal measures included a 'don't know' category, data were treated as missing.

(percentages and medians where appropriate) are provided for the total sample and by site, to complement site-related information presented in Chap. 7.

A total of 305 eligible people consented to participate and subsequently completed the interview questionnaire. Approximately equal numbers of people completed the survey at each site: Site A, $n=84$; Site B, $n=70$; Site C, $n=78$; and Site D, $n=73$.

9.3.1 *Socio-demographics*

Table 9.1 presents the socio-demographic characteristics of the sample. Approximately half of the sample were aged 64 to 74 years (42.6%), were male (54.1%), and had not continued their education past primary school level (47.5%). Most respondents were married (82.3%) and, consistent with the ethnic profile of the general Singapore population, the respondents were disproportionately Chinese (82.6%). Nearly half of the sample has been resident at their present home for more than 20 years ($n=135$, 44.3%). Almost all respondents resided in public ($n=296$, 97.3%) rather than private housing. Of those respondents living in public housing, just over half are housed in 4-room flats ($n=168$, 56.8%).

9.3.2 *Social Capital*

9.3.2.1 Neighbourhood Connections

Respondents were asked how many people they know in their neighbourhood and response options were no one (0), a few (1), 1–3 people (2), 4–5 people (3), 6–10 people, and many, >10 people (4). Most respondents indicated they know many people in their neighbourhood (median = 4). As shown in Table 9.2, a small number of participants indicated they know no one in their neighbourhood.

Participants were asked about the places where they first met those people in their neighbourhoods that they now consider their close friends. People first met their closest neighbourhood friends in a diverse range of places. The most typical places were the common corridor (30.5%), local coffee shop/eating house/hawker centre (28.5%), void deck (24.6%), lift lobby (21.3%), footpath/linkway (15.4%), and neighbourhood market (14.4%). The least common places where only a small number of respondents had met friends were: gym/sports stadium (0.3%), speciality cafes/restaurants (0.3%), library (0.3%), school (1.0%), car park (1.6%), shopping/entertainment centre (2.6%), work (3.0%), playground (8.2%), park/garden (8.2%), and community centre (9.2%).

Regarding site differences, the two most common places where people had met their close friends for the first time were: at Site A, the coffee shop and market (65.5%); at Site B, the coffee shop or void deck (77.4%); at Site C, the common corridor and lift lobby (74.3%); and at Site D, the void deck or common corridor (65.7%). A higher proportion of respondents at Site A (28.6%) compared to respon-

Table 9.1 Socio-demographic characteristics

		Total (N = 305)		Interview site: Site A (n = 84)		Site B (n = 70)		Site C (n = 78)		Site D (n = 73)	
		n	%	n	%	n	%	n	%	n	%
Age group	55–64 years	120	39.3	33	39.3	22	31.4	31	39.7	34	46.6
	65–74 years	130	42.6	37	44.0	33	47.1	34	43.6	26	35.6
	≥75 years	55	18.0	14	16.7	15	21.4	13	16.7	13	17.8
Gender	Female	140	45.9	36	42.9	23	32.9	47	60.3	34	46.6
	Male	165	54.1	48	57.1	47	67.1	31	39.7	39	53.4
Ethnicity	Chinese	252	82.6	68	81.0	59	84.3	69	88.5	56	76.7
	Malay	32	10.5	11	13.1	6	8.6	6	7.7	9	12.3
	Indian	20	6.6	5	6.0	5	7.1	3	3.8	7	9.6
	Others[a]	1	0.3	0	0.0	0	0.0	0	0.0	1	1.4
Marital status	Married	251	82.3	75	89.3	60	85.7	63	80.8	53	72.6
	Not married	54	17.7	9	10.7	10	14.3	15	19.2	20	27.4
Education[b]	Primary or lower	145	47.5	48	57.1	36	51.4	33	42.3	28	38.4
	Secondary or higher	148	48.5	34	40.5	31	44.3	40	51.3	43	58.9
Public housing type	1 and 2-room	5	1.9	3	3.8	2	2.9	0	0.0	0	0.0
	3-room	32	10.8	13	16.3	5	7.2	7	9.0	7	9.6
	4-room	168	56.8	39	48.8	43	62.3	47	60.3	39	53.4

(continued)

Table 9.1 (continued)

		Total (N = 305)		Interview site: Site A (n = 84)		Site B (n = 70)		Site C (n = 78)		Site D (n = 73)	
		n	%	n	%	n	%	n	%	n	%
	Others[c]	91	30.7	25	31.3	19	27.5	24	30.8	23	31.5
	Total	296	100	80	100	69	100	78	100	69	100
Living arrangement	Living alone	14	4.6	2	2.4	4	5.7	1	1.3	7	9.6
	Not living alone	291	95.4	82	97.6	66	94.3	77	98.7	66	90.4
Years at current address	≤10 years	61	20.0	17	20.2	20	28.6	7	9.0	17	23.3
	11–20 years	109	35.7	35	41.7	20	28.6	28	35.9	26	35.6
	>20 years	135	44.3	32	38.1	30	42.9	43	55.1	30	41.1

Note [a]Others include Indonesian (n = 1); [b]Cells do not sum to 100 due to missing data (n = 12); [c]Others include elderly studio apartment (n = 1); maisonette (n = 5); and executive condominium (n = 5); chi square test not performed for ethnicity, marital status, public housing type and living arrangements due to cells with expected count less than 5

Table 9.2 Number of neighbourhood acquaintances

	Site A		Site B		Site C		Site D		Total	
	n	%	n	%	n	%	n	%	n	%
No one	3	3.6	1	1.4	0	0.0	1	1.4	5	1.6
A few, 1–3	3	3.6	5	7.1	4	5.1	15	20.5	27	8.9
4–5	8	9.5	5	7.1	6	7.7	7	9.6	26	8.5
6–10	20	23.8	15	21.4	19	24.4	9	12.3	63	20.7
Many (>10)	50	59.5	43	61.4	49	62.8	41	56.2	183	60.0
Missing	0	0	1	1.4	0	0	0	0	1	0.3
Median		4		4		4		4		4

Table 9.3 Places where close friends meet up

	Site A		Site B		Site C		Site D		Total	
	n	%	n	%	n	%	n	%	n	%
Coffee shop/eating house/hawker centre	41	48.8	31	37.0	30	38.5	14	19.2	116	38.0
Void deck	11	13.1	12	17.1	12	15.5	19	26.0	54	17.7
Market	17	20.2	3	4.3	7	9.0	9	12.3	36	11.8
Park/garden	4	4.8	6	8.6	5	6.4	15	20.5	30	9.8
Community centre	1	1.2	7	10.0	12	15.4	9	12.3	29	9.5
Shopping/entertainment centre	8	9.5	0	0	4	5.1	1	1.4	13	4.3
Specialty café/restaurant	1	1.2	1	1.4	1	1.3	4	5.5	7	2.3
Fitness gym/sports stadium	0	0	1	1.4	1	1.3	2	2.7	4	1.3
Library	0	0	1	1.4	0	0	0	0	1	0.3
Others[a]	7	8.3	8	11.4	7	9.0	3	4.1	25	8.2

Note [a]Other locations included mosque, church, temple, homes of friends

dents at Sites B (10%), C (11.5%) and D (5.5%) indicated the market had been the place where they had met their close friends for the first time.

Respondents were also asked where they would normally meet up with friends for social activities these days. Most respondents meet up for social activities at the coffee shop/eating house/hawker centre (38%) and the void deck (17.7%). The market (11.8%), parks and gardens (9.8%), and community centres (9.5%) were also places were respondents socialise with friends these days. The library (0.3%), fitness gym/sports stadium (1.3%), specialty café/restaurants (2.3%), and shopping/entertainment centre (1.3%) were the least frequented social spaces.

As shown in Table 9.3, this general pattern of results was replicated at each site. The coffee shop was the most common place to meet–up and socialise with friends at Sites A (48.8%), B (37.0%) and C (38.5%). At Site D, however, the most common spaces for socialising were the void deck (26.0%) and the garden/park (20.5%). The void deck was also a popular space for social activity at Sites A (13.1%), B (17.1%) and C (15.5%).

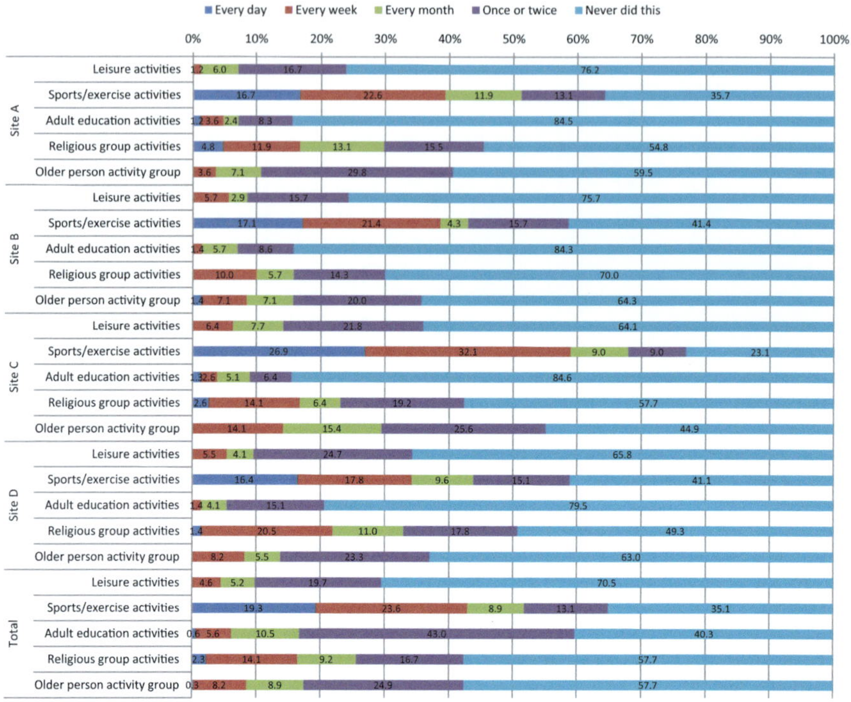

Fig. 9.1 Frequency of social participation by activity

9.3.2.2 Social Participation

All respondents were asked how many times in the past six months they have participated in a range of activities, including: leisure activities, sports and exercise; religious group activities; activities/groups for older people; and, adult education activities. Response options were never (1), once or twice (2), every month (3), every week (4) and every day (5).

About 83% of people had not participated in adult education and 58% had not participated in activities or groups for older people in the past six months. About 42% of respondents had engaged in a religious group activity. On average, people participated in these activities only once or twice during the past six months (median = 1).

Sports and physical exercise was the most frequently cited type of activity in which respondents participated. About 65% of all respondents participated in sports or exercise activities. On average, respondents engaged in sports or exercise every month (median = 3). On the other hand, 35.1% of respondents had never participated in sport or exercise during the past six months.

Descriptive statistics for frequency of participation in social activities are illustrated by site in Fig. 9.1.

Table 9.4 Main reasons for not participating in social activities

	Site A		Site B		Site C		Site D		Total	
	n	%	n	%	n	%	n	%	n	%
Not interested	28	33.3	40	57.1	32	41.0	47	64.4	147	48.2
No time	34	40.5	28	40.0	31	39.7	24	32.9	117	38.4
No information	13	15.5	7	10.0	7	9.0	10	13.7	37	12.1
No company	17	20.2	7	10.0	1	1.3	4	5.5	29	9.5
No transport	4	4.8	0	0	2	2.6	2	2.7	8	2.6
Don't like the place	0	0	3	4.3	1	1.3	1	1.4	5	1.6
Don't like the facilities	3	3.6	1	1.4	1	1.3	0	0	5	1.6
Don't like the people there	1	1.2	1	1.4	0	0	0	0	2	0.7
Others[a]	8	9.5	13	18.6	11	14.1	11	15.1	43	14.1

Note [a]Other reasons included: too busy with work; caring responsibilities; personal preference; activities are mixed gender; age; medical conditions

A lack of interest and a lack of time were the main reasons that respondents gave for not participating in the listed activities, as shown in Table 9.4. About 48.2% of all respondents indicated that a lack of interest was their main reason for non-participation and 38.4% suggested they had no time for such activities. A lack of information about activities also influenced activity engagement for some respondents (12.1%).

It is noteworthy that 9.5% of respondents indicated that not having someone to accompany them to an activity was a barrier to participation. At Site A, 'no company' was the third most common reason for non-participation. About 20.2% at Site A, 10% at Site B, 1.3% at Site C, and 5.5% at Site D cited no company as a reason they did not engage in any, or all, of the listed activity types.

9.3.2.3 Civic Participation

Respondents were asked how many times in the past six months they have participated in voluntary community work and response options were never (1), once or twice (2), every month (3), every week (4) and every day (5). As shown in Table 9.5, 64.9% respondents had not participated in voluntary community work over the past six months.

Respondents were asked whether they would contribute time, money or knowledge to a community project that would not directly benefit them but would benefit many others in their neighbourhood. Descriptive statistics for this item are presented by site in Table 9.6 and show that almost half of all respondents would contribute their time ($n=152$, 49.8%) or knowledge ($n=156$, 48.9%) to a community project in circumstances where the expected benefits are likely to be experienced at the neighbourhood rather than at the individual level. Fewer respondents were prepared to contribute money to a community project ($n=101$, 33.1%).

Table 9.5 Frequency of participation in voluntary community work

	Site A		Site B		Site C		Site D		Total	
	n	%	n	%	n	%	n	%	n	%
Never	57	67.9	47	67.1	46	59.0	48	65.8	198	64.9
Once or twice	16	19.0	11	15.7	11	14.1	18	24.7	56	18.4
Every month	7	8.3	8	11.4	10	12.8	4	5.5	29	9.5
Every week	0	0	3	4.3	7	9.0	3	4.1	13	4.3
Every day	4	4.8	1	1.4	4	5.1	0	0	9	3.0
Median		1		1		1		1		1

Table 9.6 Contribution to community project

		Site A		Site B		Site C		Site D	
		n	%	n	%	n	%	n	%
Time	No	44	52.4	36	51.4	41	52.6	32	43.8
	Yes	40	47.6	34	48.6	37	47.4	41	56.2
Money	No	56	66.7	51	72.9	54	69.2	43	58.9
	Yes	28	33.3	19	27.1	24	30.8	30	41.1
Knowledge	No	43	51.2	34	48.6	40	51.3	32	43.8
	Yes	41	48.8	36	51.4	38	48.7	41	56.2

Table 9.7 Impact in making neighbourhood a better place

	Site A		Site B		Site C		Site D		Total	
	n	%	n	%	n	%	n	%	n	%
No impact	16	19.0	23	32.9	26	33.3	22	30.1	87	28.5
Small impact	33	39.3	29	41.4	25	32.1	16	21.9	103	33.8
Moderate impact	29	34.5	14	20.0	23	29.5	28	38.4	94	30.8
Big impact	6	7.1	4	5.7	4	5.1	7	9.6	21	6.9
Median		2		2		2		2		2

Participants were asked how much impact ‘people like you’ can have in making your neighbourhood a better place in which to live and response options were no impact (1), small impact (2), moderate impact (3) and big impact (4). As shown in Table 9.7, only 7% of respondents believed people like them can have a big impact on their neighbourhood. Many (68%) respondents suggested that they can have a small or moderate impact on the liveability of their neighbourhood whilst 26% thought they can have no impact.

9.3.2.4 Helping Behaviour

Respondents were asked about how well people in their neighbourhood help each other out these days and response options were: never (1), rarely (2), sometimes, (3), most of the time (4) and always (5). Most respondents felt that people in their

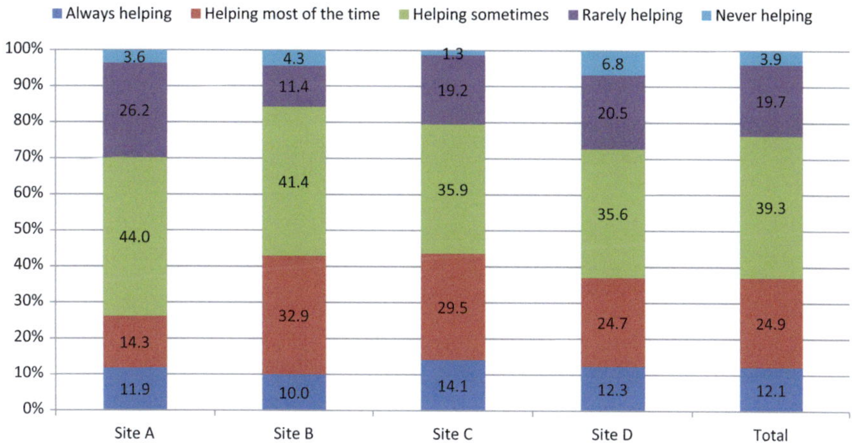

Fig. 9.2 How well people help each other out

neighbourhood were good at helping each other some or most of the time (64%). About 12% of all respondents suggested that people in their neighbourhood were always helping each other out, compared to 4% who thought people were never helping. This general pattern of responses was mirrored at the sites, as illustrated in Fig. 9.2.

9.3.2.5 Trust

Respondents were asked to rate their trust for different groups of people in their neighbourhood and response options were: don't trust them, trust them little, trust them sometimes, trust them very much and don't know. Descriptive statistics for this question are provided in Table 9.8.

Most respondents felt the neighbourhood was a place where many groups of people can be trusted. The extent to which different groups can be trust varied somewhat, with a tendency for more people to have trust for nurses, doctors and health workers, with about 36% of all respondents trusting this group very much. Approximately 28% of respondents trust the local constituency/town council/grassroots organisations very much, 20% trust their neighbours very much, 18% trust the police very much, 10% trust storekeepers very much, and 8% trust people of other races/nationalities very much.

Small proportions of respondents said they did not trust people who live near them (5%), the local police (9%), people who work in the stores where they shop (11%), people of other races/nationalities (12%), local constituency/town council/grassroots organisations (11%), and nurses, doctors and health workers (3%). Most respondents did not trust strangers (68%). This general pattern of responses was mirrored at the sites as illustrated in Fig. 9.3.

Table 9.8 Trust for different groups of people in neighbourhood

	People in neighbourhood		Local police		Local store keepers		People of other races/nationalities		Local groups		Health workers		Strangers	
	n	%	n	%	n	%	n	%	n	%	n	%	n	%
Don't trust them	14	4.6	26	8.5	32	10.5	35	11.5	34	11.1	8	2.6	208	68.2
Trust them little	81	26.6	71	23.3	95	31.1	78	25.6	56	18.4	29	9.5	21	6.9
Trust them sometimes	109	35.7	84	27.5	104	34.1	89	29.2	75	24.6	122	40.0	20	6.6
Trust them very much	61	20.0	85	27.9	31	10.2	25	8.2	56	18.4	110	36.1	4	1.3
Don't know	40	13.1	39	12.8	43	14.1	78	25.6	84	27.5	36	11.8	52	17.0

Note Response options are: People in neighbourhood = People in the neighbourhood (they live near you, on same floor/block/street/estate); Local police = The police in your local community; Local store keepers = People who work in the stores where you shop (food, groceries, etc.); People of other races/nationalities = People of other races/nationalities different to you; Local groups = Your local constituency, town council, or grassroots organisations; Health workers = Nurses, doctors and health workers; Strangers = Strangers

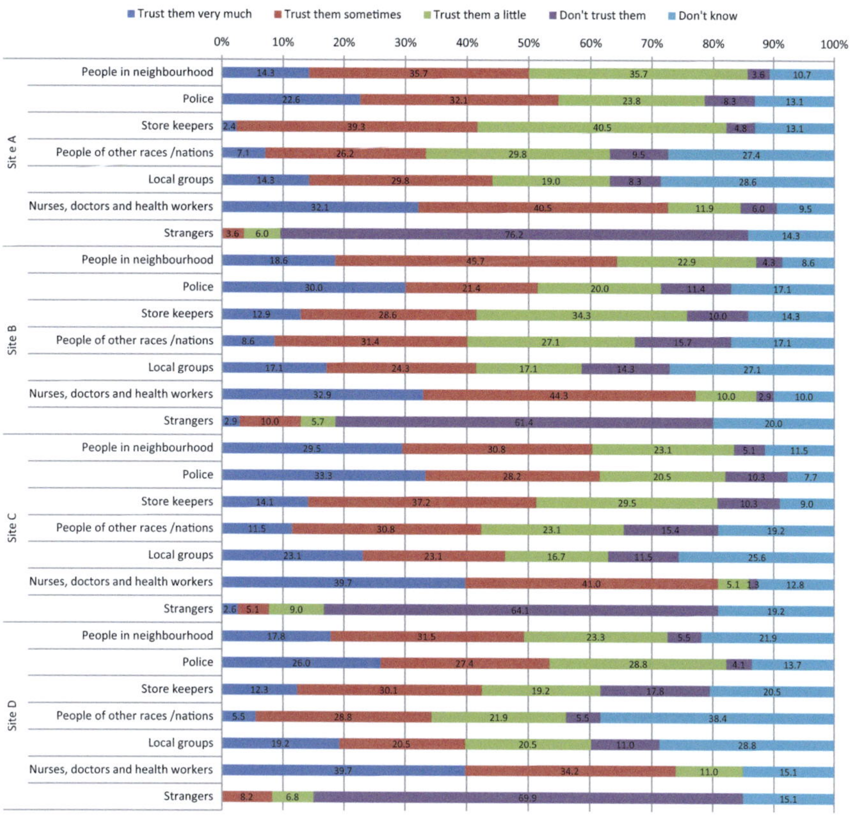

Fig. 9.3 Trust for groups of people in your neighbourhood by site

9.3.2.6 Neighbourhood Facilities

All respondents were asked about their experiences of the following facilities and services in their neighbourhood: social/leisure facilities for people like you, for young children (<12 years old), and for teenagers (age 13–17 years); local health services; and, local public transport. Response options were very poor, poor, average, good, very good and don't know and descriptive statistics to these questions are provided in Table 9.9.

Experiences relating to local public facilities were generally highly rated, particularly local transport services, which 73% regarded as good or very good, and health services, which 56% rated as good or very good. More people regarded facilities for children to be good or very good, compared to facilities for people of other ages. About 48% rated facilities for children as good or very good, whilst 40% rated facilities for people like them as good or very good and 32% rated facilities for teenagers as good or very good. This general pattern of responses was mirrored at the sites, as illustrated in Fig. 9.4.

Table 9.9 Quality of local facilities

	Total									
	Social/leisure facilities for people like yourself		Facilities for young children up to age of 12		Facilities for teenagers (aged 13-17)		Local health services		Local public transport	
	n	%	n	%	n	%	n	%	n	%
Very poor	6	2.0	4	1.3	9	3.0	4	1.3	6	2.0
Poor	25	8.2	23	7.5	33	10.8	21	6.9	12	3.9
Average	128	42.0	68	22.3	85	27.9	85	27.9	59	19.3
Good	87	28.5	109	35.7	83	27.2	118	38.7	136	44.6
Very good	34	11.1	36	11.8	14	4.6	52	17.0	86	28.2
Don't know/no experience	25	8.2	65	21.3	81	26.6	25	8.2	6	2.0

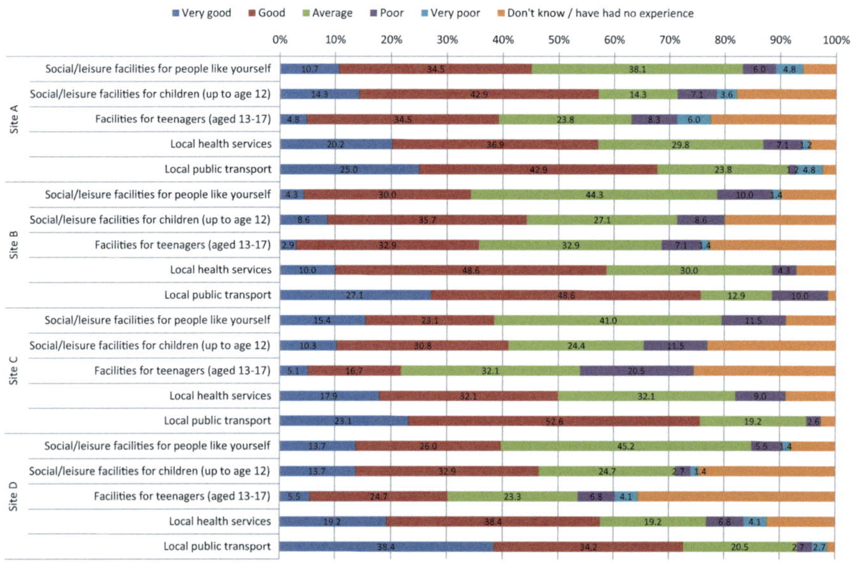

Fig. 9.4 Perceptions of local facilities/services

9.3.2.7 Neighbourhood Problems

Respondents were asked to consider how problematic a number of common neighbourhood issues are in their own neighbourhood, including: road traffic volume and speed, rubbish and litter in the streets, mess from dogs and cats, graffiti and vandalism, noise, and alcohol and drugs. Response options were: not a problem, minor problem, fairly big problem, very big problem, and don't know. As illustrated in Fig. 9.5, most respondents indicated that the listed issues were not a problem in the

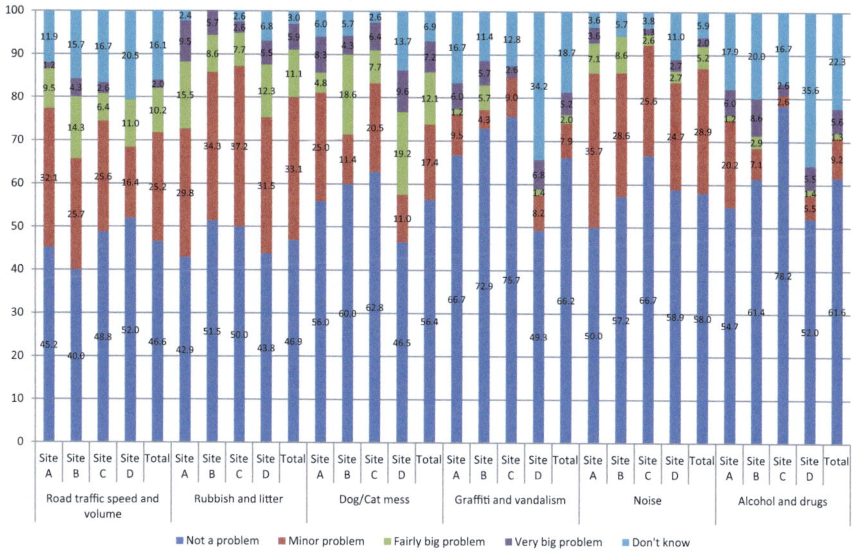

Fig. 9.5 Neighbourhood problems

neighbourhood: traffic speed and volume, 47%; rubbish and litter, 47%; cat and dog mess, 56%; graffiti and vandalism, 56%; noise, 58%; and alcohol and drugs, 62%. Few felt any issue was a very or fairly big problem: traffic speed and volume, 12%; rubbish and litter, 17%; cat and dog mess, 19%; graffiti and vandalism, 7%; noise, 7%; and alcohol and drugs, 7%.

9.4 Conclusion

This survey aimed to reveal older people's perceptions about their neighbourhood-based social capital and to explore whether perceptions regarding social capital were higher among users of spaces that have purportedly been successful in bringing older people in the community together. Towards this goal, 305 people aged 55 years and over were asked a series of questions about their perceptions of their social connections, social and civic participation, helping behaviour, trust, and neighbourhood facilities and problems.

Contrary to expectations, findings indicated that perceptions of older people regarding their social capital were approximately consistent across each of the four sites. This suggests that the sites at which people were interviewed were not associated with social capital as measured in this study. Other contextual factors not measured in this survey that shape older people's perceptions about their neighbourhood and exert their influence over a wider spatial scale may account for the null findings.

Our findings revealed that most people have many acquaintances in their neighbourhood. Nevertheless, some older people felt that they had few or no acquaintances. These people may be at risk of social isolation and poorer health than their more socially connected age counterparts. As the absence of social relationships is a major cause of feelings of loneliness and experience of depression, the findings provide cursory support for ageing in place initiatives that target mental health.

Informal places such as common corridors, void decks, lift lobbies and footpaths/linkways around buildings, were found to be places where many older people had first met their closest neighbourhood friends and also places where they meet up their friends these days. Informal spaces such as these are located immediately outside the home and are spaces where people are most likely to encounter others in their neighbourhood. Therefore, they are critical spaces where social networks are not only formed but also cultivated and may represent points of intervention for urban planners and designers wishing to influence social health for older residents.

Sports and physical exercise was a key form of activity in which older people participated on a weekly or a daily basis. Nevertheless, our analysis also revealed that some people do not currently engage in physical activity and may lead sedentary lifestyles. Given the well documented physical and mental health benefits that accrue from regular exercise and physical activity in daily life (including walking) (World Health Organisation 1998), the findings suggest there is a need in Bukit Panjang Town for further strategies to increase the number of older residents who exercise daily. Such strategies may include improving the walkability of neighbourhoods, establishing walking groups and/or companions, and reducing the fear that many older people have about falling whilst outdoors by making age-friendly modifications to footpaths (e.g. clearly marked grade changes, pedestrian only footpaths, etc.).

Adult education activities, leisure activities, and activity groups for older people were not common activities in which older adults engaged. A lack of interest and a lack of time were the two main reasons why respondents chose not to participate in these activities. A lack of interest suggests that greater consultation must be undertaken with older people to ascertain the types of activities that they enjoy and would like to have provided in the neighbourhood. A lack of time to participate in leisure-time activities might be attributed to work-related commitments. Many people, particularly those actively engaged in paid work, may spend a substantial portion of their day at formal work places outside their neighbourhoods and have limited time to participate in organised, leisure-time activities in their neighbourhoods.

A lack of information about activities or not having someone to accompany to social activities was also identified by some respondents as barriers to participation. Other reasons for non-participation included health and medical conditions, fear of falling, caring commitments (grandchildren, ageing parents), and old age. It was not possible in this cross-sectional study to determine if the presence of few or no close social ties inhibits opportunities to be socially active or whether having no company is the consequence of a lack of social activity engagement. Investments in longitudinal research are needed to establish the temporal association between social ties and social activity.

Our findings also showed that there are some people who through personal preference simply do not wish to join in with organised activities. This highlights the role of psychological factors in determining people's behaviours and explaining individual differences in experiences of neighbourhood life and social capital.

We found low levels of civic participation among respondents. Only a third of people surveyed had ever done voluntary community work and about half expected that they would be likely to contribute time and knowledge to a community project. Only about one third would contribute money to a community project in circumstances where the expected benefits were likely to be felt at the neighbourhood rather than at the individual level. Nevertheless, people surveyed felt that people in their neighbourhood were good at helping each other some of the time. Thus, low rates of community work may simply reflect a preference for more informal modes of helping (Chong et al. 2015).

Additionally, most people surveyed suggested that they can make a small to moderate impact on the liveability of their neighbourhood. This suggests that older people may not be sufficiently empowered to steer and lead change in their neighbourhoods. This reasoning may also account for the finding that only a small number of people participated in community work in their neighbourhoods.

Crucially, most people felt their neighbourhood was a place where many groups of people can be trusted. The extent to which different groups could be trust varied somewhat, with a tendency for more people to have highest levels of trust for nurses, doctors and health workers and the police. Evidence suggests that trusting relationships across power or authority gradients (referred to as linking capital), such as relationships with doctors, police, and political institutions, play a critical role in building community resilience and health (Szreter and Woolcock 2004). The role of linking social capital for older persons health represents an area warranting further investigation.

Our findings showed that older adults regarded health and transport services available to them in Bukit Panjang as good. As older people tend to rely increasingly on public transport to access social activities as they withdraw from driving personal automobiles, the quality of the transport system is germane. However, respondents considered social facilities and services for 'people like them' to be relatively poor in quality. By comparison, they considered the quality of services available to children to be of better quality. This highlights the need for a shift in focus among urban planners and designers towards creating neighbourhoods for all ages.

We found that most older people experienced very few problematic, neighbourhood issues. Many regarded their neighbourhood as relatively free from major issues such as drugs and alcohol, and graffiti and vandalism, which may plague other areas or other densely populated cities. Among a comprehensive list of potential issues that might diminish older people's experiences of their neighbourhoods, rubbish and litter left lying around in the streets as well as speed and volume of road traffic were highlighted as a issues as potentially problematic.

Acknowledgements This project was led by Principal Investigator, Dr. Belinda Yuen. Colleagues, Parvathi Nair, Penny Kong and Adithi Moogoor assisted with designing the study protocol and instruments, data collection and/or analysis. Dr. Anna Lane contributed to writing the study report. We acknowledge the survey respondents who generously gave their time and shared their experience with the research team.

References

Burton, E. J., Mitchell, L., & Stride, C. B. (2011). Good places for ageing in place: Development of objective built environment measures for investigating links with older people's wellbeing. *BMC Public Health, 11*(1), 839.

Carmona, M., Tiesdell, S., Heath, T., & Oc, T. (2010). *Public places public spaces: The dimensions of urban design*. UK: Architectural Press.

Chong, K. H., Yow, W. Q., Loo, D., & Patrycia, F. (2015). Psychosocial well-being of the elderly and their perception of matured estate in Singapore. *Journal of Housing for the Elderly, 29*(3), 259–297.

Hawe, P., & Shiell, A. (2000). Social capital and health promotion: A review. *Social Science and Medicine, 51*(6), 871–885.

Lim, L. L., & Ng, T. P. (2010). Living alone, lack of a confidant and psychological well-being of elderly women in Singapore: The mediating role of loneliness. *Asia-Pacific Psychiatry, 2,* 33–40.

Nyqvist, F., & Forsman, A. K. (2015). Healthy ageing: Focus on social capital. In F. Nyqvist & A. K. Forsman (Eds.), *Social capital as a health resource in later life: The relevance of context* (Vol. 11, p. 2). Springer, p. 2.

Putnam, R. D., Leonardi, R., & Nanetti, R. Y. (1994). *Making democracy work: Civic traditions in modern Italy*. Princeton University Press.

Szreter, S., & Woolcock, M. (2004). Health by association? Social capital, social theory, and the political economy of public health. *International Journal of Epidemiology, 33*(4), 650–667.

Tabachnick, B. G., & Fidell, L. S. (2013). *Using multivariate statistics* (6th ed.). Pearson.

World Health Organisation. (1998). The role of physical activity in healthy ageing. http://www.who.int/ageing/publications/physical_activity/en/. Accessed January 3, 2017.

Chapter 10
Designing with Older Residents

Belinda Yuen and Sue Wei Cheong

Abstract The aim of conducting a series of community design workshops was to understand the unique needs and desires of the community, gather older residents' ideas, suggestions and solutions for their community's public space and urban problems so as to aid the design and placemaking of neighbourhood public space for an ageing community. The eight workshops were conducted in Bukit Panjang Town, specifically on Bukit Panjang Neighbourhood 2 Park. The themes included: Needs, Experiences and Vision; Accessibility and Connectivity; Park Identity and Activity; Safety and Security; Walk and Talk Urban Audit Session; Park Spirit and Sense of Place; and Community and Ownership. While participants expressed a general satisfaction with the park, they had concerns mostly with tidiness of the park, maintenance of greenery and managing the presence of undesirables.

10.1 Introduction

The Community Design Workshop is one of nine survey instruments employed in the study, Understanding the Changing Needs of Singapore's Older Adults. The workshops were conducted with the aim of listening and understanding the unique needs and desires of the community, gathering residents' ideas, suggestions and solutions for their community public space and urban problems in order to aid the design and placemaking of neighbourhood public space for an ageing community. The workshops were focused on a local community space: the Bukit Panjang Neighbourhood 2 (BPN2) Park in Bukit Panjang Town.

This Chapter outlines the study methodology, key findings and recommendations from the Community Design Workshops.

B. Yuen (✉) · S. W. Cheong
Singapore University of Technology and Design, Singapore, Singapore
e-mail: belinda_yuen@sutd.edu.sg

S. W. Cheong
e-mail: suewcheong@gmail.com

B. Yuen (ed.), *Ageing and the Built Environment in Singapore*,
https://doi.org/10.1007/978-3-319-92444-1_10

10.2 Methodology

The workshop discussion model is designed to allow researchers and residents to work creatively together to develop planning and design ideas. It has the possibility to promote and increase community knowledge, engagement and commitment, providing an opportunity for residents including older residents to interact and participate in conceptualising the vision, programme, activity and design of a public space in their community. In this method, the design is research and its results are co-interpreted by the researchers and the participants working together.

The first two sessions were conducted on a monthly basis while the remaining six sessions were conducted twice weekly over a subsequent three-week period (June 2015) to build on the momentum of participation interest.

The workshops were conducted at the Bukit Panjang Zone 4 (BPZ4) Residents' Committee (RC) Centre, Block 213 Petir Road. The RC Centre was located at the north-western periphery of the study community space—Bukit Panjang Neighbourhood 2 (BPN2) Park. With the help of BPZ4 RC, residents living around the study site were invited to join the workshops. The workshops covered a range of discussion topics concerning physical and programmatic factors such as park design, space, amenities and activities as well as other social and less tangible considerations such as community and individual experiences, concerns and aspirations. The workshop themes covered:

Workshop 1: Needs, Experiences and Vision;

Workshop 2: Accessibility and Connectivity;

Workshop 3: Park Identity and Activity Part 1;

Workshop 4: Park Identity and Activity Part 2;

Workshop 5: Safety and Security;

Workshop 6: Walk and Talk Urban Audit Session;

Workshop 7: Park Spirit and Sense of Place;

Workshop 8: Community and Ownership.

The activities conducted during the workshops included presentations, breakout group discussions, team-building games, design charrettes and individual presentations. Each session implemented two or more of these activities and generated quantitative and qualitative data, which were transcribed, analysed and distilled into eight different workshop summaries. The aim was to draw out commonalities, key findings and finally, a set of recommendations from these reports to inform the refurbishment and revitalisation of BPN2 Park.

There were a total of 30 unique participants who attended the eight workshops. An average of 20 participants attended each workshop. The participants included male and female, aged 21+ to 80+ (90% were aged 50 and older) of Chinese and Malay ethnicities living in the BPZ4 neighbourhood. All participants were Singaporeans.

Study Site BPN2 is a hill park and one of four larger neighbourhood parks in Bukit Panjang Town—the others being Bukit Panjang Park (with Pang Sua Pond), Bukit Panjang Neighbourhood Park 5 and Zheng Hua Park. BPN2 Park is connected to these other parks by the Pang Sua and Bukit Panjang Park Connectors (PCN), which aim to link new and existing green spaces in Singapore's North-West District. BPN2 Park is connected to the Central Water Catchment and Bukit Timah Nature Reserve through a walking trail.

BPN2 begins its gain in gradient from its north-western and western periphery and rises to a peak in its south-end, facing Petir Road. Vegetation also increases with increase in elevation; the bottom of the hill is grassy with little or no trees, at the halfway point, there are more trees and the top of the hill culminates as a forest. Trees are largely mature and planted at the inception of the Park about 30 years ago when the surrounding HDB[1] blocks were constructed.

BPN2 Park can be divided into three main areas—the Exercise Area, Forested Area/Trails, and Community Gardens (Fig. 10.1).

The Exercise Area is about two-thirds way up the hill and consists of a concrete-finish hard court where fitness classes (e.g. qigong, tai chi) are conducted, an IPPT[2] Fitness Area and Seniors Exercise Equipment Area. There is seating provision at either short ends of the hard court. At one end, the seats are located under a trellis (open to the sky) shelter, and at the other end, there are two small pavilions and a portable toilet. West of the hard court are two more pavilions holding three benches each.

Further up the hill, the Forest Trails begin and lead into a thickly vegetated forest with tall, mature trees and dense undergrowth. There are several trails in this area, providing different options for moving through the forest.

Then, there are two community gardens, one located at the northern foothill of the Park and the other at the eastern corner of the Park. Opposite the latter garden resides a private gardening shop that sells gardening provision, conducts classes and promotes new methods of planting like hydroponics.

Beyond the Park, at its north-western foothill next to BPZ4 RC at Block 213, is a Recreation Area that comprises of a sheltered gathering space, children's playground, elderly exercise equipment area, basketball court and BBQ pits. In the east is a narrow strip of parkland that provides a connection to Zheng Hua Park and Dairy Farm Nature Park further south. Finally, immediately beyond the south of the forest and community garden are the new condominium developments.

10.3 Findings

Recurrent themes from the eight workshop sessions will be discussed under seven issues:

---

[1]HDB is the abbreviation for Housing and Development Board, the public housing agency in Singapore. Public housing is popularly referred to as HDB housing/block/flat in Singapore.

[2]Refers to the Individual Physical Proficiency Test (IPPT) set by the Singapore Armed Forces, Singapore Police Force and Singapore Civil Defence Force to test the basic components of physical fitness and motor skills of their service people.

Fig. 10.1 Map of BPN2 Park

- Vision
- Favourite Spaces
- Accessibility and Connectivity
- Maintenance and Management
- Safety and Security
- Activity and Amenity
- Community Spirit and Ownership.

Participants' suggestions and recommendations are included for the recurrent themes to illustrate some of the emerging co-design ideas.

10.3.1 Vision

Aspirations for the Park Participants were generally happy with the Park. Most participants' ideal neighbourhood park was a green space, with trees and beautiful flowers, a place to walk, to do exercise and activities (e.g. meet people, talk to friends, growing vegetable). Several participants shared that they had met firm friends through the Park, continuing to meet for regular activities and outings to the Park and elsewhere. BPN2 Park presented a welcoming neighbourhood park because of its unique landscape, natural terrain and mature trees. Participants felt themselves blessed to have this unique place as their neighbourhood park when compared to the ubiquitous neighbourhood park provision in many other HDB estates/towns.

They generally did not want too much to be done to the Park, just for the Park to be kept tidy, clean and regularly maintained. Participants highlighted several, sometimes crosscutting issues of concern and suggestions for change. Some hoped for the planting of flowers (introduce some colours, not just green even though green is beautiful), and to have more chairs and shelters for older people to rest. Others talked about accessibility, safety and maintenance (see subsequent discussion on these recurrent themes).

When asked to summarise their vision for BPN2 Park in one word, the most common wish was for the Park to remain beautiful (Table 10.1).

Table 10.1 Vision for BPN2 Park in one word

Word	No. of mentions
Beautiful	9
Clean	3
Flowers	1
Nice	1
Good	1
Kind	1

N = 18

10.3.2 Favourite Spaces

Great Places in the Park When asked to list and then vote for their favourite places in the Park, 'nature' emerged on top with 23 votes, followed by 'exercise area' with 18 votes. Participants identified many favourite spots in the Park. The top 14 places are listed in Table 10.2. They include nature in the Park, exercise area, particular trees they had remembered or discovered from the past, commonly used access-points, footpaths to and around the Park, shelters and spaces where they frequently spent time. The activity spaces in the Park, e.g. community gardens, hard court and exercise area are among the liveliest places in the Park while the forested area on top of the Park with its dense undergrowth and overgrown vegetation was perceived as creepy and unsafe, to be avoided at all times.

10.3.3 Accessibility and Connectivity

Entry Nearly all participants lived within a 2.5–5 min walking distance from the Park and entered the Park from the north or north-east of the Park, with the majority from Block 210 and a smaller number from Block 213 and Block 203 (Fig. 10.2). Entering from Block 210 was popular as there were two options for entry; both were the shorter and easier routes into the Park. These were: a service driveway (suitable for cyclists and wheelchair users) and a sloped footpath to the hard court and exercise area. The other two entries were at Block 213, a gradually sloped path but longer

Table 10.2 Participants' favourite places in the park

Places	No. of votes
Nature (including grassy sloped area)	23
Exercise area	18
Large trees	14
Stairs to park at Blk 201	13
Pavilion	11
Hard court	10
Community garden	9
Stairs to park at Block 203	9
Path with 'footprint'—designated footpath	9
Feature rock-lined footpaths	8
Mangosteen trees	8
To breath fresh air	8
Toilet	7
Large tree with 'lips'	6

N = 22

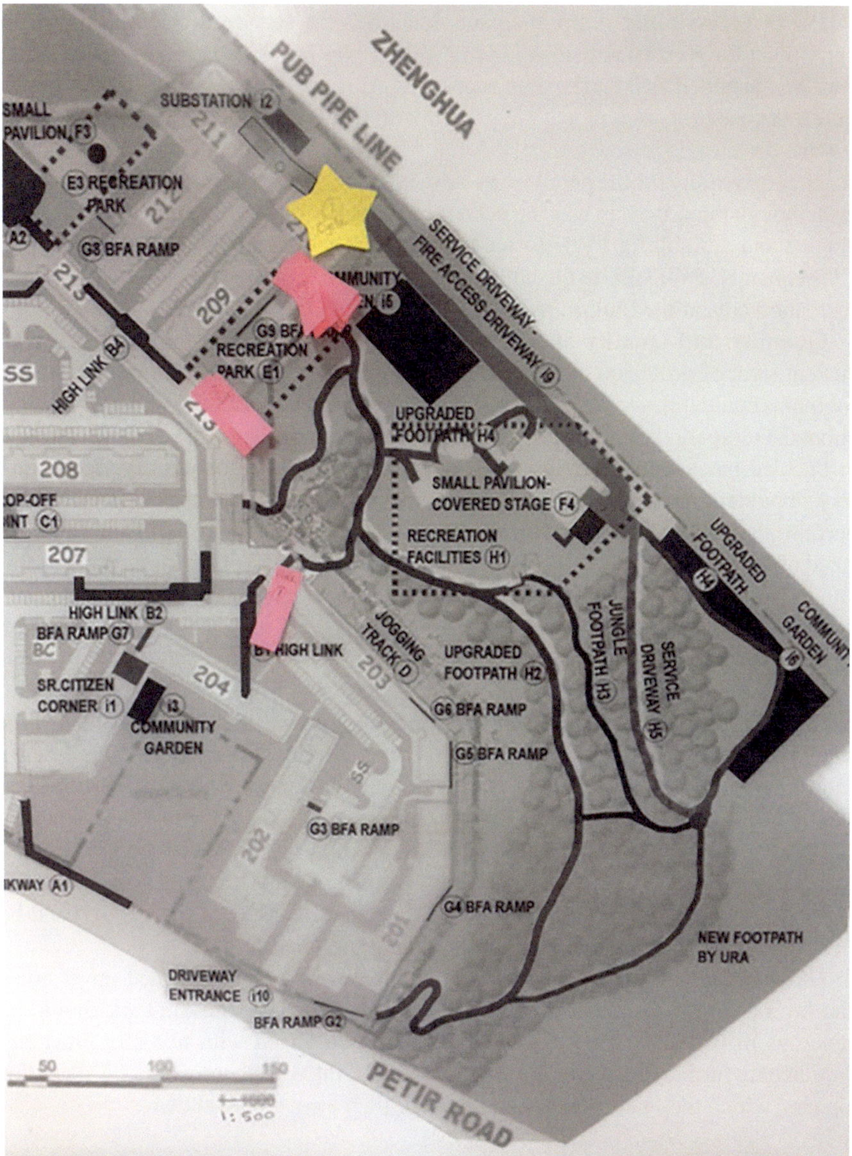

Fig. 10.2 Participants entry to the park. *Notes* Yellow Star: Enter by cycling; Pink Tags: Enter by walking

than that from Block 210, and at Block 203, a flight of steps with shallow risers. The access point at Block 201 was described as too steep; not preferred/used by participants, only used by one participant during a morning walk around the Park.

PWD Accessibility All participants identified the service driveway as the best entry point for persons with disability (PWD). It was observed that the service driveway was intended as a barrier-free access (BFA) route into the Park after upgrading in recent years; the jogging path upgraded with BFA ramps lead to the bottom of the service driveway. However, this BFA route did not really meet the requirements of comfort and safety for the people they were intended for. Participants pointed out that even though the driveway was the easiest path for PWD, the driveway was still steep at parts and difficult for PWD to access without a helper or a motorised wheelchair. Consequently, PWD did not usually go up to the Park but preferred to gather at the recreation area at the foothill near Block 209.

Quantity and Quality of Footpaths Participants were generally content with current level of provision and did not wish to see more footpaths introduced. More footpaths could destroy the park environment. However, participants raised concerns about the footpaths being slippery due to mosses and lichen and uneven due to broken or buckled pavements, potholes or level changes. Some suggested that the feature rocks/boulders along the old footpaths (developed at Park inception) be removed because they believed that the rocks posed a danger when people slip and fall—instead of falling on to grass, people might hurt themselves badly falling on the rocks. Other participants were in favour of retaining these feature rocks as they are because the rocks would serve as seating for resting joggers and walkers. These paths were reported to be too narrow for people to walk side-by-side.

Participants went on to make recommendations, covering, for example,

Accessibility Flat rest areas are recommended along the service pathway for wheelchair access, with grab bar provided. The legibility of barrier-free access ways and other pathways is viewed as vital for circulation in/to the park and should be clearly visible and properly maintained. Legibility promotes a sense of ease of access and clarity of direction. Importantly, it helps wayfinding. Another recommendation is to introduce benches at convenient intervals along the flight of stairs at Block 201 to provide rest points for the older adults. The one toilet in the Park should be PWD-friendly to ensure that the Park is truly accessible for all.

Footpath Improvement and Maintenance Paths need to be patched up of holes and level changes. Participants suggested that the paths around the Park should be generous in width, provide clear sightlines and be paved with non-slip finishing. Participants further suggested for paths to be tree-lined so there would be shade for joggers, which they believed would be better than a covered walkway.

10.3.4 Maintenance and Management

Greenery General upkeep of the greenery (i.e. trimming, pruning and sweeping) was an issue that was repeatedly red-flagged. Nearly all participants complained about the unkempt, dense undergrowth of creepers and grasses in the forested area at top of the Park. Participants opined that the dense vegetation provided refuge for ‘undesirables’ such as strangers, pests and wild animals, which could be a disturbance to the peace of the Park or a threat to park users' safety. The vegetation obscured direct line of sight to other parts of the Park. According to the participants, the authorities had left the forested area as it is in its 'natural' state. However, this should not mean leaving it in neglect. Trees were old and falling branches were observed to pose a danger to park users. In the months of August and September, trees would shed their leaves the most and without regular sweeping, the build-up of dead leaves was unsightly and a potential inconvenience to walking.

Pests and Wildlife Participants reported the presence of stray and wild animals such as dogs, snakes and monkeys. Dogs defecated in the Park and could intimidate Park users. The problem had been reported to the authorities but remained unresolved. There had been difficulty finding the dogs when the Agri-Food and Veterinary Authority of Singapore made their visit to the Park. Some felt the problem could be better resolved if authorities worked together, rather than 'passing the buck' between authorities on whose responsibility it was to remove the stray dogs.

Monkeys were a new presence in the Park, which the participants opined was due to the Eco-Link Bridge[3] over the Bukit Timah Expressway. The monkeys did not reside in the Park but came through the Eco-Link Bridge. When they appeared, they disturbed the vegetable gardens (dug up the plant, ate the fruits, etc.). Participants expressed their displeasure at people feeding the monkeys. However, it was noted that the monkeys presented an attraction to the Park. Participants reported that visitors from other neighbourhoods would come specially to sight these creatures during the weekend.

Mosquitoes discouraged visitors from staying too long in the Park, especially in the forested area.

Facilities A number of observations, both positive and negative, were voiced. The cloistered feeling of the forested area and the presence of 'undesirables' meant that facilities in those area such as benches were infrequently used and subjected to vandalism (e.g. bench back was flattened to aid sleeping). Wayfinding signs were vandalised (they should be replaced as they are no longer legible). Some seats were rusted and left in a bad state; they did not welcome use. Participants shared that there was nothing to see, too many mosquitoes, and therefore, no reason to stay and sit around.

Participants also spoke about the exercise equipment being dirty and that they had to clean them before use. They felt that the equipment should be sheltered. Pavilions had not been fixed of their damages and therefore, looked dilapidated and uninviting (during the workshop period, the pavilions were refitted, new seats were installed).

The participants observed that the one portable toilet was generally well maintained and a much-welcomed facility in the Park, especially among the older participants. Some suggested having separate toilets for each gender.

---

[3]This is a 62 m long purpose-built wildlife bridge across the Bukit Timah Expressway. Built some 30 years after the expressway, the ecolink bridge seeks to reconnect Bukit Timah Nature Reserve to Central Catchment Nature Reserve, Singapore's largest nature reserve. These two nature reserves were connected before the construction of the expressway.

Others Some respondents wished the Park to be smoke-free and smoke-free signs to be put up. Several had negative sentiments about the surrounding condominium as the new development apparently reduced the size of the Park (this could be perceived/psychological impact of the imposing heights of the condominium towers). Participants reported that the hard court was used to store building materials during estate upgrading. This was dangerous and many stayed away (reduced the number of people coming to exercise).

Participants shared various recommendations relating to park maintenance including,

Park Management Responsibility There is a need for clear designation of park management responsibility. It is not immediately obvious to the participants which authority oversees the park jurisdiction—whose duty it is to manage the Park and they believe this lack of clarity prevents scheduled maintenance from happening. Neither are the participants aware of who to inform when there is a problem in the Park. Some would approach the RC Centre when they encountered problems in the Park. Others felt that the problem would be picked up by other park users or RC members eventually. It is recommended that residents be informed of the authority(s)-in-charge and of their possible roles in taking care of the Park.

Regular Maintenance It is vital that park maintenance is kept up regularly and not a one-off effort when complaints are made or when a Member of Parliament visits. There is a need for regular inspection, pruning or planting of new trees when old trees die and rot. Falling branches pose a great danger to park users; they should be removed without delay in a well-maintained welcoming park. Clearings should be planted with new trees/vegetation, especially the blooming kind.

Participants emphasised the need for the undergrowth to be tidied up, which would solve multiple problems such as the presence of stray dogs, snakes, monkeys, mosquitoes, and poor sightlines. Vandalised benches would have to be replaced and seat dividers were suggested to prevent people from sleeping on them. Some said that it was vital to consider the location and design of these seats to ensure that they were inviting and conducive for frequent usage and under the surveillance of park users.

10.3.5 Safety and Security

Participants identified the community garden (in lower part of the Park) and exercise areas as the safest and liveliest parts of the Park (Fig. 10.3). The upper part of the Park, in the forest trail and the stairs from Block 201 were identified as the most unsafe and quietest parts of the Park.

A combination of physical and psychological factors was offered for why the Park was perceived as unsafe (Table 10.3).

Area of Isolation The forested area at the top of the Park was unanimously perceived to be unsafe as it was concealed by dense vegetation and appeared 'wilder' and darker than the rest of the Park. The forest undergrowth and the winding paths

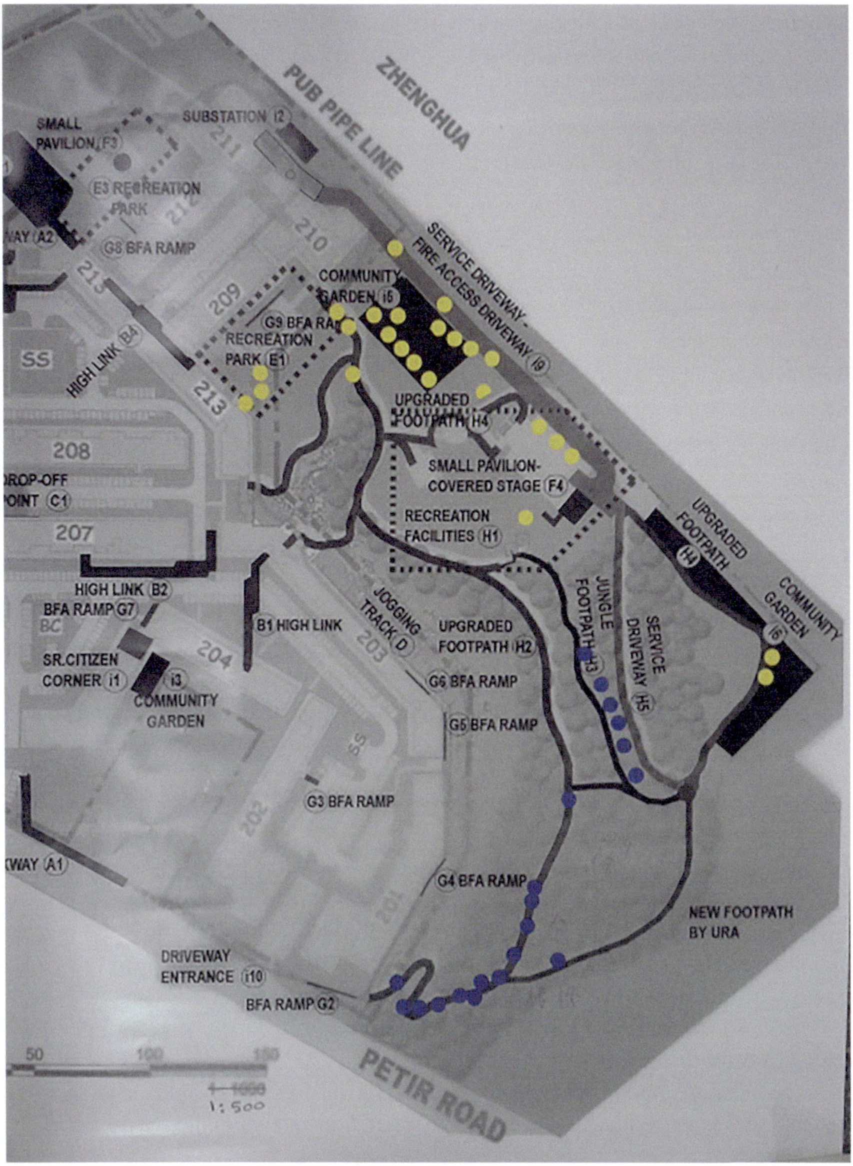

Fig. 10.3 Places where participants felt safest (yellow) and most unsafe (blue) in the park

prevented any direct lines of sight to the active parts of the Park and created a feeling of isolation. The footpaths in this area were also channelised and therefore, a user's movements were predictable, which could be a real danger or add to the feeling of perceived danger.

Table 10.3 Participants rate what makes them feel unsafe in the park

Design, maintenance and surveillance issues	Participant counts
Social and behavioural	
Undesirables (e.g. behaviour, animals)	11
Petty crimes (e.g. theft in garden, littering)	7
Vandalism	0
Physical design	
Poor visibility	15
Unkempt grass and falling branches	14
Areas of concealment	12
Poor lighting	10
Accessibility—steep slopes	8
Confusing lighting	1
Others	
No access to help	10

N = 21

Presence of Undesirables Participants attributed the sense of danger to the presence of 'undesirables' such as people who sleep on park benches, foreign workers or students who engage in rowdy behaviour, the presence of stray dogs and other wild animals. Vandalised wayfinding signs and damaged park benches further add to the participants' unease.

Insufficient Lighting Participants pointed to the entry route from Block 201 as a very dark area, lacking sufficient lighting after dark. These areas were reported to feel most unsafe at night (past 7 pm) or in the early morning before sunrise (4–5 am). They believed that dark areas encouraged 'bad people' to lurk—there was a perceived danger when there was insufficient lighting. Some participants highlighted that the Park should not need to be too bright as this may cause the Park not to feel like a park. Lights are needed when there is activity or for movement.

Others Participants also spoke of a homicide that happened 3–4 years ago, which seemed to add to their negative impression of that part of the Park. A number of participants spoke about personal responsibility for one's own safety and security. That is, there is a need to know where and when you can walk and where and when you cannot.

The following recommendations were offered.

Informal Surveillance and Intensifying Activity To assure better visibility, the dense undergrowth of the forested area should be cleared away and tidied up. In so doing, this area would come within the line of sight of nearby blocks and areas of the Park where it is more active. Tackling the issue of vegetation would reduce areas of concealment and a sense of isolation, which would encourage more users and thereby increase surveillance in the area. 'Park Patrol'/Park Watch groups could be considered to encourage surveillance by local residents and regular park users.

Creating multiple activity nodes; ensuring that activity in the Park is not just focused in one area can increase movement from one cluster of activity to the next, thereby reducing isolated areas. For example, the community garden located at the top of the Park increases visitation to the area and movement between lower (exercise equipment) and upper parts of the Park.

Appropriate Lighting Lighting needs to be positioned to illuminate areas of potential concealment such as in the forested part of the Park with its winding paths. Lighting should also be used to illuminate facilities such as exercise equipment, hard court and playground to make the Park safer by encouraging evening use. However, this has to be balanced with creating a false sense of security created by lighting areas that are inappropriate for night-time use (Project for Public Spaces 2009a).

Signage A comprehensive system of wayfinding signs will increase a sense of safety, clarity and order when people know where things are and how to get to where they want to go. Participants also suggested sizeable, eye-catching signage (displaying the name of the Park) at the fringes of the Park for easy recognition and a greater sense of ownership. It can serve as a landmark and is a part of place management. Participants claimed that visitors to the Park would not necessarily know where they are/the name of the Park until they see the only signboard of the name of the Park located next to the community garden. Wayfinding is an important aspect of the imageable or memorable features of the Park.

Informative and positive signage (such as information about the origins, flora and fauna of the park, etc.) can be used to encourage a sense of ownership and pride among users. Educational signs can inform users about management and safety issues and encourage them to have a more direct involvement in keeping the Park tidy and safe, like knowing who to approach about problems in the Park or reporting a crime they witness.

Telephone or Call Button Introducing an emergency telephone or a safety call button is not only for convenience but a symbol of safety and security in the Park. These installations need to be easy to locate and accessible.

10.3.6 Activities and Amenities

Participants were encouraged to talk about their experiences with Park amenities and activities. Various activities, both existing and new, were listed and discussed.

Community Gardens All but one participant rated the community gardens as of high importance (rating 4–5 out of 5 points where 5 is the most important). The one that rated it low is not interested in gardening. Participants believed that without these gardens, less people would use the Park. They opined that the community gardens were a place for residents to gather, bond and build neighbourly relations. Gardening offered a form of exercise and therapy to keep physically and mentally active. Participants expressed a sense of satisfaction in growing their own vegetables and believed that the vegetables tasted better because they were self-grown and also because they were organic.

Weekly Farmers' Market Participants expressed mixed opinions about introducing a weekly farmers' market. Some believed that it was not necessary and that it did not concern them. Those who rated this highly thought it was a good activity to liven up the Park. However, participants who worked in the community gardens believed their harvest quantity was not large enough for market sales, especially after distributing them amongst community members and their families. Participants were also afraid that the quality of their produce was not up to sale standard. They suggested trying out the market as a monthly event and if results were favourable, its frequency could be increased.

Community Flower Garden Participants expressed a desire to see more flowers in the Park (pretty to look at). They expressed mixed opinions about having a community flower garden. Those in favour of it believed that it would be very beautiful, colourful, smell good and would keep people relaxed and happy. They suggested planting flowers from different countries and that it could serve as an additional attraction for the Park. Others who disagreed with the idea pointed out that there was limited space for planting and any available space should be used for planting vegetables (edible) instead of being 'wasted' on flowers. They also thought that 'nice things' like flowers would get stolen. One suggested that the flowers could be grown in old water bottles as shown in the presentation example, involving recycling.

Fitness Classes Four-fifths of the participants rated fitness classes as an activity of high importance. They said many fitness activities (both free and paid classes) were in place in the community (e.g. organised by RC) such as qiqong, tai chi, line dancing and brisk walking. But, they would like even more. Many participants were not able to make it to the classes or to continue with them as they were busy or the timing was not right. Classes were usually held before 8 am or at night, as it was too hot during the day. Participants were of the view that free activities would encourage more participation. One participant elaborated on the importance of holding age-appropriate fitness classes for older adults.

Age-Friendly Exercise Equipment Two-thirds of the participants rated age-friendly exercise equipment as an important amenity in the Park. Many lamented about the condition of the current equipment being tailored to a Western physique as they found the equipment difficult to reach with feet placing too far apart. Some suggested the introduction of adjustable equipment to suit all sizes. Others talked about the resistance being too heavy for the older adults. Yet, others opined that priority should be given to the old as they used the equipment more frequently. Some said more equipment was needed. But, a small number did not think everyone was interested in using these equipment as they preferred doing other forms of exercises like walking, cycling, gardening or playing badminton.

(Better designed) Children's Playground Participants felt that a well-designed children's playground was important to the community. The current playground is at the foothill of the Park next to Block 209, a distance away from the adult and elderly exercise equipment, which is halfway up the hill park. The other children's playground is located at the periphery of the Park next to the sheltered gathering space at Block 213. Participants opined that a playground in the Park would encourage more children and young families to visit, which could liven up the Park. Locating a new playground next to the existing exercise equipment could allow parents and grandparents to watch over children while they exercise.

Participants suggested that playgrounds need to be suitable for all ages (have age appropriate design) and be creative and stimulating, otherwise children would not use them. Some participants noted that there was now more exercise equipment for adults and older adults than children's equipment (in the past, there was more children's playground), indicating (there was an assumption) that the demographic in the neighbourhood was growing older.

Pavilions and Shelters Many participants opined that there were not enough shelters in the Park and more were needed. Those presently provided in the Park were too small, had seats but lacked tables, were not aesthetically pleasing, were dilapidated or vandalised and in need of repair and renewal. Participants viewed pavilions as a necessary amenity for the retired and the older adults to relax, sit and socialise. Currently, they used the void deck as a gathering space and played chess at the coffee shops.

In the Park, people tended to use the pavilions after exercise as a resting spot. Participants also talked about the importance of having shelter against the elements. The trellis shelter by the hard court did not provide protection against sun or rain. It was dangerous for the older person to run for cover down the hill when it rained. The common opinion was that shelters must be well designed and well-placed; they should facilitate cross ventilation and be free of mosquitoes.

Seating along Footpaths in Park Most participants believed that introducing more seating along footpaths in the Park was unimportant or unnecessary. Participants said that there were enough seats around (they could sit on the large feature rocks/boulders). Existing seats were unpopular for a range of reasons, e.g.,

- those not under shelter and therefore, not conducive to use;
- those located in the forested area where participants did not like to loiter;
- those that were rusted or damaged.

Participants suggested that the seats should be provided with dividers to prevent people from sleeping on them. Some also said it was vital to consider the location and design of the seats as no one would use them otherwise.

BBQ Area Participants believed that having a BBQ area in the Park was unnecessary. There were already two BBQs at the foothill of the Park, which were well used. They would cost S$10.70 a pit and required pre-booking. The pits required a lot of maintenance to ensure they did not breed vermin. Participants did not think a BBQ pit in the Park was suitable as there were many practical problems like no water point up in the Park, not enough garbage bins, monkeys and dogs may be attracted to food waste from the BBQ, etc. The Park also was not as accessible as the blocks. Some participants believed that BBQ was an activity for young people and not the older adults; family gatherings were usually over meals at home and not BBQ.

Café in the Park Some participants preferred having a vending machine instead. They opined that there was not enough foot traffic to sustain a café business. Participants pointed out that they would not have time to visit the Park cafe themselves, let alone outsiders.

Table 10.4 Participants' activity and amenity recommendations for BPN2 Park

Activity/Amenity	Recommendation
Community garden	Due to popularity, queues for allotment and availability of space in the park, expansion of community gardens can be considered
Farmers' market	Organising a pilot run with the help of residents can be considered
(Community) Flower garden	Can be introduced in the area surrounding existing pavilions to beautify area or in the forested area should some of it be cleared
Fitness classes	Currently satisfactory though more fitness classes for older adults are welcome
Age-friendly exercise equipment	To consider implementation of adjustable equipment and resistance levels to suit different heights and fitness levels
Children's playground	To consider introducing children's playground closer to exercise equipment area to bring in younger users and families, and expanding demographic range in the Park. Also, to allow for easy surveillance of children while adults worked out in the Park
Pavilions and shelters	To consider designing and locating sheltered pavilions as a focal point in the Park. This would increase interest in usage, thereby increasing surveillance and reducing chances for vandalism or misuse
Seating	No need for more seating; rusted and damaged seating should be replaced. To consider seating location carefully and not be arbitrarily placed
Wild nature	A study can be made of the type of flora to be maintained or introduced to encourage desirable/native flora and fauna in the park
Wayfinding signage	To consider improving wayfinding signage in the Park; vandalised and illegible signage to be replaced
Park signboard	To consider installing a Park signboard (Park name) at key access points to improve place branding and sense of identity
Lighting	To consider appropriate park lighting to illuminate dark and deserted areas would improve sense of safety in the Park
Emergency/safety call point	Both for convenience and a symbol of safety and security in the Park; installation need to be easy to locate and accessible

Pet Grooming Area Participants were unanimous in rating the introduction of a pet grooming area lowly. They thought it was unnecessary, as dog owners would take care of their pets at home. They did not see this as a plausible solution for dealing with the stray dog problem. In regard to training the stray dogs in the Park into pets and security dogs of the community, they believed that the stray dogs were too old to be trained and not worth the costs; professionals need to be hired to train them and this would be very expensive.

Wild Nature Participants were mixed in their opinion of the presence of 'wild nature'. Some participants complained about birds being noisy and birds that fly into homes and defecate on clothes hung out to dry, about the stray dogs and monkeys that had come to dominate parts of the Park, scaring off park users. Participants were selective with the kind of nature they want around them; they did not want dogs, monkeys or snakes, which would scare the children; they would like nature that is not dangerous to the residents. Others viewed 'wild nature' as very important, part of the park-like environment and supportive of its preservation. They talked about spotting many birds in the community gardens, native birds of the Park making nests in the potted plants grown by residents, taking photos of the birds and recognising the unique natural environment that they have in the Park.

Table 10.4 summarises the activity and amenity suggestions from the participants.

10.3.7 Community Spirit and Ownership

For suggestions to further enhance the Park's sense of place, facilitators sketched ideas and took notes on paper while participants contributed their ideas (Fig. 10.4).

Building a sense of community spirit and ownership is vital in creating a better park for all. Participants expressed a willingness to contribute to this community space and to volunteer their time if there was leadership, guidance and an initiative put in place. The community garden where a large number of participants worked in and contributed to demonstrates community spirit and sense of ownership, which has resulted in its upkeep and success.

To gauge their willingness to contribute to the Park and the type of contribution, participants were asked to indicate their willingness to be involved in a number of community park activities.

Community Flower Garden Many were in favour of introducing flowers to the Park. Some participants already spent one to three hours a day gardening, so they did not mind spending a little more time looking after flowers, especially if there was a management system, facilities and provisions (such as a water point) in place.

Farmers' Market Some participants suggested that the profit made could be donated to charity or could be channelled back into the upkeep of the Park or funding the community gardens. Participants would like to prepare food for the market but were aware that they would not be able to sell food in the markets due to health and safety regulations.

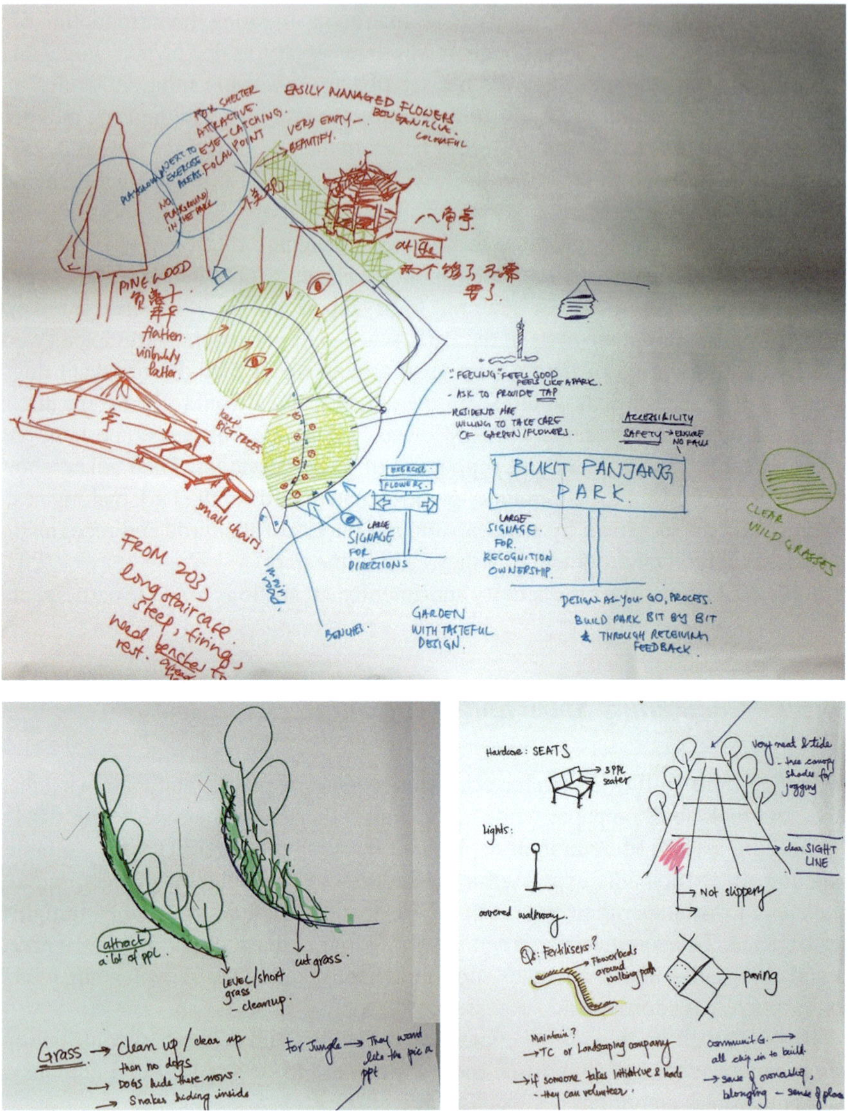

Fig. 10.4 Participants' design ideas

Plant-a-Tree A small number of participants were willing to participate in tree-planting. One participant was highly in favour of the activity and said she did not mind planting more than one, even 5 or 10 trees. Some participants opined that new trees did not have to be planted, only flowering plants.

(Volunteer) Nature Guide There was generally a positive response to this activity. A number of participants were familiar with tree species in the Park and believed they could take guests around the Park on a tour like they did during the 'Walk & Talk' survey [4] (another data collection instrument).

Adopt-a-Bench A number of people were interested in this activity. One even joked about implementing 'adopt-a-toilet', showing how interested they were in volunteering their time and contributing back to the Park.

They offered several recommendations.

On Strategy for Efficient Volunteerism, e.g. 'Adopt-a-Park' and 'Divide & Conquer' Nearly all participants were enthusiastic and supportive of taking ownership for a part of the Park and being responsible for its upkeep. According to them, assigning a group to a part of the Park can encourage a sense of ownership and because of that, they will show up for duty. However, they emphasised that it could not be a lone effort and would require leadership, that a group needs to be formed with leaders, a division of workload, a roster of different duties and enough people within a group to cover those that could not make it for duty. Work hours should not be too long, 1 or 2 h a week is good enough.

On Sense of Ownership and Participation Most participants showed great willingness to join in these park-building activities. But, a small number acknowledged that it is human nature to be lazy and uninitiated (wait for others to start/lead). One participant shared that women were more interested in volunteering and contributing, more than men, as evident in the number of women present during the community design workshops. Another participant identified the need to survey and get feedback from the wider neighbourhood community, not only the small group of participants in the workshop who are highly motivated individuals already willing to participate in most activities. Some reminded that, for volunteering to take place, there would need to be interest, young age (not just old people) and time to volunteer. Others talked about the philosophy of contribution—a little goes a long way. Finally, participants felt that if the garden were beautiful, people would be happy and feel a further sense of responsibility and ownership. This could help strengthen the spirit of caring and looking out for the Park.

In general, there was positive feedback for participation in community and volunteer activities to create a great community park. Community education and communication ought to be considered integral to participation. Residents could be educated on a range of ways in which they could contribute to the Park, from small and easy steps like bringing their own trash bags to collect trash on the way (which some participants already said they do) to bigger initiatives like adopt-a-park/tree and others mentioned above.

10.4 Conclusion

Participants are generally happy with the Park. Participants do not want too much to be done, just for the Park to be kept tidy, clean and regularly maintained. They

---

[4]See Chap. 5

recognised the unique qualities of their neighbourhood park—its hilly terrain, natural forest and mature trees, and felt themselves blessed to have this Park when compared to the prosaic neighbourhood park provision in many other HDB towns.

The Park is seen as a place for the enjoyment of natural environment (e.g. greenery, trees and fresh air), exercise and fitness (e.g. walking and group exercise classes) and activities (e.g. meeting people, talking to friends and gardening). The Park is important also as a place for neighbourly gathering and bonding. Nearly all participants shared that they had met their friends through the Park, continuing to meet for regular activities and outings to the Park and elsewhere. A fine example of such an opportunity is community gardening. The neighbourly bonds and the proactive behaviour that was encouraged amongst the community gardeners were apparent during our workshops, when participants invited and reminded their gardening friends to attend the workshops, and when they showed willingness and eagerness in the participation of other community or survey activities (such as See and Snap,[5] another data collection method of this project).

While participants have expressed an overall contentment with the Park, where they have concerns it is primarily to do with management issues: tidiness of the park, maintenance of greenery and managing the presence of undesirables.[6] This lends support to findings reported in high-income developed countries such as North America, Europe and Australia. In a study on the influence of age on recreational trail preferences in public green spaces in Vienna, consideration of management and social attributes (e.g. litter, dog mess, aggressive visitor behaviour and other incivilities) appeared more important with increasing age (Arnberger and Eder 2011). Arnberger and Eder interpreted these results in terms of the older person's (aged above 60) perceived vulnerability.

Others have found that deficits in recreational settings and park environments that are less well maintained, look too wild and overgrown are disliked and can discourage usage and physical activity (see e.g. Godbey and Blazey 1983; Gobster and Westphal 2003; Mowan et al. 2008; Bird et al. 2010). Older persons seem to have a more positive attitude towards the removal of undergrowth than the young and middle-aged person (Arnberger and Eder 2011). It has been found that limited visual surveillance caused by impermeability of vegetation can induce the fear of crime as well as accidents by blocking the field of vision (Arnberger and Eder 2011). The implication is that areas in BPN2 Park with perceived lack of maintenance and management (e.g. forested area with its dense vegetation, stray dogs, over-stayers and damaged benches) and repeatedly identified as places where participants feel unsafe or vulnerable and to avoid at all cost after dark, should be reviewed and made as unthreatening as possible.

Creating and improving recreational (park) spaces bring many benefits including social, health and wellbeing (Kaplan and Kaplan 1989; Bartolomei et al. 2003; Berman et al. 2008; Campbell and Wiesen 2009; Canadian Parks Council 2014). In

[5]See Chap. 6.

[6]Succeeding the report submission to the local town council, many of the recommendations for BPN2 Park were implemented.

many cases, as Kent and Madden (1998) reminded, 'We have not even scratched the surface of the impact parks could have on the revitalization of communities'. Studies have shown that having a park within walking distance of one's home is the strongest predictor that an older person would use a park (Goodbey et al. 2004). Neighbourhood park offers an opportunity to connect older folks to nature locally.

In conclusion, the community design workshop yielded several lessons and insights.

Management is a key concern and should not be neglected as this would cause a domino effect on all other aspects of the park, its maintenance, tidiness, perceived safety and security, which would result in disuse and the attraction of undesirables and anti-social behaviour.

Recognising parks as vital community spaces and a place of sociability (especially for the elderly) and to design for it. Project for Public Spaces (2009b) characterises a good park as a place that supports a variety of activities and programmes for everybody—different age groups and different kinds of people. A good park is easily accessible and well linked to the surrounding neighbourhood. It is a comfortable, safe, tidy and inviting place. Most crucially, a good park has sociability; it should be a place to meet other people, an integral part of community life. A good park is also an impetus for community development and enrichment of its surrounding neighbourhood. These characteristics also coincide with a space for supporting continued community participation, physical and psychological health and wellbeing of the ageing population. Now (not later), we have to design to facilitate and perpetuate this community space.

Placemaking or Responsive Design as a means to create Great Parks. Project for Public Spaces describes the steps involved in creating parks as vital community spaces as beginning with '*understanding a community's concerns about a particular space… [Determining] how the assets of the community can be used to develop both the plan and the programming for the park*' (Project for Public Spaces 2009b). The process involves a collaborative method known as 'placemaking' where community-based participation is at its core. Place making reinforces the relationship between places and the people that use the space, take advantage of the community's resources, aspirations and potentials, and it eventuates in the generation of great public spaces that contribute to people's health, happiness, and wellbeing and nourish communities (Crowhurst Lennard and Lennard 1995; Walljasper 2007; Silberberg et al. 2013).

Community Design Workshop as a Placemaking Tool—The workshop model has been deployed as one of nine instruments in our study on Singapore's ageing population. Activities such as group discussions, design charrettes and individual presentations provided a platform and opportunity for residents to interact and participate in conceptualising the vision, programme, activity and design of their BPN2 Park. It shows that residents, even the shiest, would come forward after a time and share their thoughts and ideas.

One activity that was particularly telling was '*The Power of 10+*' (borrowed from Project for Public Spaces), where participants were asked to identify their 10+ Great Places in the Park or the 10+ reasons to be there. This encouraged the participants to consider what is unique in their community space and if there were other unique places that should be identified but have not been. The list also reflected what is special in BPN2 Park, its community and social culture (nature was top on the list followed by exercise area, large trees, stairs entry at Block 201, pavilion, hard court, community garden, etc.).

Our study testifies to the importance of engaging the residents. Their involvement can assist the community, designers and other stakeholders to identify the places they would focus their efforts and strategize for future plans and changes.

Acknowledgements We gratefully acknowledge the following for their generous support and assistance with the community design workshops: Dr. Teo Ho Pin, Mayor, North-west Community Development Council for his advice and guidance; Lawrence Toh, Senior Property Manager, Holland-Bukit Panjang Town Council, Lin Shuang Fu, Deputy Constituency Director, Bukit Panjang Community Centre, Lim Yong Moi, Bukit Panjang Zone 4 Residents Committee Chairman, K. C. Tan, Bukit Panjang Zone 4 Residents Committee Manager for their facilitation and outreach to residents and for allowing us the use of Bukit Panjang Zone 4 Residents Committee premises for the workshops; Rita Soh, Advisor, Lee Li Ming Programme for Ageing Urbanism (2014–16), Director, RDC Architects Pte Ltd and President, Board of Architects Singapore (2010–15), for working alongside us during the workshops; our colleagues, Josephine Ng, Parvathi S Nair and our student intern, Lim Aunn Ning for helping us with the workshop organisation, facilitation, translation and report on workshop proceedings; residents of Bukit Panjang Zone 4 who participated in our workshops and generously shared with us their ideas, solutions as well as their own grown and baked food.

References

Arnberger, A., & Eder, R. (2011). The influence of age on recreational trail preferences of urban green-space visitors: A discrete choice experiment with digitally calibrated images. *Journal of Environmental Planning and Management, 54*(7), 891–908.

Bartolomei, L., Corkery, L., Judd, B., & Thompson, S. (2003). *A bountiful harvest: Community gardens and neighbourhood renewal in Waterloo*. New South Wales Government: Department of Housing and University of New South Wales.

Berman, M., Jonides, J., & Kaplan, S. (2008). The cognitive benefits of interacting with nature. *Psychological Science, 19*(12), 1207–1212.

Bird, S. R., Radermacher, H., Sims, J., Feldman, S., Browning, C., & Thomas, S. (2010). Factors affecting walking activity of older people from culturally diverse groups: An Australian experience. *Journal of Science and Medicine in Sport, 13*(4), 417–423.

Campbell, L., & Wiesen, A. (Eds.). (2009). *Restorative commons: Creating health and wellbeing through urban landscapes*. USDA Forest Service.

Canadian Parks Council. (2014). *Connecting Canadians with nature: An investment in the wellbeing of our citizens*. Canadian Parks Council.

Crowhurst Lennard, S. H., & Lennard, H. L. (1995). *Livable cities observed*. Gondolier Press.

Godbey, G., & Blazey, M. (1983). Old people in urban parks: An exploratory investigation. *Journal of Leisure Research, 15,* 229–244.

Goodbey, G., Payne, L., & Orsega-Smith, B. (2004). *Examining the relationship of local government recreation and park services to the health of older adults*. Robert Wood Johnson Foundation Grant Research Results.

Gobster, P., & Westphal, L. (2003). The human dimensions of urban greenways: Planning for recreation and related experiences. *Landscape and Urban Planning, 68,* 147–165.

Kaplan, R., & Kaplan, S. (1989). *The experience of nature: A psychological perspective*. Cambridge, NJ: Cambridge University Press.
Kent, F., & Madden, K. (1998). *Great parks/Great cities: Seattle, Urban Parks Institute regional workshop publication*. http://www.pps.org/reference/creating-great-urban-parks/. Accessed July 27, 2015.
Mowan, A., Kaczynsk, A., & Cohen, D. (2008). The potential of parks and recreation in addressing physical activity and fitness. *Research Digest*, *9*(1).
Project for Public Spaces. (2009a). *Designing and maintaining safer parks*. http://www.pps.org/reference/what-role-can-design-play-in-creating-safer-parks/. Accessed July 27, 2015.
Project for Public Spaces. (2009b). *What is placemaking?* https://www.pps.org/reference/what_is_placemaking/. Accessed July 27, 2015.
Silberberg, S., Lorah, K., Disbrow, R., & Muessig, A. (2013). *Places in the making*. MIT.
Walljasper, J. (2007). *The great neighbourhood book*. BC, Canada: New Society Publishers.

Index

A
Accessibility, 48, 49, 53, 56, 73, 88, 99–101, 104, 105, 107–110, 117–123, 141, 143, 147, 154, 168, 171–173, 209, 235, 259, 260, 263, 266, 270
Activity nodes, 152, 158, 165, 197, 271
Adaptability, 105–107, 110, 111, 117–121, 123
Affordability, 66, 76, 84, 105–107, 112, 117–121, 171
Age-friendly city, 73, 90, 95, 101
Age-friendly design, 75, 76, 87, 95, 99, 100
Age-friendly housing, 46, 47, 49, 60, 73, 100, 105, 106, 110, 121
Age-friendly neighbourhoods, 47, 53
Ageing in place, 49, 50, 71, 73, 78, 80, 86, 105, 106, 113, 256
Amenities, 45, 47–49, 53–56, 67, 68, 74, 77, 81, 88, 101, 129, 135, 139, 147, 149, 152–155, 157, 161, 162, 164, 170–172, 177, 179, 184, 192, 209, 211, 218, 235, 260, 271
Assisted living, 79, 80, 82, 99, 122
Attitudes, 46, 65, 71, 87, 95, 101

B
Barrier-free access, 47, 100, 135, 162, 168, 171, 173, 177, 208, 266
Benches, 145, 161, 163, 177, 184–188, 195, 196, 198, 200, 209, 217, 261, 266, 268, 270, 278
Big data, 83, 223, 238, 239
Built environment, 99, 100, 121, 149–151, 162, 172, 209, 217

C
Caregiver, 54, 111
Community centres, 56, 155–157, 171, 172, 223, 235, 247
Community design, 64, 68, 128, 130, 175, 179, 192, 259, 277, 279, 280
Community gardens, 56, 58, 130, 139, 160, 161, 170–172, 179, 205–208, 213, 214, 218, 261, 264, 271, 272, 275
Connectivity, 49, 54, 67, 87–90, 101, 169, 186, 209, 259, 260, 263
Convenience, 53, 67, 72, 81, 101, 149, 154, 161, 164, 170, 172, 235, 271

D
Dementia care, 73, 85, 86
Disability, 48, 105, 110–112, 121, 123

E
Elderly studio apartment, 51, 246
Emerging old, 83, 106, 109, 110, 112, 114, 115, 117
Employment, 81, 98, 102, 151, 152, 236, 238
Exercise, 56, 59, 129, 130, 133, 134, 138, 145, 146, 155–158, 161, 171, 172, 177–179, 182, 185, 186, 189, 193, 195–197, 201, 205–209, 213, 226, 243, 248, 256, 261, 264, 267, 268, 271–273, 278, 280

F
Favourite places, 150, 170, 264
Flexibility, 61, 105–107, 110, 111, 117–121, 158

B. Yuen (ed.), *Ageing and the Built Environment in Singapore*,
https://doi.org/10.1007/978-3-319-92444-1

G

Gardening, 81, 133, 134, 139, 146, 160, 161, 170, 205–207, 213, 261, 271, 272, 275, 278
Greenery, 67, 88, 101, 130, 158, 170, 214, 259, 278
Growing old, 6

H

Hawker centre, 57, 67, 153, 154, 209, 234–236, 244, 247
Healthcare, 72, 73, 79, 80, 85, 92, 95, 99, 101, 105, 112, 135, 146, 157, 165, 171, 221, 234, 239
Heartware, 73, 100
Home improvement, 66, 78
Home ownership, 4, 60, 161, 202, 275
Housing audit, 103, 104, 121, 124
Housing option, 3, 74, 79, 82, 83, 162, 225

I

Identity, 49, 217, 225, 259, 260
Inclusivity, 65, 95, 101, 105, 107, 109, 110, 117–121
Independence, 51, 56, 61, 65, 66, 121–123, 192
Infrastructure, 54, 55, 72, 73, 87, 88, 92, 101, 104, 124, 147, 152, 162, 163, 169, 171–173, 209, 214, 241
Innovations, 46, 73

K

Kopitiams, 153, 154, 170

L

Landmarks, 177, 179, 217
Lifetime homes, 79
Lighting, 88, 101, 108, 109, 270, 271
Likert scales, 105
Liveability, 60, 105, 107, 115–120, 123, 250, 257
Living arrangements, 246

M

Maintenance, 48, 66, 104–107, 112, 113, 116–123, 147, 160, 162, 163, 171–173, 191, 192, 195, 209, 216, 259, 263, 266, 268, 270, 273, 278, 279
Markets, 53, 149, 154, 158, 166, 168, 189, 275
Methodology, 45, 46, 72, 104, 127, 128, 150, 175, 176, 221, 222, 242, 259
Middle old, 106, 110, 112, 114, 115, 117
Mindset, 95, 99–101
Mobility, 47, 54, 55, 59, 65, 66, 68, 81, 87, 88, 101, 122, 131, 135, 143, 146, 150, 151, 168, 173, 176, 177, 222, 239

N

Neighbourhood park, 56, 128, 130, 175, 179, 214, 242, 260, 278, 279
Neighbourhood spaces, 86, 99, 101, 127
Nursing homes, 82, 84, 85, 217

O

Observation, 58, 64, 114, 127, 175, 176, 178–182, 184, 185, 188, 189, 191–193, 195, 196, 198, 201, 202, 209, 216, 218, 223, 238, 241
Oldest old, 106, 109, 110, 112, 114, 115, 117
Outdoor activities, 181, 216, 221
Outdoor spaces, 56, 57, 73, 88, 89, 101, 152, 155, 175, 176, 217, 241, 242

P

Parks, 56, 57, 67, 68, 89, 92, 101, 136, 158, 160, 161, 164, 166, 170–172, 209, 217, 247, 260, 261, 278, 279
Perceptions, 71, 151, 176, 216, 218, 241, 242, 254, 255
Photo-taking, 149, 150, 170
Placemaking, 65, 67, 104, 217, 259, 279
Places of worship, 154, 157, 171
Playground, 129, 134, 137, 138, 145, 147, 161, 163, 172, 177, 184–186, 188, 212, 216, 226, 244, 261, 271–273
Preferences, 86, 100, 149, 202, 216, 221, 278
Privacy, 52, 66, 105–107, 113, 114, 117–121, 123, 209, 225, 226, 233
Private housing, 48, 74, 75, 79, 106, 108, 109, 112, 113, 115, 117, 118, 123, 244
Public housing, 47–49, 53, 56, 59, 66–68, 75, 78, 79, 82, 87, 93, 106, 108, 109, 112, 113, 115, 117, 118, 122, 123, 128, 130, 135, 155, 165, 177, 179, 202, 234, 242, 244–246, 261
Public spaces, 53, 56, 64, 65, 67, 81, 101, 127, 128, 131, 134, 135, 138, 146, 147, 158, 164, 172, 173, 175, 176, 180–182, 188, 196, 208, 214, 216–218, 241, 271, 279

Q

Quality of life, 46, 49, 65, 66, 83, 100, 103, 104, 123, 127
Questionnaire, 45, 105, 242–244

R
Recreational activities, 127, 133–135, 146, 171, 172, 216
Renting, 4, 76
Retirement villages, 51, 61, 83, 84
Retrofitting, 50

S
Safety, 57, 68, 73, 85, 88, 89, 99–101, 104–107, 113, 114, 117–121, 123, 164, 172, 173, 177, 209, 222, 259, 260, 263, 268, 270, 271, 275, 279
Satisfaction, 48, 54, 172, 259, 271
Security, 52, 68, 105, 147, 171, 177, 259, 260, 263, 268, 270, 271, 275, 279
Senior activity centres, 155, 157, 172, 173
Senior citizen corners, 56, 58, 68, 157, 192, 202, 210, 211, 213, 214
Sense of place, 259, 260, 275
Services, 51, 53–55, 66–68, 72, 82, 83, 85, 86, 88, 99, 101, 104–107, 112, 113, 117–122, 153, 157, 166, 169, 171, 192, 241, 243, 253, 254, 257
Shopping, 133, 134, 137, 149, 152–156, 161, 166, 182, 184, 189, 195, 200, 221, 226, 234–238, 244, 247
Social activities, 56, 133, 134, 146, 181, 182, 184, 186, 189, 191, 192, 195, 196, 200, 201, 206, 207, 214, 216, 217, 247–249, 256, 257
Social capital, 241–244, 255, 257
Social interaction, 65, 67, 68, 77, 81, 137, 161, 180, 188
Social spaces, 45, 47, 56, 60, 68, 138, 149, 152, 154–159, 170–172, 247
Survey, 45, 46, 66, 105, 113, 124, 127, 128, 130, 131, 133, 144, 146, 150, 151, 173, 176, 218, 223, 225–227, 235, 241–244, 255, 258, 259, 277, 278
Sustainability, 60, 65, 97, 105, 107, 115–120

T
Technology, 47, 55, 61, 62, 65, 82, 83, 86, 100, 122, 123, 218, 224
Toilets, 47, 48, 56, 68, 74, 88, 100, 101, 103, 110, 135, 146, 161, 164, 172, 192, 267
Transportation, 45, 47, 54, 68, 149, 152, 162, 167, 169, 171–173, 209, 235, 239
Trust, 115, 116, 241, 243, 251–253, 255, 257

U
Universal design, 47, 66, 75, 87, 95, 99, 100, 105, 135, 147
Urban audit, 127, 149–151, 173, 175, 259, 260

V
Void decks, 49, 53, 56, 58, 66, 67, 82, 85, 99, 128, 155, 157, 158, 170, 171, 179, 217, 242, 256

W
Walkability, 49, 104, 256
Wayfinding, 177, 266, 270, 271
Wellbeing, 49, 56, 65, 67, 92, 99, 100, 103, 104, 122, 241, 242, 278, 279

V
Void decks, 2, 52, 53, 56–58, 66, 67, 85, 99, 155, 157, 158, 170, 171, 179, 217, 256

Y
Young old, 46, 47, 49–51, 53–58, 61, 62, 64, 106, 109, 110, 112, 114, 115, 117, 131, 134, 135, 138, 141, 146

MIX
Papier aus verantwortungsvollen Quellen
Paper from responsible sources
FSC® C105338

If you have any concerns about our products,
you can contact us on
ProductSafety@springernature.com

In case Publisher is established outside the EU,
the EU authorized representative is:
Springer Nature Customer Service Center GmbH
Europaplatz 3, 69115 Heidelberg, Germany

Printed by Libri Plureos GmbH
in Hamburg, Germany